Economic Policy and Sustainable Land Use

Contributions to Economics

Giuseppe Gaburro (Ed.)
Ethics and Economics
1997. ISBN 3-7908-0986-1

Frank Hoster/Heinz Welsch/
Christoph Böhringer
CO_2 Abatement and Economic Structural Change in the European Internal Market
1997. ISBN 3-7908-1020-7

Christian M. Hafner
Nonlinear Time Series Analysis with Applications to Foreign Exchange Rate Volatility
1997. ISBN 3-7908-1041-X

Sardar M.N. Islam
Mathematical Economics of Multi-Level Optimisation
1998. ISBN 3-7908-1050-9

Sven-Morten Mentzel
Real Exchange Rate Movements
1998. ISBN 3-7908-1081-9

Lei Delsen/Eelke de Jong (Eds.)
The German and Dutch Economies
1998. ISBN 3-7908-1064-9

Mark Weder
Business Cycle Models with Indeterminacy
1998. ISBN 3-7908-1078-9

Tor Rødseth (Ed.)
Models for Multispecies Management
1998. ISBN 3-7908-1001-0

Michael Carlberg
Intertemporal Macroeconomics
1998. ISBN 3-7908-1096-7

Sabine Spangenberg
The Institutionalised Transformation of the East German Economy
1998. ISBN 3-7908-1103-3

Hagen Bobzin
Indivisibilities
1998. ISBN 3-7908-1123-8

Helmut Wagner (Ed.)
Current Issues in Monetary Economics
1998. ISBN 3-7908-1127-0

Peter Michaelis/Frank Stähler
(Eds.)
Recent Policy Issues in Environmental and Resource Economics
1998. ISBN 3-7908-1137-8

Jessica de Wolff
The Political Economy of Fiscal Decisions
1998. ISBN 3-7908-1130-0

Georg Bol/Gholamreza Nakhaeizadeh/Karl-Heinz Vollmer (Eds.)
Risk Measurements, Econometrics and Neural Networks
1998. ISBN 3-7908-1152-1

Joachim Winter
Investment and Exit Decisions at the Plant Level
1998. ISBN 3-7908-1154-8

Bernd Meyer
Intertemporal Asset Pricing
1999. ISBN 3-7908-1159-9

Uwe Walz
Dynamics of Regional Integration
1999. ISBN 3-7908-1185-8

Michael Carlberg
European Monetary Union
1999. ISBN 3-7908-1191-2

Giovanni Galizzi/
Luciano Venturini (Eds.)
Vertical Relationships and Coordination in the Food System
1999. ISBN 3-7908-1192-0

Gustav A. Horn/
Wolfgang Scheremet/
Rudolf Zwiener
Wages and the Euro
1999. ISBN 3-7908-1199-8

Dirk Willer
The Development of Equity Capital Markets in Transition Economies
1999. ISBN 3-7908-1198-X

Karl Matthias Weber
Innovation Diffusion and Political Control of Energy Technologies
1999. ISBN 3-7908-1205-6

Heike Link et al.
The Costs of Road Infrastructure and Congestion in Europe
1999. ISBN 3-7908-1201-3

continued on page 377

Nico Heerink · Herman van Keulen
Marijke Kuiper (Editors)

Economic Policy and Sustainable Land Use

Recent Advances in Quantitative Analysis for Developing Countries

With 44 Figures and 41 Tables

Physica-Verlag
A Springer-Verlag Company

Series Editors
Werner A. Müller
Martina Bihn

Editors

Dr. Nico Heerink
Marijke Kuiper
Wageningen University
Department of Economics and Management
Hollandseweg 1
6706 KN Wageningen
The Netherlands
E-mail: nico.heerink@alg.oe.wag-ur.nl
marijke.kuiper@alg.oe.wag-ur.nl

Professor Herman van Keulen
Plant Research International
P.O. Box 16
6700 AA Wageningen
The Netherlands
E-mail: h.vankeulen@plant.wag-ur.nl

ISSN 1431-1933
ISBN 3-7908-1351-6 Physica-Verlag Heidelberg New York

Cataloging-in-Publication Data applied for
Die Deutsche Bibliothek – CIP-Einheitsaufnahme
Economic policy and sustainable land use: recent advances in quantitative analysis for developing countries; with 41 tables / Nico Heerink ... (ed.). – Heidelberg; New York: Physica-Verl., 2001
(Contributions to economics)
ISBN 3-7908-1351-6

Physica-Verlag Heidelberg New York
a member of BertelsmannSpringer Science+Business Media GmbH

Softcover Design: Erich Kirchner, Heidelberg

SPIN 10784884 88/2202-5 4 3 2 1 0 – Printed on acid-free paper

Preface

Since the 1980s many developing countries have implemented macro-economic policy reforms to curb inflation, reduce fiscal deficits and control foreign debt. The policy instruments used, such as exchange rate adjustment, budget cuts, trade policy reforms, public expenditure reviews and privatisation, have different and sometimes opposite consequences for agricultural land use. During the same period awareness was growing that deteriorating soil quality could become a limiting factor to increase or even sustain agricultural production. As a result, food availability and even accessibility for large population groups in developing countries may be jeopardised in the near future.

Recently, quantitative models have made useful contributions to understanding the impact of economic policy reforms on the sustainability of land use. They provide a consistent analytical framework to deal with complex issues such as the direct and indirect effects of economic, agricultural, environmental and population policies, the role of market imperfections in transmitting economic policy signals, and the interactions between soil quality, agricultural production and household economic decision making. Different types of models can be distinguished: bio-economic models, focussing on the link between farm household decisions and the agricultural resource base, household and village models, examining the impact of the socio-economic environment on farm household decisions, and more aggregate models, analysing interactions between sectors and their implications for sustainable land use.

The international workshop "Economic Policy Reforms and Sustainable Land Use in LCDs - Recent Advances in Quantitative Analysis", brought together academics working on these types of models in Wageningen, the Netherlands from June 30 to July 2, 1999. The objective of the workshop was to review recent advances in quantitative analyses of economic factors in sustainable land use at the micro, meso and macro level. In addition possibilities were discussed to integrate elements of these quantitative methodologies into a consistent framework to analyse effects of economic policy reforms on land degradation.

This volume combines the papers presented during the workshop to provide an overview of recent advances in the quantitative analyses of the link between economic policy reforms and sustainable land use in developing countries. In addition it identifies strengths and weaknesses of different approaches, areas in which the approaches are complementary, and fruitful directions for future research.

Reflecting the organisation of the workshop, the chapters in the book are organised in four groups: (1) an overview of the issues and of ways to model the link between agriculture en soil quality, (2) soil quality and household decisions, (3) soil quality at aggregate levels, and (4) soil quality and institutional changes. Most chapters describe the key features of a model and an example of its application to illustrate the type of analysis possible. The last chapter compares the contributions of the different types of models, from the point of view of directions for future research.

The combination of an overview of the current state of quantitative modelling of sustainable land use in developing countries with applications of different model types should appeal to academics and students from both the social and the biophysical sciences. Policy makers interested in different ways of analysing the impact of alternative economic policy instruments on sustainable land use may also find this book a useful source of reference.

We thank all the participants in the workshop for their contributions and discussions, that added to the quality of the present volume. We are also much indebted to Peter Uithol for his expertise in transforming a stack of papers into a book. Organisation of the workshop and this publication have been made possible by financial support from the Netherlands Directorate-General for International Co-operation (SAIL programme), the European Union (INCO-DC programme) and the Netherlands Ministry of Agriculture, Nature Management and Fisheries (North-South programme). Their support is gratefully acknowledged.

NICO HEERINK
HERMAN VAN KEULEN
MARIJKE KUIPER

JULY 2000

Table of Contents

1 Economic Policy Reforms and Sustainable Land Use in Developing Countries: Issues and Approaches

Nico Heerink, Arie Kuyvenhoven & Maarten S. van Wijk

Development Economics Group, Department of Economics and Management, Wageningen University and Research centre, Wageningen, the Netherlands

Abstract

An important factor in land degradation and farmers' investment in soil conservation in developing countries are the changes in the socio-economic environment of farm households resulting from economic policy reform measures taken at higher levels. This chapter provides an overview of the mechanisms through which economic reforms may affect soil degradation and conservation investments, and discusses some recent developments in quantitative approaches that can be used for analysing these relationships. It starts with a discussion of the soil degradation problem and its effects on agricultural production. Next, the potential implications of economic policy reforms for the sustainability of agricultural land use are discussed. Relevant factors at the macro, meso, household and plot level, and the potential relationships between them, are reviewed. The resulting analytical framework takes into account the role of market imperfections and institutional factors in shaping the relation between economic policy instruments and farm household decision making. Recent developments in quantitative modelling may contribute to improving our understanding of the complex effects of economic policies on changes in soil quality. Major approaches in this respect, and their potential contributions, are briefly reviewed. The chapter ends with a discussion of the objectives of the book and a brief overview of its structure.

1.1 Introduction

Soil degradation is increasingly regarded as a major, and perhaps the most threatening, environmental problem in developing countries. Large areas of land are facing reduced productivity or are abandoned as a result of water erosion, wind erosion, nutrient depletion, salinisation or waterlogging. If these trends continue, food availability and accessibility of large population groups may be severely reduced in the near future (World Bank, 1992).

Soil degradation is affected by many factors, such as the quality of the land resource, available technology, poverty, security of land use rights, ignorance of soil mining processes, population growth, and land use and environmental policy. Economic policy reforms in developing countries may strongly modify the socio-economic environment of farm households, and can have a major impact on the sustainability of land use and soil conservation decisions.

This chapter provides an overview of the mechanisms through which economic policy reform measures affect soil degradation and conservation investments, and discusses some recent developments in quantitative approaches that can be used for analysing these relationships. Section 1.2 provides a brief discussion of the available evidence of the soil degradation problem and its effects on agricultural production. Factors affecting soil degradation and soil conservation investments, and the influence of economic policy reforms on some of these factors, are discussed in Section 1.3. In Section 1.4 an analytical framework for the major relationships between economic policy reforms and soil degradation is presented. This analytical framework can serve as a basis for developing quantitative models to assess the impact on soil quality change of major economic policy instruments. Section 1.5 considers recent developments in bio-economic models, farm household and village models, and models at the macro and regional level that can be used to this purpose, and their relations to the analytical framework presented in Section 1.4. Finally, Section 1.6 presents the objectives of the present volume and discusses its structure.

1.2 Soil Degradation and Agricultural Production

World-wide there is a growing concern that negative changes in soil quality[1] in developing countries are becoming a limiting factor in increasing or even sustaining agricultural production. Based on an extensive review of soil degradation literature, Scherr (1999, 16) concludes that soil degradation is widespread on about one quarter of the world's agricultural land and that the pace of degradation has accelerated in the past 50 years. For some regions the picture is even more alarming. According to the GLASOD survey[2], 38% of the world's agricultural land and 23% of all used land is degraded, while in Africa and Central America, the share of degraded land in total agricultural land is as high as 65 and 74%, respectively (see Fig. 1.1). Preventing further degradation of agricultural

[1] Defined by Scherr (1999) as the inherent capability of the soil to perform a range of productive, environmental and habitat functions.

[2] The Global Assessment of Soil Degradation project, based on a formal survey of more then 250 regional experts (Oldeman *et al.* 1991)

land is therefore of fundamental importance for promoting sustainable land use in LDCs.

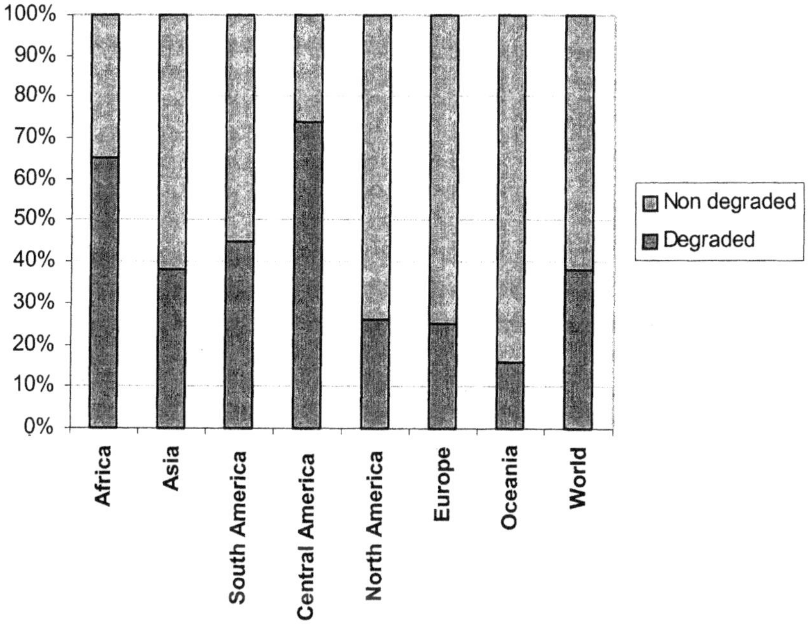

Figure 1.1: Global estimates of soil degradation of agricultural land, by region (compiled from Scherr (1999, 18)).

The main negative consequences of soil degradation are on-farm decline of crop production, and off-farm damages as a result of the siltation of reservoirs (reducing the capacity to store water and the availability of water for irrigation, and increasing the production costs of irrigation schemes) and riverbeds (increasing the frequency of flooding and the costs of flood prevention), and the destruction of the ecological environment (reducing the environmental absorption capacity and increasing the frequency of natural calamities such as droughts).

The magnitude of on-farm effects of soil degradation is under dispute. Estimates of the annual rate of loss of agricultural land through degradation range from 0.3 to 1.0% of the world's arable land. Using GLASOD data, Crosson (1995, 1997) argues that land degradation has not seriously jeopardised agricultural productivity on a global scale. He estimates the global cumulative production loss due to past and present soil degradation at 12–13%; by the year 2030, the cumulative production loss may increase to 17%. Estimates by Brown, Pimentel and others, however, suggest a more serious effect on global agricultural production (Crosson & Anderson, 1999).

Agricultural production losses strongly vary among regions. Crosson & Anderson (1992) estimate that if the 1982 rates of erosion in the United States would continue for 100 years, national average yields would decline by 2–3%. In tropical areas, soil erosion may result in more strongly reduced yields than in temperate regions, because soil organic matter in tropical soils is more concentrated in the topsoil. Erosion-induced production decline is probably a serious problem in large parts of especially Central America and Africa, because of the relatively high degree of land degradation in these two regions. Another problem region is South and East Asia. Crosson & Anderson (1999) present evidence that salinisation and waterlogging in irrigated rice production significantly impaired the productivity of the land in this region. Moreover, they argue that the degradation-induced loss of a unit of irrigated land imposes much higher social costs than that of a unit of rainfed cropland or pasture land.

In general, the relation between soil degradation and yields is poorly documented and quantitative empirical work is scarce (Mitchell *et al.*, 1997), because it requires long-term, longitudinal research. The key soil characteristics that affect yield are organic matter content, nutrient contents, waterholding capacity, soil acidity, topsoil depth and salinity. Yields may decline if one or more of these characteristics is unfavourably affected (depending on which factors are limiting yields in a certain region). Substitution (trade-offs) between these characteristics affects productivity, as do trade-offs between soil quality and external inputs such as fertiliser that can compensate for the loss of soil quality. Another complication is that soil stocks are often large enough to buffer soil degradation for many years. But after a critical level of *e.g.* soil depth has been reached, a rapid productivity decline takes place (*e.g.* Antle & Stoorvogel, this volume).

A well-documented long-term trial in Kabete, Kenya (Sanchez *et al.*, 1997) shows the effect of soil fertility depletion on yields. A continuous maize-common bean rotation with no external inputs was monitored for 18 years. During this period, the fertile red soil (Oxic Rhodudalf) lost about one ton per ha of soil organic N and 100 kg per ha of soil organic P, resulting in maize yields declining from 3.8 to 0.9 tonnes per ha, a reduction of more than 70% in 17 years.

Empirical work of Pimentel *et al.* (1995) shows that in the United States a soil erosion rate of 17 tons per ha per year leads to a reduction in soil depth of 1.4 mm, a loss of 15 kg of available nitrogen, and a decline in water-holding capacity of less than 0.1 mm. These and other effects like the destruction of soil biota translate into an 8% reduction in crop productivity of maize after one year and a reduction of 20% after 20 years when reduced nutrient and water availability is not compensated.

The effects of soil degradation can be temporarily offset by the use of fertilisers, irrigation, improved varieties and/or labour (making it difficult to separate the effects of soil degradation from other factors affecting agricultural productivity in research among farm households). However, the long-term cumulative loss of soil organic matter, soil depth and waterholding capacity is difficult to replace. In their state-of-the-art review of research on nutrient flows and balances, both Nandwa & Bekunda (1998) and Bationo *et al.* (1998) conclude that mineral fertilisers can be effective in increasing yields, but cannot sustain yields in the long run. Productive and sustainable production systems require a combination of mineral fertilisers with technologies using crop residues and manure to maintain soil organic matter levels.

Off-side effects of soil degradation are even more difficult to quantify. Erosion results in displacement of topsoil, causing lower topsoil depth in the erosion area but a thicker topsoil in sedimentation areas. According to some studies, only about 5% of the soil eroded during the past 100 years has been lost for use in agriculture (Mitchell *et al.*, 1997, 55). Farmers may react to the shift of topsoil by migrating from erosion areas to sedimentation areas or areas not affected by erosion. Plot-level yield declines in erosion areas may therefore overestimate the total impact on agricultural production. On the other hand, erosion can cause considerable damage by siltation of waterways, loss of reservoir storage, disruption of ecology, flooding, and so on. Pimentel *et al.* (1995) estimate the on-farm cost of soil degradation for the USA at $27 billion and the off-farm cost at $17 billion per year.

Although the empirical evidence on the magnitude of the soil degradation problem and its impact on global agricultural production is mixed, agricultural production in some regions, particularly Africa, South and East Asia and Central America, appears to be severely affected. The short-term effect of soil degradation on agricultural yields may be limited, continued degradation can cause substantial yield declines when critical thresholds are exceeded, and result in irreversible damage. Positive off-site effects may result from sedimentation in other areas, but these are often offset by the economic damage caused by siltation of reservoirs, waterways and irrigation systems, increased risks of flooding, and destruction of the ecological environment. The flooding of rural and urban land by the Yangtze river in China in 1998 provides an (extreme) example of the damage that may result from continued soil degradation. Various policy options are available to reduce the on-site and off-site impact of soil degradation. These include policies directly focusing on changing land use patterns, adoption of soil conserving technologies, and promotion of conservation investment, as well as population or other policies indirectly affecting land use in developing countries. Reforms of macro-economic and agricultural sector policies may play an important role as well. They are the topic of the next section.

1.3 Economic Policy Reforms and Soil Degradation

Soil degradation in developing countries is affected by many natural and human-induced factors. Natural factors include geology (particularly properties of parent material, topsoil depth, natural soil fertility, and slope length and steepness), climate (especially rainfall and its seasonal distribution), vegetation cover density. Human activity may affect soil quality in many ways. For example, the density of the vegetation cover is affected by farm household decisions on activities (arable farming, animal husbandry, forestry, or off-farm employment) and land use (choice of crops, livestock types); the depth and fertility of the topsoil is affected by technology selection (use of tillage, fertilisers, crop residues) by farmers; and length and steepness of slopes and the impact of the climate may be modified through investments in terraces and irrigation systems.

In most countries, farm households are the main micro-level decision making units on human activities that affect soil quality. Their choices of land use activities and technology and (conservation) investment decisions are influenced by various socio-economic factors. Prices of outputs, inputs, and production factors play a prominent role, although their impact is not clear-cut. Literature suggests that higher agricultural output prices would encourage soil conservation practices of farmers (Repetto, 1989), but also that higher prices would stimulate soil mining by stimulating production of commercial crops (Lipton, 1987). Barrett (1991) argues that price changes will not have much effect on soil conservation, because price increases not only increase instantaneous benefits of over-exploitation, but also future benefits when the soil is conserved. According to Lutz (1992), the negative effect of more intensive resource use may partially be offset by the income effect of higher prices. The higher income of farmers would allow them to use production techniques that are more environmentally friendly and to make larger investments in conservation. Input prices directly affect input demand and its composition through the choice of production technology. In areas where more nutrients are removed from the soil than are added, higher fertiliser prices are likely to aggravate the problem of soil mining. Conservation investments (such as terraces or stone bunds) depend crucially on factor prices. An increase in land prices will generally stimulate such land improvement investments. An important precondition for making conservation investments is the availability of sufficient labour and working capital. High costs of credit and labour, or reduced access to these factors, will limit the extent of these investments.

The system of land use rights plays a crucial role in soil conservation investment decisions. Security of land use is generally seen as an important pre-condition for soil conservation investments by farmers (von Benda Beckmann, this volume). When transfer rights are absent (or land markets are missing), farmers will not be able to recover land improvement investments and therefore have less incentives to make such investment even when relative prices are favourable (Carter & Olinto, 1998). Another key element is farmers' awareness of soil degradation, in which information and extension services can play a crucial role.

At the macro level, population growth and policy, and agricultural and environmental policies are important factors in soil degradation. Population growth can lead to increased cultivation of marginal land or to shorter fallow periods (*i.e.* a reduction in the recovery period of the soil). On the other hand, it will tend to increase land prices, and hence stimulate soil conservation investments. Prevailing environmental policies also play a major role. Measures to promote soil conservation practices constitute a major element of environmental policies in some developing countries. However, the importance attached to sustainability issues and the type and implementation of environmental policies strongly varies among countries. Agricultural policies affect soil degradation by influencing the level of output, input and factor prices farm households are confronted with, and the variation in these prices. Moreover, public investment in rural infrastructure affects the effective prices faced by farmers by reducing the transaction costs made in using output and input markets. When agricultural markets develop as a result of successful agricultural policy and an overall climate of growth is created, higher marketed surpluses and incomes allow farmers to invest more in soil conservation measures. Effective extension services can contribute to more sustainable land use by inducing farmers to grow different crops, use more environment-friendly technologies, or invest in soil conservation measures.

Macro-economic policies can have important implications for agricultural production and for the sustainability of land use that may, in fact, dominate the effects of agricultural, environmental and population policies. Quantitative analyses of the effects of macro-economic and sectoral policies during the post-war period in developing countries show that (1) exchange rate, trade and tariff policies rather than agricultural sector policies have been the major determinants of agricultural prices and farmers' real incomes; and (2) agricultural price support, if any, has been insufficient to compensate for the strong price-depressing effect of macro-policies (World Bank, 1986; Krueger, 1995; Sadoulet & De Janvry, 1995). In the 1980s and 1990s, macro-economic policy reforms were introduced in many low- and middle-income countries to deal with increasing and unsustainable deficits on the balance of payments and government account, rising inflation, and rapidly increasing foreign debts. Major instruments include exchange rate, money supply, government budget, import and export tariffs, price controls, and privatisation of public enterprises. The new agricultural sector policy orientation implied a change from a controlled but unprotected sector to a more free but moderately protected sector in line with other activities, without major claims on the budget (Norton, 1992, 52). Economic policy reforms sought to achieve this ambitious turn-around toward restoring equilibrium and enhancing efficient supply response of agricultural production. Most reforms therefore stressed the need to increase production and net export of agricultural products, to adjust the real exchange rate, and to increase agriculture's productivity and efficiency.

The potential environmental impact of economic policy reforms affecting agriculture has attracted increasing attention (*e.g.* Sebastian & Alicbusan, 1989; Repetto, 1989; Pearce & Warford, 1993; Reed, 1993, 1996; Munasinghe & Cruz,

1995; Munasinghe 1996; Taylor, 1996). There is a wide consensus that macro-economic policy reforms constitute by no means sufficient conditions to induce *e.g.* sustainable land use and conservation measures, for many reasons: (1) persistent poverty (of groups that do not benefit from the reforms) and few opportunities to transfer resources to the poor, (2) structural imperfections in rural markets (high transportation and other transaction costs, access and information problems, absence of capital and insurance markets, lack of secure land use rights), (3) environmental 'goods' and 'bads' are not signalled in the market, (4) many environmental goals have a trade-off in terms of income, and (5) environmental concerns are usually subordinated to growth and distributional issues in low-income countries. As a result, macro and sector policies usually show mixed effects on sustainable land use (and natural resource use in general). Different policy measures have different and often opposite consequences (Kuyvenhoven *et al.*, 1999).

Modelling techniques can be applied to analyse the effects of economic policy reforms on sustainable land use in developing countries. They can be used to examine the various direct and indirect effects of different policy measures on the socio-economic environment of agricultural households, and the implications for their crop choice, technology choice and conservation investments. Simulations with alternative policy options can show the potential off-setting effects on soil degradation of measures that are part of such policy packages. The modelling of relations between economic policy measures and soil quality is still in its infancy. In the next section an analytical framework is developed, as a basis for such models.

1.4 An Analytical Framework

Fig. 1.2 gives a schematic representation of major relationships between economic policy reforms and soil degradation, based on the discussion in the previous section. Four levels of analysis are distinguished: the macro level (where economic policy decisions are taken), the socio-economic environment of farm households (where the impact of policies appears), the farm household (where agricultural production choices and investment decisions are made), and the field or plot level (where soil degradation processes occur). Feedback mechanisms (for example, the impact of soil degradation on household decisions) and exogenous factors (for example, household characteristics) are not shown in this scheme.

A common distinction of soil degradation processes is between water erosion, wind erosion, chemical deterioration (loss of nutrients and organic matter, salinisation, acidification, and so on), and physical deterioration (compaction, crust formation, sealing, and so on). The outcomes of these processes may be expressed in terms of indicators, such as topsoil depths, nutrient balances, organic matter balances, and other indicators.

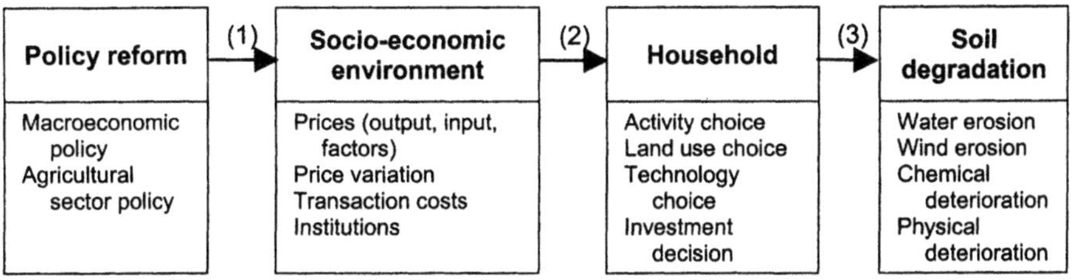

Figure 1.2: Analytical framework.

Farm household decisions on activities (arable farming, animal husbandry, forestry, off-farm employment), land use choice (food crops, cash crops, green manure, livestock type, timber, fuelwood), technology choice (use of fertiliser, manure, tillage, crop residues) and investments (terraces, ditches, stone rows, trees) affect the vegetation cover, topsoil depth and fertility, slope length and steepness, and the impact of climate, and therefore play an important role in aggravating or reducing soil degradation processes[3]. Farm household decisions on these issues are dependent on relative prices of agricultural products, inputs and production factors, and on the variation in these prices. The responsiveness of farm household decisions to such economic incentives depends on institutions such as the existing land tenure system, the available social capital, prevalent customs and norms, and availability and interlinkages of markets. Transportation, information, and other transaction costs involved in using output, input and factor markets affect the effective prices received and paid by farm households, as well as (according to new institutional economic theory) the institutions faced by farm households (Hoff *et al.*, 1993).

Reforms of macro-economic and agricultural policies may fundamentally change the socio-economic environment of farm households. Agricultural market liberalisation and price and input policies have a direct (usually positive) impact on the prices received and paid by farmers. For traded products, these effects are reinforced when combined with international trade liberalisation and exchange

[3] The concepts used to classify farm household decisions in this chapter differ from those used by Fresco *et al.* (1992) and Bouman *et al.* (2000). In their studies, the concept of a land use system is used to reflect differences in activity, land use, technology and investment choices in a given physical environment. In our analytical framework, we prefer to distinguish four categories of farm household decisions, since their impact on soil quality change and their relation with the socio-economic environment of households differ fundamentally between these types of decisions. For example, higher wages may lead to an increase in off-farm employment, cultivation of crops that require less labour, substitution of chemical fertilisers for manure, and less investments in terraces or other soil conservation structures. All these changes are likely to have important, sometimes off-setting, effects on soil quality.

rate adjustments. As a result, price ratios of previously subsidised vs. unsubsidised and traded vs. non-traded products (and inputs) will change considerably. When moving from controlled to market-determined prices, the temporal variation in prices may increase. On the other hand, private traders may be more efficient in carrying over supplies in time, thereby reducing variation in prices (Badiane, 2000). Other elements of policy reform, such as cuts in public investment in rural infrastructure, privatisation of extension services, and adjustments in laws and regulations on, for example, land ownership can have major effects on transaction costs and institutions facing farm households. For public investment cuts in rural infrastructure, for example, the impact on prices received and paid by farm households may be opposite to the effect of output subsidy reduction and exchange rate devaluation.

The framework in Fig. 1.2 can serve as a basis for developing models to simulate the impact of economic policy reform measures on soil degradation. Depending on the purpose of the model, a number of elements may be added to gain in realism. First, the impact of soil degradation on yields and farm household responses to such production declines can be added. The need for inclusion of these effects depends on the sensitivity of yields to soil degradation (see above) and on the extent to which farm households react to such production declines. Secondly, in models that concentrate on the regional or watershed level spatial factors may be added, such as off-site erosion effects and clearance of new land areas. The next section will discuss some recent advances in modelling approaches that provide important building blocks for models examining the impact of economic policy measures on soil degradation.

1.5 Modelling Approaches

Various aspects of the analytical framework presented in the previous section have been implemented in quantitative models. Firstly, several bio-economic models have been proposed in recent years that integrate biophysical processes and economic decision making at the household level (and more aggregate levels), that can be used to analyse the impact of economic incentives on land use decisions and sustainability indicators. Secondly, a number of household and village models have been developed to examine the impact of market imperfections, missing markets and institutional change on household behaviour. Finally, macro- and regional level models are increasingly applied, either to analyse the impact of macro-economic and sectoral policies on different economic sectors and socio-economic groups (taking into account their interactions), or to aggregate the results of models at the household level to higher levels of analysis. We will discuss these approaches in turn.

1.5.1 Bio-economic Models[4]

It is increasingly recognised that disciplinary research on agro-ecological and socio-economic issues is not very fruitful. Bio-economic modelling approaches have been developed to address the growing mismatch between agro-ecological research focusing on technological options to enhance food security and sustainable land use, and socio-economic research focused on the analysis of production efficiency and the (non-)adoption of technologies (Anderson, 1994; Ruben *et al.*, 1998). Their purpose is to enable a better assessment of the impact of economic incentives on the adjustment in agricultural technology choice and land use practices. They rely on quantitative procedures that specify the interaction between bio-physical and socio-economic processes, including feedback mechanisms between farm and regional levels. Bio-economic models represent an interdisciplinary approach, that specifies both agro-ecological and behavioural mechanisms and their interactions. Socio-economic aspects deal mainly with farm household responses to different policy instruments and the market and institutional arrangements that influence supply response. Agro-ecological aspects refer to biophysical processes, such as water and nutrient regimes, plant and animal growth, accumulation and leaching of pollutants, and soil erosion, that influence the quality of the resource base.

Bio-economic modelling has been applied at the level of the household (Dalton, 1996; Kruseman & Bade, 1998), village (Barbier, 1994) and watershed (Barbier & Bergeron, 1999) and for the agricultural sector (Deybe, 1994; Schipper 1996). These models generally consist of separate modules that specify farm household preferences, technological choice and sustainability parameters. Their common characteristics include: (i) econometric specification of consumptive choice and labour allocation, (ii) discrete definition of technical production coefficients, (iii) a linear programming framework for optimisation of activity choice, (iv) specific indicators for sustainability (*e.g.* nutrient and carbon balances; soil erosion; pesticide emissions), and (v) aggregation through market clearance procedures. Linkages with spatial, inter-temporal and aggregate aspects are sometimes incorporated, but still require further elaboration. Intra-household allocation procedures are mostly neglected.

Bio-economic models at the farm level incorporate farm household modelling (FHM) procedures as developed by Singh *et al.* (1986). These models explicitly account for natural resource endowments, input and labour allocation decisions, output choice and consumption preferences under different conditions of market development. Biophysical information can be linked to the production side of the farm household model (Altieri *et al.*, 1993; Dalton, 1996) by making use of mathematical programming procedures. Integrated bio-economic household models rely on technical production coefficients derived from production ecology

---

[4] This section draws on Ruben *et al.* (2000)

and land evaluation (Kruseman *et al.*, 1995; Kruseman & Bade, 1998). Econometric techniques are applied to specify farm household behaviour with respect to consumption and risks. For purposes of model calibration, information from farming systems research is used. These models enable an appraisal of supply responsiveness of farm households to different policy incentives, taking into account different household objectives (*e.g.* income, sustainability).

Bio-economic modelling approaches provide an important contribution to understanding the environmental implications of economic policy reforms at the micro level. Consequences of market reforms (*i.e.* exchange rate adjustment, price liberalisation) and institutional reforms (*i.e.* public budget restrictions, land rights regulations) for technological choice, land use adjustment, labour allocation and agro-ecological sustainability can be examined. By distinguishing different types of representative farm households, they take into account that generic policy instruments may well yield different reactions from different types of farmers.

Still, fundamental research on the specification of model components and procedures needs to continue, since some major interactions are incompletely understood. At the biophysical level, relationships between soil erosion and yield potential are still subject to debate. Moreover, the dynamics of multiple cropping systems cannot be captured adequately in available crop growth models. Farm household behaviour can be better specified when procedures for modelling risk and adoption behaviour are explicitly incorporated. Finally, the economic dimension of bio-economic modelling can be improved through incorporation of multi-market analysis and linkages with non-agricultural activities and other regions.

1.5.2 Household and Village Models

Farm households in developing countries are typically located in an environment characterised by market failures for some of its outputs (*e.g.* food), inputs (*e.g.* manure) and factors (*e.g.* capital, labour). A market is considered to fail for a particular household, when it faces a wide price band between the (low) price at which it could sell the good or factor, and the (high) price at which it could buy it (De Janvry & Sadoulet 1994; Sadoulet & De Janvry, 1995, Chapter 6). Faced with a wide price band - as a result of high transaction costs, shallow markets and high risks - the household may choose self-sufficiency when its subjective price (equating supply and demand by the household) falls within the price band. This good or factor can then be considered a non-tradable for that household. Markets may therefore fail for some households, but not for others. An extreme case of market failure is the non-existence of a market due to legal prohibition.

The frequency of market failures, *e.g.* in credit, insurance, labour and land markets in rural economies in developing countries, points to non-separable production and consumption decisions, as a rule rather than as an exception (De Janvry *et al.*, 1991; Hoff *et al.*, 1993). In economies with significant transaction costs and missing or imperfect markets, few policy conclusions can be drawn from conventional economic models. Some of the standard results from production and consumption theory may be reversed when production and consumption decisions are interdependent (De Janvry *et al.*, 1991; Finkelshtain & Chalfant, 1991; Singh *et al.*, 1986). Responses to price changes and to risk may be opposite to those predicted by partial analyses of production or consumption. A more empirical approach, capable of capturing complex production and consumption linkages within households, is therefore required for policy analysis to derive reliable policy conclusions. Non-separable farm household models (see *e.g.* De Janvry *et al.*, 1992; Singh *et al.*, 1986) have been developed in recent years to deal with these linkages.

De Janvry *et al.* (1991), on the basis of results of a stylised non-separable model of an African farm household, show the impact of a 10% increase in the price of cash crops on production and consumption for a situation with no market failures and a situation with failing markets for food and/or labour. When both markets fail, the elasticity of supply response of cash crops drops from 0.99 (in the case of no market failures) to 0.18, due to the inability to use more labour in production and to significantly increase food production (since the family needs to feed itself). The output response mainly originates from increased fertiliser use. The resulting higher income leads to a 15.8 % higher consumption of manufactured goods, compared to 5.6 % without market failures, at the expense of the consumption of food and leisure.

Economic linkages and feedback among village households may affect the impacts of policy, market and environmental changes (Taylor & Adelman, 1996). If all agricultural households would be self-sufficient and supply their own inputs and factors, income and expenditure linkages among village households would be absent. At the other extreme, if all village households would be perfectly integrated into markets outside the village – all goods and factors are tradables for the village – production and expenditure linkages among village households would also be absent. However, most villages in developing countries are situated somewhere in between. Households typically are self-sufficient in some goods or factors –'household non-tradables'–, other goods or factors are only exchanged within the village –'village non-tradables'–, while still others are 'village tradables', implying that their prices are determined outside the village. Costs of transactions with markets outside the village, and local demand and supply determine whether a village is self-sufficient for, or a net importer or exporter of a good or factor.

Taylor and Adelman (1996) argue that modelling local income and general equilibrium feedbacks in different economic, social and cultural settings is critical for understanding the likely impacts of government policies. Farm household models, focusing on first-round effects of economic policy measures, including those that take into account market imperfections, lack the second- and higher-round income and expenditures linkages and general equilibrium effects. Village Social Accounting Matrix (SAM) multiplier models can be used to analyse income and expenditure linkages within villages (assuming production capacity limitations are absent and all relationships are linear), but not the price changes in village non-tradables.

Village (or micro) computable general equilibrium (CGE) models are designed to analyse income as well as general equilibrium feedback effects. When the SAM at the basis of the CGE-model adequately reflects the socio-economic structure of the village to which the model is applied, some of the local institutions that set the rules for production, consumption, and exchange relations can be taken into account.

The advantages of a village CGE-model are demonstrated in a comparison of the simulated impact of a 10% increase in migrant remittances, with a CGE model with missing markets for labour and staples[5], with a separable household model, and a village SAM multiplier model, for a Mexican village (Taylor & Adelman, 1996, 27-29). In the (neo-classical) household model, local prices do not change so that production remains unaffected despite the increase in migrants' household incomes. In the SAM model, remittances stimulate the demand for goods and services inside and outside the village. Production in all sectors increases, creating new rounds of income increases in the village. In the village CGE model, migration competes with local production activities for scarce family time. Increased remittances result in lower village production levels, except for livestock (which uses very little labour) and staples (which is a non-tradable).

The results for the farm household and village CGE models presented in De Janvry *et al.* (1991) and Taylor & Adelman (1996) illustrate the impact of market imperfections on household responsiveness to policy incentives and external shocks. To analyse the impact of economic policy reforms on farm household production decisions, a careful assessment of the presence of such imperfections is required and should be reflected in the choice of the modelling technique.

1.5.3 Macro- and Regional Models

Macro CGE models and multi-market models can be used to analyse the impact of economic policy reforms on prices faced by farm households. Macro CGE models simulate the interaction between the agricultural sector and the other sectors in the economy. They can examine the price effects for the agricultural sector (and sub-sectors, if specified) of policies oriented at that sector, as well as of exchange rate adjustments, foreign trade and tariff policies, industrialisation policies and other policy measures focusing on other sectors or the economy as a whole. In multi-market models, the analysis can be extended to deal with more disaggregated levels within the agricultural sector (and sub-sectors), by examining the interactions among markets at the product and factor sides. An example is provided in De Janvry *et al.* (1992), who use a combination of a macro CGE model and a multi-market model for the livestock sector in Morocco, to examine

[5] For simplicity sake, the results of the CGE model with a missing market for family labour only is not discussed here.

the impact of a 20.1% price increase[6] in cereals. The results indicate substantial price effects on other agricultural products (meat: 12.8%, milk: 8.3%, fruits and vegetables: 8.7%), other consumption goods (5%), inputs (forage: 24%, machinery and fertiliser: 1.5%), and production factors (wages: 5%). Evidently, the implications for farm household decision making will be quite different from that of an analysis based on cereal price changes only. To examine the impact of economic policy reforms on soil degradation for an entire country, some studies have extended a macro CGE model with some econometrically-specified linkages with agro-technical modules (Alfsen *et al.*, 1997; Aune *et al.*, 1997; Glomsrød, this volume).

These macro approaches focus in particular on linkages between (sub-)sectors. Other studies have used bottom-up approaches to aggregate models for individual households or typical farm types to higher levels of analysis such as watersheds, regions, or the agricultural sector. Such models may take into account spatial aspects, such as off-site erosion effects, geological and/or climatological differentiation, population dynamics, or availability of infrastructure. Market clearing procedures are often included to adjust prices of non-tradables that are exchanged within the spatial unit. Models in this category were mainly built in the 'French tradition' of analysing watersheds or comparable micro-regions (*Modèles Villageoises*; Barbier, 1994; Deybe, 1994; Reyniers & Benoît-Cattin, 1995; Deybe & Robilliard, 1996; Barbier & Bergeron, 1999).

1.5.4 Potential Contributions

The model approaches discussed above provide important contributions to the analysis of the consequences of economic policy reforms in developing countries for sustainable land use in agriculture. Aggregate general equilibrium models provide insights in the impact of macro-economic and agricultural sector policies on the socio-economic environment of households (relation (1) in Fig. 1.2). Economic incentives for the agricultural sector, originating from outside the sector, and feedback mechanisms through production and income linkages, can be examined by such models. Farm household and village economy models are suitable tools for analysing the impact of the market environment on farm household decision making in developing countries (relation (2) in Fig. 1.2). They take into account the impact of market imperfections on household behaviour, and the possibility of feedback mechanisms (through income and expenditure effects, and local price formation processes) at the village level. Bio-economic models are especially suited for analysing the links between socio-economic and bio-physical processes at the household level (relation (3) in Fig. 1.2). In these models, the complexity of bio-physical processes, including feedback effects on agricultural

[6] This price increase would result, if pricing rules were implemented that would closely relate domestic prices to international prices.

productivity, are integrated with socio-economic analyses of farm household decision making.

In addition, there is a need for aggregating the results for (representative) households to higher levels of analysis. The main reason is that economic policies focus on population groups - either at the macro, regional or other aggregate levels - not on individual households. Results pertaining to individual households may be difficult to aggregate because of geographical diversity, interactions among households, aggregation biases, and economic and biophysical processes operating at higher levels. Regional models that explicitly focus on aggregating micro-level relations to higher levels of analysis can play an important role in this respect.

1.6 Purpose and Structure of the Book

The increased interest in recent years in the potential environmental impact of economic policy reforms in developing countries has also led to a proliferation of quantitative models that can be used to analyse such effects. These models vary widely in various aspects, such as coverage of biophysical and/or economic relationships, level of analysis, length of time period considered. To guide future quantitative modelling work on the effects of economic policy reforms on sustainable land use, a detailed assessment is required of the state-of-the-art, the strengths and weaknesses of current approaches, and identification of issues on which future modelling work should concentrate.

The main objectives of the present volume are therefore:

1. to assess recent advances in quantitative analyses of economic policy reforms and sustainable land use in developing countries,
2. to identify key issues in the modelling of the effects of economic policies on sustainable land us in developing countries, and
3. to formulate recommendations for future research directions that may improve our understanding of the potential effects of economic policy instruments on soil quality changes in developing countries.

The book is structured as follows. Part I provides an overview of issues and approaches. The brief review given in the current chapter is followed by an overview of available approaches for quantifying the links between soil quality and agricultural production in Chapter 2. Quantitative methods to assess interactions between household decision making and soil quality changes are presented in Part II. This part consists of eight chapters describing the essential elements of the model approaches applied, and (in most cases) providing an illustration of a model application for households in a specific country or region. Part III consists of six chapters presenting and illustrating model approaches at the (sub-)regional and country level. Part IV deals with a crucial issue in any quantitative analysis of economic reforms and sustainable land use, *viz.* the role of changes in institutions. The chapters in this part focus on two major institutions that are likely to play an important role, social capital and land property rights. In the final chapter, a number of key issues in the modelling of economic policy reforms and sustainable land use in developing countries are identified. In addition, the way in which these key issues are incorporated in the various model approaches presented in the book is assessed, and recommendations are made for future research.

1.7 References

Alfsen, K.H., T. Bye, S. Glomsrød & H. Wiig (1997): Soil degradation and economic development in Ghana. Environment and Development Economics 2, 119-143

Altieri, M., C. Benito, A. Gomez-Lobo, P. Faeth, T. Tomic & J. Valenzuela (1993): Wheat farming in central Chile. In: P. Feath (Ed.) Agricultural policy and sustainability: case studies from India, Chile, The Philippines and the United States. World Resources Institute, Washington, D.C.

Anderson, J.R. (Ed.) (1994): Agricultural technology: Policy issues for the International Community. CAB International, Wallingford, UK

Antle, J.M. & J. Stoorvogel (2000): Integrating site-specific biophysical and economic models to assess trade-offs in sustainable land use and soil quality. This volume

Aune, J., S. Glomsrød, V. Igersen & H. Wiig (1997): Structural adjustment and soil degradation in Tanzania: A CGE-model approach with endogenous soil productivity. Discussion Paper 189, Statistics Norway, Research Department, Oslo

Badiane, O. (2000): The effects of liberalization on food markets in Africa. In: A. van Tilburg, H.A.J. Moll & A. Kuyvenhoven (Eds.) Agricultural markets beyond liberalization. Kluwer., Boston

Barbier, B. (1994): Modélisation agronomique et économique de la durabilité d'un système agraire villageois. Ph.D. thesis. CIRAD, Montpellier

Barbier, B. & G. Bergeron (1999): Impact of policy intervention on land management in Honduras: Results of a bio-economic model. Agricultural Systems 59, 1-16

Barrett, S. (1991): Optimal soil conservation and the reform of agricultural pricing policies. Journal of Development Economics 36, 167-187

Bationo, A., F. Lompo & S. Koala (1998): Research on nutrient flows and balances in West Africa: state of the art. Agriculture, Ecosystems and Environment 71, 19-36

Benda-Beckmann, F. von (2000): Between free riders and free raiders: property rights and soil degradation in context. This volume

Bouman, B.A.M., H.G.P. Jansen, R.A. Schipper, J. Bouma, A. Kuyvenhoven & M.K. van Ittersum (2000): A toolbox for land use analysis. In: Bouman, B.A.M., H.G.P. Jansen,

R.A. Schipper, H. Hengsdijk & A. Nieuwenhuyse (Eds.) Tools for land use analysis at different scales – with case studies from Costa Rica. Boston: Kluwer

Carter, M. & P. Olinto (1998): Getting institutions right for whom: The wealth-differentiated impact of property rights reform. Paper prepared for the XXI International Congress of the Latin American Studies Association, Chicago, 24-26 September, 1998

Crosson, P. (1995): Soil erosion and its on-farm productivity consequences: What do we know. Discussion Paper 95-29. Resources for the Future, Washington, D.C.

Crosson, P. (1997): Will erosion threaten agricultural productivity? Environment 39, 4-9, 29-31

Crosson, P. & J.R. Anderson (1992): Resources and global food prospects: Supply and demand for cereals to 2030. World Bank Technical Paper no. 184. World Bank, Washington, D.C.

Crosson, P. & J.R. Anderson (1999): Technologies for meeting future global demands for food. Paper prepared for Bellagio Conference on Sustainable Agriculture

Dalton, T.J. (1996): Soil degradation and technical change in Southern Mali. Ph.D. thesis. Purdue University, West Lafayette

De Janvry, A. & E. Sadoulet (1994): Structural adjustment under transaction costs. In: F. Heidhues & B. Kerr (Eds.) Food and agricultural policies under structural adjustment. Peter Lang, Frankfurt

De Janvry, A., M. Fafchamps & E. Sadoulet (1991): Peasant household behavior with missing markets: some paradoxes explained. Economic Journal 101, 1400-1417

De Janvry, A., E. Sadoulet, M. Fafchamps & M. Raki (1992): Structural adjustment and the peasantry in Morocco: a computable household model. European Review of Agricultural Economics 19, 427-453

Deybe, D. (1994): Vers une agriculture durable. Ph.D. thesis. CIRAD, Montpellier

Deybe D. & A.S. Robilliard (1996): MATA: an integrated micro, meso and macro-economic model for decision making in agriculture, some applications for Burkina Faso. Notes et Documents URPA no. 57. CIRAD, Paris

Finkelshtain, I. & J.A. Chalfant (1991): Market surplus under risk: do peasants agree with Sandmo? American Journal of Agricultural Economics 73, 557-567

Fresco, L., H. Huizing, H. van Keulen, H. Luning & R. Schipper (1992): Land evaluation and farming systems analysis for land use planning. FAO working document, FAO, Rome

Glomsrød, S. (2000): Soil degradation in macro CGE models. This volume

Hoff, K., A. Braverman & J.E Stiglitz (1993): The economics of rural organization: theory, practice and policy. Oxford University Press, New York

Krueger, A. (1995): Policy lessons from development experience since the Second World war. In: J. Behrman & T.N. Srinivasan (Eds.) Handbook of Development Economics. Volume 3B. North-Holland, Amsterdam, 2497-2550

Kruseman, G. & J. Bade (1998): Agrarian policies for sustainable land use: bio-economic modelling to assess the effectiveness of policy instruments. Agricultural Systems 58, 465-481

Kruseman, G., R. Ruben, H. Hengsdijk & M. van Ittersum (1995): Farm household modelling for estimating the effectiveness of price instruments in land use policy. Netherlands Journal of Agricultural Science 43, 111-123

Kuyvenhoven, A., N. Heerink & R. Ruben (1999): Economic policy in support of soil fertility: which interventions after structural adjustment? In: E.M.A. Smaling, O. Oenema & L.O. Fresco (Ed.): Nutrient disequilibria in agrosystems. CAB International, Wallingford

Lipton, M. (1987): Limits of price policy for agriculture: Which way for the World Bank? Policy Development Review 5, 197-215

Lutz, E. (1992): Agricultural trade liberalization, price changes, and environmental effects. Environmental and Resource Economics 2, 79-89

Mitchell, D.O., M.D. Ingo & R.C. Duncan (1997): The world food outlook. Cambridge University Press, Cambridge

Munasinghe, M. (Ed.) (1996): Environmental impacts of macro-economic and sectoral policies. World Bank, Washington, D.C.

Munasinghe, M. & W. Cruz (1995): Economywide policies and the environment. World Bank Environment Paper no. 10. World Bank, Washington, D.C.

Nandwa, S.M. & M.A. Bekunda (1998): Research on nutrient flows and balances in East and Southern Africa: state-of-the-art. Agriculture, Ecosystems and Environment 71, 5-18

Norton, R.D. (1992): Integration of Food and Agricultural Policy with Macro-economic Policy: Methodological Considerations in a Latin American Perspective. ESD Paper 111, FAO, Rome

Oldeman, L.R., T.A. Hakkeling & W.G. Sombroek (1991): World map of the status of human-induced soil degradation. ISRIC, Wageningen; UNEP, New York

Pearce, D.W. & J.J. Warford (1993): World without end. World Bank and Oxford University Press, Oxford

Pimentel, D., C. Harvey, P. Resosudarmo, K. Sinclair, D. Kurz, M. McNair, S.Crist, L. Shpritz, L. Fitton, R. Saffouri & R. Blair (1995): Environmental and economic costs of soil erosion and conservation benefits. Science 267, 1117-1123

Reed, D. (1993): Structural adjustment and the environment. World Wide Fund for Nature, Earthscan Publications, London

Reed, D. (1996): Structural adjustment, the environment, and sustainable development. World Wide Fund for Nature, Earthscan Publications, London

Repetto, R. (1989) Economic incentives for sustainable production. In: G. Schramm & J.J. Warford (Eds.) Environmental Management and Economic Development. Johns Hopkins University Press, Baltimore

Reyniers, F.N. & M. Benoît-Cattin (1995): Couplage de modèles en agriculture. Collection colloques, CIRAD, Montpellier

Ruben, R., H.A.J. Moll & A. Kuyvenhoven (1998) Integrating agricultural research and policy analysis: analytical framework and policy applications for bio-economic modelling. Agricultural Systems 58, 331-349

Ruben, R., A. Kuyvenhoven & G. Kruseman (2000): Bio-economic models for eco-regional development: Policy instruments for sustainable intensification. In: D.R. Lee, D.R. & C.B. Barrett (Eds.) Critical trade-offs: Agricultural intensification and the environment in developing countries. CAB International, London (in press)

Sadoulet, E. & A. De Janvry (1995): Quantitative development policy analysis. Johns Hopkins University Press, Baltimore

Sanchez, P.A., K.D. Shepherd, M.J. Soule, F.M. Place, R.J. Buresh, A.M.N. Izac, A.U. Mokwunye, F.R. Kwesiga, C.G. Ndiritu & P.L. Whoomer (1997): Soil fertility replenishment in Africa: An investment in natural resource capital. In: R.J. Buresh , P.A. Sanchez & F. Calhoun (Eds.) Replenishing soil fertility in Africa. Special Publication 51, Soil Science Society of America Inc., Madison, WI

Scherr, S.J. (1999): Soil degradation: A threat to developing-country food security by 2020? Food, Agriculture and the Environment Discussion Paper 27, IFPRI, Washington, D.C.

Schipper, R.A. (1996): Farming in a fragile future. Economics of land use with applications in the Atlantic Zone of Costa Rica. Ph.D. thesis. Wageningen Agricultural University, Wageningen

Sebastian, I. & A. Alicbusan (1989): Sustainable development: issues in adjustment lending policies. Environmental Department Divisional Paper 1989-6, World Bank, Washington, D.C.

Singh, I., L. Squire & J. Strauss (1986): Agricultural household models: extensions, applications, and policy. Johns Hopkins University Press, Baltimore

Taylor, L. (1996): Sustainable development: an introduction. World Development 24, 215-225

Taylor, J.E. & I. Adelman (1996): Village economies: the design estimation and use of villagewide economic models. Cambridge University Press, Cambridge

World Bank (1986): World Development Report 1986. Oxford University Press, New York

World Bank (1992): World Development Report 1992: Development and the environment. World Bank, Washington, D.C.

2 Soil Degradation and Agricultural Production: Economic and Biophysical Approaches

Gideon Kruseman[1] & Herman van Keulen[2]

[1] Development Economics Group, Department of Economics and Management, Wageningen University and Research centre, Wageningen, the Netherlands

[2] Plant Research International, Wageningen University and Research centre, Wageningen, the Netherlands

Abstract

In this chapter we give an overview of a number of approaches used in the biophysical sciences to quantify the relationship between agricultural production and soil degradation. First the concept of soil degradation is discussed, and the various components detailed. Then the approaches are assessed in the context of bio-economic models in which biophysical processes are linked to decision making processes at household, village and regional levels. The main emphasis of the chapter is on the household level where the direct interaction between decision makers and the biophysical processes takes place. Different biophysical approaches are presented, used for generation of data for incorporation in bio-economic models.

2.1 Introduction

The crucial point in bio-economic modelling is the interface between the social sciences and the biophysical sciences. The key issue is the formulation of production functions that capture on one hand the interactions between biophysical processes and the environment, and on the other hand agricultural technology choice and allocation of production factors, as dictated by economic factors. As illustrated in this volume, recent developments in bio-economic modelling have resulted in a variety of methods to combine biophysical and economic knowledge at varying degrees of integration (from *multi*-disciplinary to *inter*-disciplinary). The main emphasis has been on the formulation of economic models incorporating information from the biophysical sciences.

Two aspects are fundamental to the discussion: (i) the traditional relationship between inputs and yield, where the inputs include soil quality, (ii) the relationship between agricultural management practices and soil quality. In this context, soil degradation can be viewed as both *cause* of yield decline and *effect* of agricultural practices.

Economists and biophysical scientists have different paradigms when looking at the issue of soil degradation. Biophysical scientists concentrate on the processes with respect to crop-soil interactions. Economists focus on the decision-making behaviour of agricultural households, which are assumed to maximise a utility function. This utility function may explicitly contain preferences related to soil quality, or it may implicitly be affected through different technical coefficients of production activities.

This chapter discusses the way in which biophysical and socio-economic sciences deal with so-called production functions, the quantification of agricultural production systems and sustainability indicators. It treats the way in which effects of biophysical processes, related on one hand to agricultural production, with special emphasis on intensification and technology choice, and on the other hand soil degradation, can be incorporated in economic models. Economic and sociological factors can be used to specify boundary conditions and management rules for biophysical models, but are not incorporated in a dynamic fashion (Penning de Vries, 1990).

In Section 2.2, soil degradation is discussed in more detail. Sections 2.3 to 2.5 deal with the main issues related to specification of production functions and the way economists and biophysical scientists treat them. Section 2.3 reviews a number of ways in which biophysical processes are described for use in combination with economic models. Section 2.4 discusses the paradigmatic difference in the use of the term efficiency, which is crucial in understanding the differences in approaching the problem of agricultural production and soil degradation by different sciences, and the difference in approaching soil degradation as a cause or effect of agricultural production. In Section 2.5 the biophysical approach to the production function is highlighted, and in Section 2.6 different ways of quantifying biophysical processes are described. Section 2.7 summarises the discussion and explores the possibilities to build interfaces between biophysical and economic models. Issues for future research are briefly touched upon.

2.2 Soil Degradation: What Are We Talking About?

To further elucidate the term 'soil degradation' and set the stage for its treatment in bio-economic modelling, various characteristics associated with the definition of soil quality will be examined. In connection with agricultural use, 'soil' is being defined as 'loose surface material of the earth, in which plants grow'. The term 'soil degradation' has been defined, qualitatively, as 'the state where the balance between the attacking forces of climate and the natural resistance of the terrain against these forces, has been broken ...by human intervention... leading to decreased current and/or future capacity to support life' (Oldeman *et al.*, 1991). In more tangible terms, soil degradation is associated with either displacement of soil material (erosion under the influence of wind and/or water) or changes in properties *in situ*. The latter can refer to physical properties (surface crusting, compaction) or chemical properties (nutrient mining, acidification, salinisation, alkalinisation).

In functional terms, these processes, and the properties they affect, should be related to the soil as a medium for the growth of crops (or vegetations). These functions include anchorage for the plants and supply of water, oxygen and nutrients. The reliability of the soil in providing anchorage for the plants is related to soil depth and soil density. When the 'loose top soil' is not 'loose' enough, root growth is constrained, or when it is too shallow, plant roots will be restricted to the upper layer of the soil, and relatively mild forces, such as exerted by wind, may lead to uprooting of the plants. Very little information appears available for quantitative description of the relations between soil depth and the risk for uprooting and the associated effects on yield. Therefore, incorporation of these effects in bio-physical crop growth models has not been attempted.

Soil depth also affects the capacity of the soil to supply water, nutrients and oxygen to the growing crop. In a simulation study on the interactions of soil depth, rainfall and nitrogen supply on dry matter production of a Mediterranean pasture, Seligman & van Keulen (1989) and Seligman *et al.* (1992) have shown that, whereas rainfall (moisture supply) and nitrogen supply strongly influence both average production and variability, soil depth had only a minor effect. Of course, shallower soils provide lower storage capacity, so that drainage (water) and leaching (nitrogen) beyond the rooting zone are higher, but that played an important role only at the higher rainfall regimes. These simulation results confirm experimental evidence on production (data on nutrient uptake have not been identified) from literature (*cf.* Engelstad *et al.*, 1961; Frye *et al.*, 1983; Pagoulatos, 1989).

Under the influence of exploitation, especially grazing livestock (Stroosnijder, 1996; Gifford & Hawkins, 1980), but also in arable farming in (semi-)arid regions when the soil at the beginning of the rainy season is bare and hence exposed to the kinetic energy of high-intensity rainfall, soil compaction may take place. The physical consequence of soil compaction is that the volume ratio of the solid fraction and the voids in the soil increases, *i.e.* both its potential oxygen and water content decrease. As a result, the infiltration capacity of the soil surface decreases, increasing the risk for run-off, and hence for soil erosion (Hoogmoed & Stroosnijder, 1984; Stroosnijder & Hoogmoed, 1984). Under high exploitation pressure, a negative spiral may set in, *i.e.* compaction (crust formation at the soil surface) leads to reduced infiltration, resulting in lower moisture availability for the vegetation, and lower production. When the grazing pressure is not adapted, a larger proportion of the vegetation will be grazed, leading to reduced protection at the beginning of the rainy season, accelerated compaction, *etc.* This eventually may lead to complete 'destruction' of the soil as a medium for plant growth (Rietkerk, 1998). In compacted soils, root development may be hampered, resulting in impaired uptake of water and nutrients (Schuurman, 1965). This effect may be aggravated by lack of oxygen that hampers root functioning. Predicting the degree of soil compaction under specified exploitation pressure is difficult, hence incorporation of these effects in cropping systems model is speculative (*cf.* Stroosnijder, 1996).

Soil erosion, as one of the components of soil degradation, did receive ample attention in the 1930's, when especially wind erosion threatened and actually destroyed extensive agricultural areas in the Great Plains of the US ('dust bowl', *cf.* Oveson & Besse, 1967). Intensification of agriculture in the Western world after World War II largely halted the process, and attention only increased again in the 1970's when, as a result of population growth in the developing countries, more and more fragile soils were brought under cultivation (Erenstein, 1999; Lal, 1984; Stocking, 1984). The attention for soil erosion remained in the 1980's when it became increasingly clear that low soil fertility was a major constraint to crop production and hence to agricultural development in developing countries (Penning de Vries & Djitèye, 1982). The attention then shifted from soil loss to the loss of soil nutrients through erosion (*cf.* de Jager *et al.*, 1998). This was also the time that modelling of soil erosion received attention. Various attempts have been made to develop physically based soil erosion models (Williams *et al.*, 1989; Littleboy *et al.*, 1992a; 1992b; Williams, 1994). However, until today, the Universal Soil Loss Equation (USLE), developed by Wischmeier and co-workers (Wischmeier & Smith, 1978), is still widely used for estimation of soil loss and analysis of the effects of alternative technologies. Soil loss through erosion is still difficult to estimate accurately (Wischmeier, 1976), and further research in this area is absolutely necessary. However, as other components in calculations of dynamics of soil qualities suffer from similar inaccuracies, the USLE for calculating soil loss at least provides a consistent picture.

In terms of soil quality, the 'indicator' that has received by far the greatest attention in the last two decades is 'soil nutrient content'. It is often further subdivided in the macro-nutrients nitrogen (N), phosphorus (P) and potassium (K), and soil organic matter (or soil carbon) (*cf.* Stoorvogel & Smaling, 1998; Breman, 1998). It might be questioned whether the process of nutrient dynamics should be equated to 'soil degradation', as the extraction of nutrients from the soil in harvested products is the very essence of agro-ecosystems (Conway, 1987). Since the decisive work of von Liebig (1855), it has unequivocally been established that inorganic nutrients are necessary for the production of organic components by plants. As nutrient dynamics play such an important role in determining the production capacity of the soil, the soil quality indicator 'nutrient content' has gained a rightful place of its own in land use descriptions.

In practically all situations the 'natural supply' of (especially the macro) nutrients is insufficient for realisation of the production potential of agricultural crops.

Hence, fertilisers have to be applied to complement these natural sources. The relation between fertiliser application and crop yield is treated in detail in Section 2.6. In relation to 'soil degradation', emphasis should then be placed on the fact that, 'in the end', crop yields will be determined by the natural supply of plant nutrients, when no nutrients are supplied from outside the system.

In conclusion, it appears that 'soil degradation' is such an elusive term from an agro-technical point of view, that for operational use, it should be explicitised in terms of soil quality indicators. Next, it is necessary to define desirable threshold values and ranges for these indicators.

2.3 Economic Interpretation of the Production Function

The debate between economists and agronomists about the production function can be clarified when looking at the mathematical representation of that function. Economists and agronomists have different paradigms and therefore have different points of departure. To make the comparison with biophysical interpretations of the production function, a basic economic interpretation is presented in this section.

In agricultural-economic models a production function is commonly postulated:

$$q = q(L, A, K, X, \tau) \tag{2.1}$$

where q is production, L labour, A land, K capital, X variable inputs and τ the technology level. The equation is continuous (quasi-)concave. The Cobb-Douglas production function satisfies this formulation (Chambers, 1988), implying substitutability of inputs.

One way of describing the change in soil quality $[\dot{s}]$ was first proposed by McConnell (1983). It basically entails describing the change in soil quality as a function $[h]$ of yield increasing[1] $[X_1]$ and soil conserving inputs $[X_2]$. Production is a function of soil quality, and yield increasing and soil conserving inputs. Analysing the first order conditions of the ensuing system of equations[2] under different assumptions about the behaviour of the partial derivatives, gives analytical insight in the relationship between household behaviour and processes of soil degradation. Examine the basic function for change in soil quality, suppressing time indices:

[1] In economic literature the yield increasing input is equated with soil quality decreases, while in fact many technologies exist that are both yield increasing and soil quality enhancing.

[2] Subject to an objective function and resource constraints. Since we only look at the production side of the model those equations are not developed here (also see Kruseman, this volume).

$$\dot{s} = h(X_1, X_2) \tag{2.2}$$

where, $h_{x1} \leq 0$, $h_{x1x1} \leq 0$, $h_{x2} > 0$, $h_{x2x2} < 0$. In this specification, h denotes a twice continuously differentiable function with homothetic properties. The production function becomes (in this simple version, production is only dependent on variable inputs):

$$q = q(s, X_1, X_2) \tag{2.3}$$

with $q_s > 0$, $q_{ss} < 0$, $q_{x1} > 0$, $q_{x1x1} < 0$, $q_{sx1} = q_{x1s} > 0$, $q_{x2x2} < 0$, and for all $s \leq s^{\#}$ $q(s,X_1,X_2) = 0$. The partial derivatives involving X_2 are indeterminate because the effect of soil conservation on short-term crop productivity is unknown. Note that a threshold level for soil quality $s^{\#}$ exists, below which production is impossible. This threshold value represents the minimum soil quality required for production. It does not, however, indicate the threshold value below which soil degradation is irreversible.

Analysing the production function under different assumptions of the effect of soil conservation measures on crop productivity and under different objective functions, the relationship between land degradation and economic behaviour is explored. In reality, however, $\boldsymbol{X_1}$ and $\boldsymbol{X_2}$ are vectors of inputs that are not independent with respect to their effect on change in soil quality, thus falsifying the basic assumptions.

An alternative way of looking at the process of soil degradation is to assume the same basic function without restrictive assumptions about functional form. Though this may seem appealing from a biophysical point of view, it renders the system of equations unsolvable analytically. This type of approach is especially appealing in situations with many crops and many alternative technologies. Such multiple land use options are difficult to fit into the simple analytical model presented in Equations (2.2) and (2.3).

The dynamic model (McConnell, 1983; Barrett, 1991; Shiferaw, 1998) in which the vector of inputs is divided in two parts $\boldsymbol{X_1}$ which is for production purposes and $\boldsymbol{X_2}$ which is for conservation purposes can be extended with a coefficient for technology level τ:

$$q = q(L, A, K, \mathbf{X}_1, \mathbf{X}_2, \mathbf{s}, \tau) \tag{2.4}$$

where, $\boldsymbol{s}$ is a vector of soil quality attributes[3]. The dynamics enter the model through the equation:

$$\dot{\mathbf{s}} = s(\mathbf{X}_1, \mathbf{X}_2, \tau) \tag{2.5}$$

[3] A subset of these attributes may contain soil organic matter content, soil nitrogen, phosphorus and potassium, soil acidity, maximum rooting depth, *etc.*

This model still implies substitutability of inputs, which is not attractive from the agro-technical point of view, since not all inputs are interchangeable. Substitutability of inputs in the production function can be avoided by using Leontief type input-output relations. The fixed input-output ratios used in this approach reduce the mathematical reliance on non-existent substitution mechanisms. However, the Leontief production function does not satisfy a number of criteria that are needed for analytical mathematical analysis. It does not satisfy strict monotonicity, nor is it twice–continuously differentiable.

The approaches presented above deal with the primary production function. In economic production function analysis the dual approach is at least as popular (Diewert, 1971; Chambers, 1988). The fundamental assumption of the dual approach is that households optimise some known objective function. Empirical evidence on technology choice and production factor allocation represents the subjective interpretation of allocative efficient points on the unknown production function. Econometric analysis of the reduced form equations does not uncover the underlying production function, but yields a functional relationship between variables and the allocative efficient points. Hence, the dual approach does not offer the required interface with biophysical sciences to analyse and assess the possible effects of (new) technology on household decisions.

2.4 Fundamental Issues Underlying the Debate

The fundamental issues underlying the debate between economists and agronomists with respect to the production functions are related to differences in paradigms. Where economists are interested in the decisions of farm households and the criteria to select among different technologies and different levels of inputs to obtain desirable outputs, agronomists are interested in the processes that determine the yields (outputs) and the externalities.

Within their own fields both types of models are useful. Difficulties arise when an attempt is made to link biophysical and economic models in an integrated framework. Two concepts are important to solve these problems: the concept of efficiency, that is used in very different ways by different scientists and the conceptualisation of cause and effect in the analysis of production systems.

2.4.1 Efficiency

Efficiency is interpreted differently by economists and biophysical scientists and therefore merits special attention. In plant production science, agriculture has been defined as the production of useful organic materials with the sun as the source of energy (de Wit, 1992), whereas by agricultural economists it is regarded as the production of desirable and undesirable outputs with scarce land, labour and capital resources. The relative scarcity of a resource determines the importance of using it efficiently. In agricultural production, resource use is directly linked to agro-ecological sustainability (Kruseman *et al.*, 1996). Sustainability implies that short-term extractions from the stocks of natural resources do not exceed regenerative capacities. Short-term fluctuations are admissible, provided that long-term stability is not threatened. Given the pressure on the acceptable level of resource use, research is directed towards finding ways to maintain or increase utility (not necessarily production) without jeopardising the available stock of natural resources. In this context, one of the key concepts is efficiency.

Sometimes efficiency is directly and positively correlated to sustainable resource use, as in ecological efficiency (Dover & Talbot, 1987) or in resource use efficiency (Wit, 1990; 1992). It may be correlated positively and indirectly to sustainable resource use, as in light use efficiency (Monteith, 1981) and water use efficiency (van Keulen & Van Laar, 1986). Sometimes it is not necessarily correlated to sustainable resource use, as in the X-efficiency concept (Leibenstein, 1979), or in the concept of economic efficiency.

In some of the discussions on agro-ecological sustainability, efficiency is used as a criterion for evaluating system performance. Efficiency concepts hinge on felt scarcity. It is related to the quantity of an input needed to obtain desirable outputs (quantity and quality), often linked to an objective of minimising scarce resource use, or its dual, maximising output. All these concepts of efficiency, although applied to diverse fields, have in common that they relate to production and productivity, expressed in comparison to the *costs* of maintaining production and/or productivity, *i.e.* a ratio.

Efficiency, production and productivity, though widely used in many disciplines, are often ill-defined, if at all, and even within disciplines their definitions are ambiguous. In this section the many definitions are treated explicitly, to provide an unambiguous framework for communication among disciplines.

Strictly speaking in an agro-ecological sense, production or output refers to total 'losses' of matter and/or energy from the system, including losses through erosion, leaching, volatilisation, theft, *etc.*, and useful losses in terms of primary product and harvested residues. Reported production data, in general, only cover the partial losses in biomass production. For economic analysis, production may be expressed in physical terms or in monetary terms. In analyses of 'sustainability', defined in terms of changes in the stock of natural resources, the output of a system includes, in addition to economically useful biomass, also changes in the stock of resources. The latter are often expressed in terms of 'sustainability indicators'.

Levels of production are related to technology, *i.e.* output is determined by input, and technology can be defined as a given combination of inputs (including factor inputs)[4]. An increase in production can be achieved either by proportionally increasing input use or by changing the technology, *i.e.* modifying the combination of inputs.

While *production* refers to physical or monetary output, without necessarily specifying the quantities of inputs used, *productivity* describes the output in terms of the inputs used. A system can be described in terms of various types of productivity, dependent on the type of output (grain, straw, manure) it produces and on the inputs it refers to (labour, land, capital, fertiliser, pesticides). The most common concept is land-time productivity, *i.e.* yield, expressed in terms of valued output during one growing season per unit of land on which the crop is grown. Note that, in addition to the ratio of output to input, also a time element is involved. Productivity[5] is a partial measure, since it refers to type of output per type of input. Increased production through technology change implies increased productivity of some of the inputs, but not necessarily of all. Technology change may even imply a productivity decrease for less scarce inputs.

While productivity is expressed in ratios of dissimilar quantities (inputs and outputs), efficiency is always expressed as a (unitless) ratio of quantities expressed in the same units. Expressing that ratio as a percentage reduces the risk of using the term efficiency to denote some other concept (usually productivity). We define efficiency as an effective operation as measured by a comparison of production with cost (as in energy, time and money). Efficiency, in general, can be defined as the ratio of an actually realised value and the maximum attainable value. The *ceteris paribus* condition commonly assumed by economists, is a prerequisite in this definition.

As a reaction to the difficulty of meeting the *ceteris paribus* condition, use of the concept of relative efficiency has become common, and thus definitions can proliferate, since the benchmark is no longer fixed. The single condition that has to be satisfied in all cases is that the measured efficiency should be unitless, and it can therefore be expressed as a percentage increase over some explicitly defined benchmark. Its value then expresses the degree of realisation of potential performance under the *ceteris paribus* condition. The term technical efficiency in the following definitions refers to comparison to the maximum attainable productivity for a given technology, and linking this to price efficiency (Farrel, 1957). Ellis (1988) defines technical efficiency as actual output (O_{act}) divided by maximum output (O_{max}) for a given level of inputs (I):

---

[4] Embodied technology in inputs fits into this definition.

[5] Often, productivity is erroneously equated to efficiency, c.f. 'Efficiency refers to the way production factors or inputs are combined to produce outputs, i.e. the conversion ratios of each input into output.' (Fresco & Kroonenberg, 1992). This is an excellent definition of productivity but not of efficiency.

$$\frac{O_{act}}{O_{max}} | I . \tag{2.6}$$

Coleman and Young (1989) define technical efficiency in terms of a given output level:

$$\frac{I_{min}}{I_{act}} | O . \tag{2.7}$$

Superficially, there is no conceptual difference between the two definitions, especially when considering the point of highest efficiency. The Ellis definition is especially applicable in analysing situations where input levels are more or less fixed (due to socio-economic or institutional constraints), the Coleman and Young definition in situations where output levels have fixed targets (due to market segmentation, imperfect demand, or inelastic household subsistence food-security requirements).

Economic efficiency is partitioned in technical efficiency, leaving determination of the benchmark values to the biophysical sciences, and allocative efficiency. Allocative efficiency determines whether, on the technically efficient frontier, outputs are obtained and resources are used in accordance with their scarcity values, *i.e.* prices.

The *ceteris paribus* condition is especially important here, as a biophysical example will clarify. In many situations, the ratio between crop nitrogen (N) uptake and N-fertiliser application exceeds 100%, because crop nitrogen includes uptake from 'natural sources', such as the organic nitrogen store in the soil. If the contribution from natural sources can not be quantified, not even by correcting for measured changes in N content in the soil, observed N uptake efficiency represents an *apparent* efficiency.

The graphical presentation (Fig. 2.1) of the relationships between fertiliser application and N-uptake, N-uptake and yield, and fertiliser application and yield (de Wit, 1992) combines the concepts of productivity (quadrants I and II) and efficiency (quadrant IV). As growing conditions improve from situation 1 via 2 to 3, maximum yield increases, N-availability from natural sources increases ($s_1 < s_2 < s_3$), and fertiliser recovery increases ($a_1 < a_2 < a_3$).

There is a clear relationship between the concept of technical efficiency used by Coleman and Young and the results presented in Fig. 2.1. In quadrant I, beyond point *a* of curve 1, no yield increase can be expected with more nitrogen uptake. Hence, it is inefficient to fertilise beyond that point.

De Wit (1992), referring to the difficulty of the *ceteris paribus* condition, indicates that (total) resource use efficiency hinges on the concept that there are no decreases and at least some increases in resource productivity associated with improved growing conditions. However, the unmeasured variability in growing conditions in empirical data makes it difficult to determine benchmark technical efficiency values that hold at farm household level.

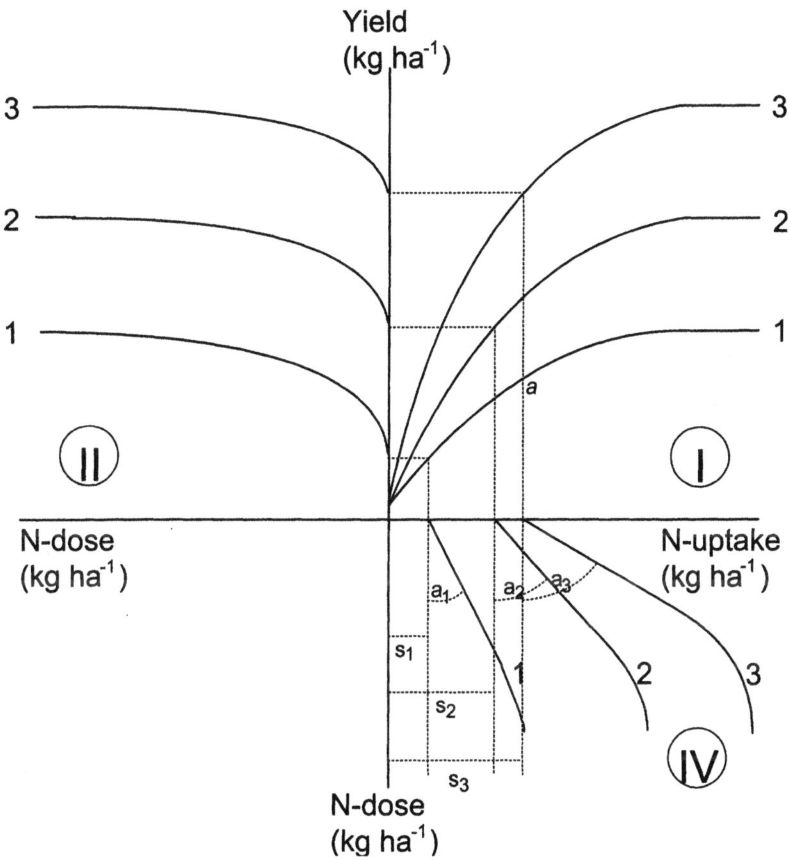

Figure 2.1: Graphical presentation of the relations between yield, nutrient uptake and nutrient application.

2.4.2 Cause and Effect

As explained in Sections 2.2 and 2.3, soil degradation can be considered as the cause of declining yields, or as the consequence of agricultural practices, or as an integral part of agricultural production. In the latter case, the question of cause and effect is no longer relevant, because the processes are endogenised. These considerations determine which relationships and feedbacks are taken into account.

Yield response functions in which soil quality is included as a determining variable, consider soil degradation as a cause of yield decline. In biophysical models, the term *growing conditions* is often used to qualify differences in the maximum attainable output level. These growing conditions include soil quality. Changes in soil quality imply changes in the growing conditions and by implication changes in the yield plateau.

Incorporating environmental information directly into production function analysis, using econometric techniques is becoming increasingly popular (Mausolff & Farber, 1995; Pattanayak & Mercer, 1998; Byiringiro & Reardon, 1998). In this approach, soil degradation is considered as a cause of declining agricultural production. The main problem in this approach is that it is very difficult to unequivocally determine the exact effects of soil degradation on yields in the short-run. Comparative studies of more or less severely eroded phases in the same locality may show the difference in yield between slightly and severely eroded soils (Weesies *et al.*, 1994; Olson *et al.*, 1994), but do not provide the information to quantify the dynamic effects of soil degradation on yields.

In the analysis of agronomists, soil degradation as a result of agricultural practices is often expressed in terms of macro-nutrient and soil organic matter balances (Section 2.2). Extracting nutrients from the soil store (soil mining) is an income-generating process. In an intertemporal framework, soil mining can be considered transfer of income from the future to the present (van der Pol, 1992).

Some economic studies also look at soil degradation as the result of land use activities. The focus in these studies is on trade-offs between income and soil loss (Veloz *et al.*, 1985; Cárcamo *et al.*, 1994). In these studies, USLE (Section 2.2) is used to quantify degradation.

Integrated approaches look at both cause and effect, without specifying *a priori* which process is most important. The analytical optimal control models (McConnell, 1983; Barrett, 1991) demonstrate an economic model in which soil quality is both cause and effect of agricultural production. Biophysical models that use an integrated approach are the technical coefficient generator (Section 2.7) and the similar, albeit more dynamic, approach used by Struif Bontkes (1999).

2.5 Biophysical Approaches to the Production Function

In the system-analytical approach to agricultural production, animal production and crop growth are calculated on the basis of explanatory models, that quantitatively describe the underlying processes (Stoorvogel *et al.*, 1998; Bouman *et al.*, 1996; Dent *et al.*, 1996). Soil quality models thus use the causal mode of explanation. The relations between the numerous input and output variables, describing the relevant soil, animal, and plant processes are, however, not undisputed

There is an ongoing debate that started in the late nineteenth century on the relation between inputs and yields (centering around plant nutrients) and to what extent synergistic effects are important (de Wit, 1992). The basic assumption in all models used to describe this relation is diminishing returns to input use. The debate revolves around the functional form of the yield function. Location-(soil and weather)specific and crop-(physiological, physical and phenological)specific characteristics and their interactions determine the maximum attainable production level, or production potential (generally this is expressed as 'yield potential', *i.e.* maximum production of economic products such as grains, tubers, *etc.*). These interactions determine such characteristics as length of the crop growth cycle, total radiation interception, total water use, distribution of total dry weight between economic product and crop residues, *etc.*

The main question for growth under sub-optimal conditions is to what extent yield potential is reduced. Schematically, three theories can be distinguished on the influence of input deficits on attainable yields, often referred to as the 'Liebig (1855)', 'Mitscherlich (1924)' and 'Liebscher (1895)' approaches, respectively (Nijland & Schouls, 1997). The main issue in the debate is how the slope changes with increasing availability of the limiting growth factor. Liebig postulated a constant initial slope, irrespective of growing conditions, with differentiation only towards the maximum yield level. Liebscher suggested a steeper slope under more favourable growing conditions. Mitscherlich hypothesised that the slope was proportional to the maximum attainable yield level.

In the hypothetical situation, illustrated in Fig. 2.2 (equivalent to Quadrant II in Fig. 2.1), curve 1 depicts the relation between nitrogen fertiliser application and yield for 'poor' soils (low in water holding capacity). The other curves referring to soils with higher soil organic matter contents, describe the response according to Liebig (curve 2), Liebscher (curve 3), and Mitscherlich (curve 4), respectively.

What are the consequences of the different functional forms illustrated here? All three represent decreasing marginal returns, *i.e.* yields tend towards an asymptotic maximum plateau level (Aune & Lal, 1995; van den Boom & Langeveld, 1992). This level is the maximum attainable production (yield) level. When all inputs are available in sufficient quantities, this maximum attainable production level is the potential yield, defined as the level determined by crop physiological and phenological properties and the environmental conditions temperature and radiation, that can not easily be modified (van Keulen, 1986; Penning de Vries, 1982).

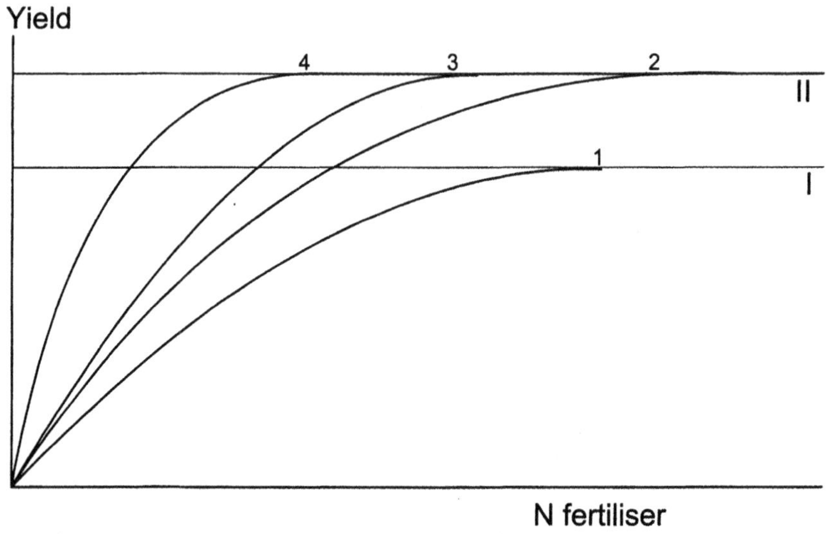

Figure 2.2: Yield response curves under two sets of growing conditions, I and II.

The Liebig approach implies absence of any substitution possibilities between, for instance, nitrogen and phosphorus, because these elements have distinctly different functions in the plant (Penning de Vries & Van Keulen, 1982). The Liebscher and Mitscherlich models allow for at least partial substitution. De Wit (1992) argues that the response curve in fact is S-shaped, and is closer to the Liebscher and Mitscherlich formulation. Substitution is not complete, hence growth factors (nutrients, water, energy) are complementary, *e.g.* the efficiency of nutrient use for a particular nutrient is affected by the level of availability of all other nutrients. The S-shaped curve reflects that agriculture is more difficult when less of the determining processes can be controlled, *i.e.* when many production factors are available in limited amounts, and timing of their availability depends on external (non-controllable environmental) conditions. Such a situation leads to both low yields and inefficient use of all production resources. This close link between yield level and efficiency of resource use is supported by the results of a comparison of pre- and post 'green revolution' production systems in the Punjab (Mundlak, 1992).

Under rain-fed conditions, water is the main limiting factor, if all other inputs are abundantly available. This situation can be adequately described in crop growth models for calculating water-limited yields (Day *et al.*, 1992; Stroosnijder & Kiepe, 1998). However, in the course of the growing season (determined by the availability of water), especially on poor soils, severe macro-nutrient shortages may occur, *i.e.* these other growth factors (inputs) may be more limiting than water (van Keulen & Breman, 1990). When addressing the agro-ecological sustainability of farming systems, it is impossible to maintain yields without balanced fertilisation as dictated by the complementarity of nutrients. This does not imply that only the law of Liebig holds and that there are no positive returns, but that input combinations are only sustainable within a limited range (Penning de Vries & Van Keulen, 1982).

Outside that range, relative abundance of one nutrient from 'external' sources, will lead to mining of another, which can be avoided only when all nutrients are abundantly supplied from external sources. The implication of this view is that, in theory, the effect of soil quality on yields and that of yields on soil quality, are tightly linked (*cf.* Müller & Reiher, 1966). In practice, however, empirical evidence to quantify these relations is missing, because, in the short-run, the effects of soil-mining are lost in the random variation of yields due to climatic variability, varying incidence of pests and diseases, and unquantifiable effects of differences in management. But ample evidence exists, that in the long-term, soil-mining leads to declining yields (*cf.* Pichot *et al.*, 1981).

Two types of biophysical models of crop growth and soil quality are introduced here. The first calculates *sets* of unique input-output coefficients. These Leontief production technologies can also be interpreted as points on an n-dimensional production function of unknown functional form. Or they can be interpreted as technically efficient production technologies, using a loose definition of *best technical means* (Section 2.7), while varying the output goals. This type of model originates in the systems analytical approach to crop growth and crop-soil interactions. The second type assumes a more narrow view on soil processes, and considers erosion as the major process contributing to soil degradation. Soil loss, reducing soil depth, used as the main soil quality indicator, is assumed the main effect. Crop yields are a function of inputs and soil depth, which in turn, is a function of soil conservation measures, crop yield in the preceding growing season and fertiliser application. This model is attractive because of its simplicity and ease of implementation in bio-economic models.

With the first type of model, a large number of point data are generated, that are introduced in the mathematical programming model as technical coefficients; in the second approach average yields are used, with standard, albeit sometimes linearly segmented, reduction and addition factors for negative and positive growth factors. Efficiency of nutrient use is independent of the level of use of other nutrients or water.

In the following sections, a number of quantified approaches to biophysical modelling of soil-plant interactions, suitable for use in economic models, is illustrated, to give insight in the type of approaches used. The first example is the Technical Coefficient Generator (TCG; Hengsdijk *et al.*, 1996) developed for the *Cercle de Koutiala*, a semi-arid region in the South of Mali (Kruseman & Bade, 1998; Kuyvenhoven *et al.*, 1998). Different models, also using a systems analytical approach, for tropical soils, are combined in the *Tropical Soil Productivity Calculator* (Aune & Lal, 1995). An example of the second type of biophysical modelling is EPIC that has been applied in household and village modelling (Barbier, 1994; Barbier & Bergeron, 1999).

2.6 Generating Biophysical Inputs in Bio-Economic Modelling

2.6.1 A Technical Coefficient Generator (TCG)

In a TCG, discrete input-output combinations are generated (Hengsdijk *et al.*, 1996, Hengsdijk *et al.*, 1998a). These points are introduced directly in a mathematical programming framework, hence, there is no need to specify a production function. To assess the strengths and weaknesses of the modelling approach, it is necessary to make the assumptions and procedures explicit.

Table 2.1: Characteristics of current and alternative crop activities for Koutiala.

Alternative crop activities	**Current crop activities**
1. Maximum yields based on water-limited production	1. Yields, nutrient and manure input based on empirical data.
2. Maximum of four yield levels per crop	2. Maximum of three yield levels per crop
3. OM- and nutrient balances ≥ 0	3. OM- and nutrient balances generally < 0
4. Crop characteristics change with higher production level towards yield increasing characteristics	4. Crop characteristics invariable with yield level
5. Soil conservation measures include: none, simple and tied ridging	5. Soil conservation measures include: none and simple ridging
6. Erosion decreases with higher yield level due to higher crop cover factors in the USLE	6. Erosion does not depend on yield level
7. A maximum of three crop residue strategies is defined (stubble grazing/burning, harvesting and ploughing in)	7. A maximum of two crop residue strategies is defined (stubble grazing/burning and harvesting)
8. Fallow can be used as part of a crop activity to maintain soil organic matter stocks	8. Fallow defined as a separate activity independent of crop activities
9. Percentage of soil mineral N lost in drain water is 70	9. Percentage of soil mineral N lost in drain water is 60

For each unique and feasible combination of the definition criteria (defined as 'environmental and management options', Table 2.1, Hengsdijk *et al.*, 1998a), inputs and outputs are determined for crop, livestock, pasture and fallow activities. These inputs and outputs of land use activities are quantified on the basis of (i) basic information on soils, climate and crops, (ii) results of documented models, and/or (iii) expert knowledge. Assumptions and knowledge of experts are used where required insight is lacking or insufficient to develop models, which introduces subjectivity in the generation of the input and output coefficients of land use activities, but allows explicit expression of the coefficients in quantitative terms.

A TCG thus is a 'set of calculation rules' that combines data, processes and relationships to calculate the required input and output coefficients of land use activities. It is flexible and allows adjustment of data and assumptions to new information and insights, and thus permits analysis of the effects of arbitrary, but verifiable, assumptions on the quantitative characteristics of land use activities.

The guiding principle in biophysical modelling according to the production-ecological approach is the *target-oriented approach* and the framework defining production levels (van Ittersum & Rabbinge, 1997). The maximum attainable (sustainable) yields are dependent on defining factors (radiation, atmospheric carbon dioxide concentration, temperature, and crop physiological and phenological characteristics). First this yield level (or a level of nutrient or organic matter losses or emissions to the environment) is determined. Subsequently, it defines the required inputs, *i.e.* fertiliser nutrients, organic matter, labour and traction, to attain and maintain this yield in the long run (van Duivenbooden *et al.*, 1991).

Current yields are (often) lower than the potential, due to the effects of yield limiting factors (water and nutrients) and yield reducing factors (weeds, diseases, pests, and pollutants). These production levels are subsequently set as targets and (efficient) input combinations are defined to reach these targets, using the concept of *best technical means*. This concept implies that for each of the defined activities, characterised by its target, technically efficient combinations of inputs are defined. This implies minimisation of input use to attain a vector of outputs. Obviously, the point data thus attained do not necessarily correspond to allocative efficient solutions, because relative and shadow prices are not taken into account. Therefore, biophysical (agro-ecological) criteria (production orientations, van Ittersum & Rabbinge, 1997) are used for selecting relevant input-output combinations. The innovation in the TCG is the link of crop-growth modelling to farming systems research (Hengsdijk *et al.*, 1998b). This approach has been widely applied in explorative land use studies (van Ittersum *et al.*, 1998).

For the definition of current crop activities, yield level *and* inputs of nutrients and manure have been derived from experimental data (DRSPR, 1992) and are used as a 'target', to determine other inputs, such as labour and traction requirements, and outputs such as crop residues and nutrient balances. These average yields are differentiated for various agro-ecological conditions, according to water availability in different types of (rainfall) years and soil types, and assumptions on feasible crop/soil type combinations.

Hence, alternative crop activities not necessarily have higher yields than current crop activities, but they can, in theory, be practised for years to come. Current crop activities, due to their soil-mining characteristics (soil organic matter and nutrients), are not sustainable in the long run, and, with invariable management will result in gradually declining yields.

For livestock activities, a similar approach has been followed. First, a feed energy intake level is defined that determines the production levels of meat and milk. Subsequently, the requirements in terms of digestible organic matter and labour to realise these levels are determined. The inputs and outputs of livestock activities refer to a stationary herd and are expressed on a Tropical Livestock Unit (TLU)[6] basis. Herd structure depends on production level, determining the average productive life span and hence the age at replacement and production goal, linked to selling strategies of offspring (male and female calves).

Quantification of pasture activities is based on the approach of Breman & de Ridder (1991) for natural vegetation in the Sudano-Sahelian region. Quantity and quality of feed from the herbaceous layer and from woody species are estimated from total available N, derived from rainfall, taking into account drainage losses, the contribution from leguminous species and grazing intensity.

Three agro-ecological sustainability criteria for crop activities are quantified: macro-nutrient balances, soil organic matter (SOM) balance and erosion. Erosion (modified USLE, Section 2.2) affects SOM and nutrient balances directly through the loss of topsoil, which is an organic matter- and nutrient-rich substrate. Organic matter inputs interact with nutrient balances: manure and crop residue applications affect both types of balances. Organic matter content in the soil affects erosion indirectly through incorporation of a sustainability criterion.

2.6.2 A Process Approach

The dynamic ecological-economic simulation model (Struif Bontkes, 1999), is based on arbitrary decision rules, and does not entail optimisation procedures. In that model, yields do not depend on externally determined fixed input ratios, but on labour inputs and the minimum of potential, nitrogen, phosphorus, and water limited yields; nutrient limited yields are defined in terms of nutrient uptake and nutrient use efficiency. Nutrient uptake depends on supply from natural sources and inputs through fertilisers and their uptake efficiency. The interactions between SOM, macro-nutrients and erosion are comparable to those utilised in the TCG.

---

[6] A Tropical Livestock Unit (TLU) is a hypothetical adult animal of 250 kg liveweight

Macro-nutrient uptake is determined dynamically, following the procedure developed by Van Keulen (1995). The result is a complex highly non-linear dynamic system, that is difficult to implement in a mathematical programming setting.

2.6.3 A Tropical Soil Productivity Calculator (TSPC)

In a number of studies (Aune *et al.*, 1997, Alfsen *et al.*, 1997), a model based on a Baule Mitscherlich crop-growth model with a nutrient cycle, is used to model soil-plant interactions as input for economic models (Aune & Lal, 1995; Aune, 1997). The model predicts changes in productivity in the long run with time steps of one year. It describes changes in soil organic carbon, total nitrogen, pH, and soil erosion. Choice variables in the model are fertiliser and manure applications and plant protection measures. The general outline of the crop growth model, which allows for substitution is:

$$\gamma = \gamma^{max}\left(\prod F_i\right)^{\beta_1}\beta_2 \tag{2.8}$$

where, γ is yield, γ^{max} is potential yield, F_i are multiplication factors ($F_i \in [0,1]$) related to different productivity parameters, such as soil characteristics and inputs, β_1 is a correction factor for sub-optimal growing conditions, and β_2 is a site-specific correction factor. If β_1 is close to 1 the model overestimates yields for low input levels, while for β_1 is 0.5 the model underestimates yields (Aune & Lal, 1995).

The indices F_i are derived from empirical regression models, linking input levels or soil characteristics to attainable yields. Like the TCG the TSPC assumes away the variation in yields due to mis-specification, random variation and inaccurate measurement.

To describe the changes in the factor relating to nitrogen (F_N), a mathematical description of the nitrogen cycle has been developed. In the TSPC approach, a comparatively static system is used. F_N is determined by fertiliser and manure applications, decomposition of crop residues, erosion and atmospheric deposition. The model is non-linear.

In model applications (Aune *et al.*, 1997), the crop-growth model and the nitrogen cycle model are used recursively. The soil productivity parameter (F_N) in the crop growth model is determined in the nitrogen cycle model.

2.6.4 Soil Depth Calculation and Changes in Yields

In a number of bio-economic models, biophysical processes are integrated in economic models through the use of soil quality indicators (Barbier, 1994; Dalton, 1996). In these models crop-growth is not modelled as such, but the effect of changes in soil quality on yields is. This permits dynamic analysis of land use. It uses the model EPIC (Williams, 1994), that comprises a number of submodules describing hydrology, weather, erosion (based on the use of the USLE), nutrients, pesticide fate, soil temperature, crop growth, tillage and plant environment control. The last includes irrigation, drainage, ridging, fertilisation and pests and disease control.

The approach of Dalton (1996) encompasses econometric estimation of yield trends, using past data and the EPIC model. The implicit soil degradation rate is fixed in the economic model and an analysis is made of changing technology choice under these circumstances. Yield decrease and erosion are different for different technological options, but there is no feedback from previous technological choice to ensuing yield in the multi-period mathematical programming model.

In the village and watershed level models in which EPIC has been included (Barbier, 1994), crop production is considered a function of a basic yield, corrected for fertilisation, tillage, soil depth and erosion. Marginal returns to inputs are positive and decreasing. The relationship between soil quality and production is aggregated over whole farms, thus ignoring possible variability in soil quality development within a farm.

2.6.5 Multi-Period Modelling

The dynamic aspects of soil degradation are sometimes ignored as in the TCG (Hengsdijk *et al.*, 1996). Sometimes they are incorporated into a myopic decision model in which the biophysical model takes into account the dynamics of soil degradation, but the economic model does not (Barbier, 1994; Aune *et al.*, 1997; Struif Bontkes, 1999). Sometimes, the dynamics of soil degradation are fixed in the model with a steady downward trend (Dalton, 1996). Incorporating the dynamics of soil degradation into mathematical programming models in a flexible way requires novel ways of looking at the issue.

One way of dealing with these dynamics in multi-period models is the use of soil fertility classes. Irrespective of the soil model used, changes in soil quality determine to what degree a plot of land remains in its quality class (Kuiper *et al.*, 1998). In their study, three fertility classes are defined. Each Leontief-type production activity vector has additional input and output elements, comprising the fertility class it starts out with and the class it ends up in. A further refinement in differentiated soil fertility without increasing the number of classes, is the use of partial switches. Each Leontief type production activity, defined for a specific soil type and fertility class, has a destination matrix that determines to what degree it shifts to a different fertility class.

An alternative approach often employed in myopic exercises (Barbier, 1994) is to determine overall soil degradation for a given soil type and transfer the whole soil type to a different fertility class accordingly. The disadvantage of this approach is that non-degrading and degrading activities can balance, assuming a linear relationship in soil degradation. A non-aggregated approach allows for more diversification in soil fertility management. It can be argued (Barbier, 1994) that the variability in outcomes justifies the use of an aggregate approach.

2.7 The Interface Between Biophysical and Economic Models

It has been illustrated that crop-soil interactions as input for economic models can be modelled in many different ways. Many relationships in biophysical models are strongly non-linear, or even discontinuous. Bio-economic models often use mathematical programming or other simulation techniques to deal with these relationships.

To incorporate non-linear and discontinuous production functions and soil quality functions in simulation models requires simplifications in descriptions of economic behaviour, the biophysical processes or both. The use of Leontief production technologies, although disregarding (dis-)economies of scale completely, offer a way of incorporating the biophysical processes in terms of point data (Hengsdijk *et al.*, 1996), while also offering the possibility for a dynamic approach (Kuiper *et al.*, 1998). Approaches in which continuous production functions are incorporated, necessarily lead to other simplifications, either in terms of household behaviour that is disregarded (Struif Bontkes, 1999) or in forcing myopia on the decision makers (Barbier, 1994; Aune *et al.*, 1997).

The graphical representation of the production function as used by economists and biophysical scientists looks very similar. This similarity is deceiving. The curves, depicting the relationship between an input and an output (See for instance Fig. 2.2) deal with different issues, when constructed by economists or biophysical scientists. The biophysical representation signifies the technically efficient yield response for a single output with all other growing conditions *ceteris paribus*. The economic representation illustrates the allocatively efficient points under varying (unknown) production conditions, and household constraints.

Main issues that need to be resolved in relation to the quantification of biophysical processes, are: improved quantification of interactions between growth factors in crop modelling, and quantification, in a more functional way, of the dynamics of soil degradation, the effects of crop growth on soil quality, and the short-term effects of soil quality on crop growth. That may require a departure from present scientific methodology, towards specification of economic boundary conditions for functional relationships between the main biotechnical variables. Application of advanced econometric estimation techniques to available empirical data sets might yield results that can be used in bio-economic modelling in the absence of full specification of all relevant biophysical processes. The use of stochastic frontier production functions (Battese & Broca, 1995; Coelli, 1995), or non-parametric efficiency analyses based on Data Envelopment Analysis (Silva, 1996;

Silva & Stefanou, 1996), are among the possibilities of combining advanced quantitative econometric techniques with quantitative production ecology.

Further development of bio-economic models faces two major challenges. On the one hand, conceptualisation of the production function and soil quality change functions used in household models evidently should reflect biophysical realities. Modelling the complex interactions and the synergy between various inputs is an ongoing challenge. On the other hand, the analyses in the biophysical sciences need to incorporate the *second-best world,* that is the reality faced by farm households in the areas for which these bio-economic models are being developed. Where economists must be more rigorous in defining the relationships between inputs and outputs, biophysical scientists need to be more rigorous in defining the boundary conditions that guide their analyses.

2.8 References

Alfsen, K.H., T. Bye, S. Glomsrød & H. Wigg (1997): Soil degradation and economic development in Ghana. Environment and Development Economics 2, 119-143

Aune, J.B. (1997): Predicting soil degradation in Tanzania - a system analysis approach. Norwegian Journal of Agricultural Sciences, supplement, 7-60

Aune, J.B. & R. Lal (1995): The tropical soil productivity calculator - a model for assessing effects of soil management on productivity. In: R. Lal & B.A. Steward (Eds.) Soil management: experimental basis for sustainability and environmental quality. CRC Lewis Publishers, Boca Ratton

Aune, J.B., S. Glomsrød, V. Iversen & H. Wiig (1997): Structural adjustment and soil degradation in Tanzania: a CGE model approach with endogenous soil productivity. Statistics Norway discussion paper no. 189. Statistics Norway, Oslo

Barbier, B. (1994): Modelisation agronomique et economique de la durabilité d'un système agraire villageois. Ph.D. thesis, CIRAD-SAR, Montpellier

Barbier, B. & G. Bergeron (1998): Natural resource management in the hillsides of Honduras: bioeconomic modeling at the micro-watershed level. EPTD discussion paper no. 32. IFPRI, Washington D.C.

Barrett, S. (1991): Optimal soil conservation and the reform of agricultural pricing policies. Journal of Development Economics 36, 167-187

Battese, G.E. & S.S. Broca (1995): Functional forms of stochastic frontier production functions and models for technical inefficiency effects: a comparative study for wheat farmers in Pakistan. Paper presented at the New England Conference on efficiency and productivity, University of New England, 23-24 November 1995

Bouman, B.A.M., H. van Keulen, H.H. van Laar & R. Rabbinge (1996) The 'School of de Wit' crop growth simulation models: A pedigree and historical overview. Agricultural Systems 52, 171-198

Breman, H. (1998): Soil fertility improvement in Africa, A tool for or a byproduct of sustainable production? African Fertilizer Market 11, 2-10

Breman, H. & N. de Riddder (1991): Manual sur les pâturages des pays sahéliens. ACCT-CTA-Karthala, Paris

Byiringiro, F. & T. Reardon (1996): Farm productivity in Rwanda: effects of farm size, erosion, and soil conservation investments. Agricultural Economics 15, 127-136

Cárcamo, J.A., J. Alwang & G.W. Norton. (1994). On-Site Economic Evaluation of Soil Conservation Practices in Honduras. Agricultural Economics 11, 257-69

Chambers, R.G. (1988): Applied production analysis. Cambridge University Press, Cambridge

Coelli, T.J. (1995): Recent developments in frontier modelling and efficiency measurement. Australian Journal of Agricultural Economics 39, 219-245

Coleman, D. & T. Young (1989): Principles of agricultural economics: markets and prices in less developed countries. Cambridge University Press, 50-59, Cambridge

Conway, G.R. (1987): The properties of agroecosystems. Agricultural Systems 24, 95-117

Dalton, T.J. (1996): Soil degradation and technical change in Southern Mali. Ph.D. thesis, Pudue University, West Lafayette

Day, J.C., D.W. Hughes & W.R. Butcher (1992): Soil, water and crop management alternatives in rainfed agriculture in the Sahel: an economic analysis. Agricultural Economics 36, 267-287

Dent, J.B., M.J. McGregor & A.R. Sibald (Eds.) (1996): Livestock farming systems: research, development, socio-economics and the land manager. EAAP Publication No. 79, Wageningen Pers, Wageningen

Diewert, W.E. (1971): An application of the Shepard Duality Theorem: a generalised Leontief production function. Journal of Political Economy 79, 481-507

Dover, M. & L.M. Talbot (1987): To feed the earth: agro-ecology for sustainable development. World Resources Institute, Washington

DRSPR (1992): Synthèse des résultats de la campagne 1991/1992. DRSPR, Sikasso

Duivenbooden, N. van, P.A. Gosseye & H. van Keulen (Eds.) (1991): Compétition pour les ressources limitées: le cas de la cinquième région du Mali. Rapport 2. Productions végétales, animales et halieutiques. CABO/ Mopti: ESPR, Wageningen

Ellis, F. (1988): Peasant economics. Cambridge University Press, Cambridge

Engelstad, O.P., Schrader, W.D. & L.C. Dumenil (1961). The effect of soil surface thickness on corn yield. I. As determined by a series of field experiments in farmer operated field. Soil Science Society of America Proceedings 25, 494-497

Erenstein, O.C.A. (1999): The economics of soil conservation in developing countries: The case of crop residue mulching. Ph. D. Thesis, Wageningen University

Farrel, M.J. (1957): The measurement of productive efficiency. Journal of the Royal Statistical Society (ser. A, General) 120, 253-281

Fresco, L.O. & S.B. Kroonenberg (1992): Time and spatial scales in ecological sustainability. Land Use Policy 9, 155-168

Frye, W., L.W. Murdock & R.L. Bevins (1983): Corn-yield fragipan depth relations on Zanesville soils. Soil Science Society of America Journal 47, 1043-1045

Gifford, G.F. & R.H. Hawkins (1980): Hydrologic impact of grazing on infiltration: A critical review. Water Resources Research 14, 305-313

Hengsdijk H., W. Quak, E.J. Bakker & J.J.M.H. Ketelaars (1996): A Land Use System and Technology Generator for the Koutiala Region in South Mali. DLV Report No. 5, AB-DLO/PSS, Wageningen:

Hengsdijk, H., A. Nieuwenhuyse & B.A.M. Bouman (1998a): LUCTOR: Land Use Crop Technical coefficient generatOR; version 2.0. A model to quantify cropping systems in the Northern Atlantic Zone of Costa Rica. Quantitative Approaches in Systems Analysis No. 17, AB-DLO/TPE, Wageningen

Hengsdijk H., M.K. van Ittersum & W.A.H. Rossing (1998b): Quantitative analysis of farming systems for policy formulation: Development of new tools. Agricultural Systems 58, 381-394

Hoogmoed, W.B. & L. Stroosnijder (1984): Crust formation on sandy soils in the Sahel. I. Rainfall and infiltration. Soil & Tillage Research 4, 5-23

Ittersum, M.K. van & R. Rabbinge (1997): Concepts in production ecology for analysis and quantification of agricultural input-output combinations. Field Crop Research 52, 309-330

Ittersum, M.K. van, R. Rabbinge & H.C. van Latesteijn (1998): Exploratory land use studies and their role in strategic policy making. Agricultural Systems 58, 309-330

Jager, A. de, S.M. Nandwa & P.F. Okoth (1998): Monitoring nutrient flows and economic performance in African farming systems (NUTMON). I. Concepts and methodologies. Agriculture, Ecosystems and Environment 71, 37-48

Keulen, H. van (1986): A simple model of potential crop production. In: H. van Keulen & J. Wolf (Eds.) Modelling agricultural production: weather, soils and crops. Simulation Monographs, Pudoc, Wageningen

Keulen, H. van (1995): Sustainability and long-term dynamics of soil organic matter and nutrients under alternative management strategies. In: J. Bouma, A. Kuyvenhoven, B.A.M. Bouman, J.C. Luyten & H.G. Zandstra (Eds.). Eco-regional approaches for sustainable land use and food production. Kluwer Academic Publishers, Dordrecht

Keulen, H. van & H. Breman (1990): Agricultural development in the West African Sahelian region: a cure against land hunger? Agriculture, ecosystems and environment 32, 177-197

Keulen, H. van & H.H. van Laar (1986): The relation between water use and crop production. In: H. van Keulen & J. Wolf (Eds.). Modelling agricultural production: weather, soils and crops. Simulation Monographs, Pudoc, Wageningen

Kruseman, G. (2000): Assessing the effects of policy measures on household welfare and agro-ecological sustainability: an overview of farm household modelling approaches. This volume

Kruseman, G. & J. Bade (1998): Agrarian policies for sustainable land use: bio-economic modelling to assess the effectiveness of policy instruments. Agricultural Systems 58, 465-481

Kruseman, G., H. Hengsdijk, R. Ruben, H. van Keulen & A. Kuyvenhoven (1996): Disentangling the concept of sustainability. Agricultural Systems 50, 191-207

Kuiper, M., R. Schipper & E. van Ierland (1998): Incorporating erosion in a multi-period model: an activity specific approach. Paper presented at the fifth biannual conference of the International Society of Ecological Economics Beyond growth: policies and institutions for sustainability, November 15-19, 1998, Santiago de Chile

Kuyvenhoven, A., R. Ruben & G. Kruseman (1998): Technology, market policies and institutional reform for sustainable land use in southern Mali. Agricultural Economics 19, 53-62

Lal, R. (1984): Soil erosion from tropical arable land and its control. Advances in Agronomy 37, 183-248

Leibenstein, H. (1979): X-efficiency: from concept to theory. In: Challenge Sept-Oct, 13-22

Liebig, J. von (1855): Die Grundsätze der Agriculturchemie mit Rücksicht auf die in England angestellten Untersuchungen. Braunschweig

Liebscher, G. (1895): Untersuchungen über die Bestimmung des Düngerbedürfnisses der Ackerböden und Kulturpflanzen. Journal für Landwirtschaft 43, 49-216

Littleboy, M., D.M. Silburn, D.M. Freebairn, D.R. Woodruff, G.L. Hammer & J.K. Leslie (1992a): Impact of soil erosion on production in cropping systems. I. Development and valdation of a simulation model. Australian Journal of Soil Research 30: 757-774

Littleboy, M., D.M. Freebairn, G.L. Hammer & D.M. Silburn (1992b): Impact of soil erosion on production in cropping systems. II. Simulation of production and erosion risks for a wheat cropping system. Australian Journal of Soil Research 30, 775-788

Mausolff, C. & S. Farber (1995): An economic analysis of ecological agricultural technologies amongst peasant farmers in Honduras. Ecological Economics 12, 237-248

McConnell, K.E. (1983): An economic model of soil conservation. American Journal of Agricultural Economics 65, 83-89

Mitscherlich, E.A. (1924): Die Bestimmung des Düngerbedürfnisses des Bodens. Paul Parey, Berlin

Monteith, J.L. (1981): Does light limit crop production? In: C.B. Johnson (Ed.) Physiological processes limiting plant productivity. Butterworths, London

Müller, P. & W. Reiher (1966): Auswertung langjahriger Ertragsfeststellung (dargestellt am Versuch 'ewiger Roggenbau' in Halle). Albrecht Thaer Archiv 10, Berlin

Mundlak, Y. (1992): Agricultural productivity and economic policies: concepts and measurements. Paris: OECD Development Centre, Technical Paper no.75

Nijland, G.O. & J. Schouls (1997): The relation between crop yield, nutrient uptake, nutrient surplus and nutrient application. Wageningen Agricultural University Papers 97-3, Wageningen

Oldeman, L.R., R.T.A. Hakkeling & W.G. Sombroek (1991): World map of the status of human-induced soil degradation. An explanatory note. Wageningen/Nairobi: ISRIC/UNEP

Olson, K.R., R. Lal & L.D. Norton (1994): Evaluation of Methods to Study Soil Erosion-Productivity Relationships. Journal of Soil and Water Conservation 49, 586-590

Oveson, M.M. & R.S. Besse (1967): The Pendleton experiment station - its development, program, and accomplishments: 1928 to 1966. Corvalis: Agricultural Experiment Station, Oregon State University, Special Report 233

Pagoulatos, A. (1989): The probability of precipitation and the crop yield soil-depth function. Journal of Soil and Water Conservation 44, 349-351

Pattanayak, S. & D.E. Mercer (1998): Valuing soil conservation benefits of agro-forestry: contour hedgerows in the eastern Visayas, Philippines. Agricultural Economics 18, 31-46

Penning de Vries, F.W.T. (1982): Systems analysis and models of crop growth. In: F.W.T. Penning de Vries & H.H. van Laar (Eds.). Simulation of plant growth and crop production. Simulation Monographs, Pudoc, Wageningen

Penning de Vries, F.W.T. (1990): Can crop models contain economic factors? In: R. Rabbinge, J. Goudriaan, H. Van Keulen, F.W.T Penning de Vries & H.H. van Laar (Eds.). Theoretical production ecology: reflections and prospects. Simulation Monographs 34, Pudoc, Wageningen

Penning de Vries, F.W.T. & M.A. Djitèye (Eds.) (1982): La productivité des pâturages Sahéliens. Une étude des sols, des végétations et de l'exploitation de cette ressource naturelle. Agricultural Research Report 918, Pudoc, Wageningen

Penning de Vries, F.W.T. & H. van Keulen (1982): La production actuelle et l'action de l'azote et du phosphore. In: F.W.T. Penning de Vries & M.A. Djitèye (Eds.). La productivité des pâturages Sahéliens. Une étude des sols, des végétations et de l'exploitation de cette ressource naturelle. Agricultural Research Report 918, Pudoc, Wageningen

Pichot, J., M.P. Sédogo, J.F. Pulain & J. Arrivets (1981): Evolution de la fertilité d'un sol ferrugineux tropical sous l'influence de fumures minérales et organiques. Agronomie Tropicale 36, 122-133

Pol, F. van der (1992): Soil mining. An unseen contributor to farm income in Southern Mali. Bulletin No. 325, Royal Tropical Institute, Amsterdam

Rietkerk, M. (1998): Catastrophic vegetation dynamics and soil degradation in semi-arid grazing systems. Tropical Resource Management Papers no. 20, Wageningen Agricultural University, Wageningen

Schuurman, J.J. (1965): Influence of soil density on root development and growth of oats. Plant and Soil 22, 352-374

Seligman, N.G. & H. van Keulen (1989): Herbage production of a Mediterranean grassland in relation to soil depth, rainfall and nitrogen nutrition: A simulation study. Ecological Modelling 47, 303-331

Seligman, N.G., H. van Keulen & C.J.T. Spitters (1992): Weather, soil conditions and the interannual variability of herbage production and nutrient uptake on annual Mediterranean pastures. Agricultural and Forest Meteorology 57, 265-279

Shiferaw, B. (1998): Peasant agriculture and sustainable land use in Ethiopia: economic analysis of constraints and incentives for soil conservation. Ph.D. thesis, Agricultural University of Norway, Aas

Silva, E. (1996): Nonparametric efficiency analysis in the context of a multioutput dynamic technology. Mimeograph. University of Porto, Porto

Silva, E. & S.E. Stefanou (1996): Nonparametric dynamic production analysis and the theory of cost. Working paper no. 304. Dept. of Agricultural Economics and Rural Sociology, Pennsylvania State University

Stocking, M. (1984): Erosion and soil productivity: a review. Working Paper no. 1, Soil Conservation Program, Land and Water Development Division, Conservation, FAO, Rome

Stoorvogel, J.J. & E.M.A. Smaling (1998): Research on soil fertility decline in tropical environments: integration of spatial scales. Nutrient Cycling in Agroecosystems 50, 151-158

Stoorvogel, J.J., J. Bouma & W.T. Bowen (Eds.) (1998): Information technology as a tool to assess land use options in time and space. Quantitative Approaches in Systems Analysis No. 16, TPE/AB-DLO, Wageningen

Stroosnijder, L. (1996): Modelling the effect of grazing on infiltration, runoff and primary production in the Sahel. Ecological Modelling 92, 79-88

Stroosnijder, L. & W.B. Hoogmoed (1984): Crust formation on sandy soils in the Sahel. II. Tillage and its effect on the water balance. Soil & Tillage Research 4, 321-337

Stroosnijder, L. & P. Kiepe (1998): Simulation of maize growth under conservation farming in tropical environments. Tropical Resource Management Papers No. 19, Wageningen Agricultural University, Wageningen:

Struif Bontkes, T. (1999): Modelling the dynamics of agricultural development: a process approach: the case of Koutiala (Mali). Ph.D. thesis, Wageningen Agricultural University, Wageningen

Veloz, A., D. Southgate, F. Hitzhusen & R. MacGregor (1985): The economics of erosion control in a subtropical watershed: A Dominican case. Land Economics 61, 145-155

Weesies, G.A., S.J. Livington, W.D. Hosteter & D.L. Schertz (1994): Effect of soil erosion on crop yield in Indiana: results of a 10 year study. Journal of Soil and Water Conservation 49, 597-600

Williams, J.R. (1994): The EPIC model. Blackland, Texas: USDA/ARS

Williams, J.R., C.A. Jones, J.R. Kiniry & D.A. Spanel (1989): The EPIC crop growth model. Transactions American Society of Agricultural Engineers 32, 497-511

Wischmeier, W.H. (1976): Use and misuse of the universal soil loss equation. Journal of Soil and Water Conservation 31, 5-9

Wischmeier, W.H. & D.D. Smith (1978): Predicting rainfall erosion losses-a guide to conservation planning. Washington: USDA Agricultural Handbook No. 537

Wit, C.T. de (1990): Understanding and managing changes in agriculture. In: J.G.W. Jones & P.R. Street (Eds.), Systems theory applied to agriculture and the food chain. Elsevier Applied Science, London

Wit, C.T. de (1992): Resource use efficiency in agriculture. Agricultural Systems 40, 125-151

3 Technical Options for Agricultural Development in the Ethiopian Highlands: A Model of Crop-Livestock Interactions

Ayele A. Abiye & Jens B. Aune

Centre for International Environment and Development Studies, Noragric, Agricultural University of Norway, Ås, Norway

Abstract

The crop and livestock sectors in Ethiopia are closely interlinked. Oxen play a dominant role as draft animals. The oxen-plough system has existed for thousands of years, but is currently under pressure due to reduced farm size and low availability and quality of fodder. There is a need therefore to develop alternative tillage systems like zero–tillage or cow traction. A model was developed to assess the resource constraints of the Ethiopian crop livestock system and to examine alternative technological options for improving the system. The model can be used to assess productive effects of fertiliser use, choice of livestock system and fodder quantity and quality. The data used to construct the model were derived from household surveys and established relationships between fodder quality and livestock production. The model results show fodder resources to be a critical factor in livestock production. An interesting option for improving the crop-livestock system is to introduce dairy goats. However, for such a change to take place, the quality of the fodder resources will have to be upgraded. Urea treatment of straw is an option.

3.1 Introduction

Cattle play a dominant role in Ethiopian livestock systems and comprise 83% of the total tropical livestock units in highland Ethiopia (FDRE, 1997). A particular feature of the livestock system in highland Ethiopia is the oxen-plough system. Ploughing with oxen in Ethiopia dates back to ancient times and is a cultivation method deeply rooted in Ethiopian culture (McCann, 1995). Milk production of cattle in Ethiopia is low: about 213 litres per lactation compared to 300 litres for sub-Saharan Africa. Major reasons for this low production are low availability and quality of fodder, poor animal health and low genetic potential of animals (FDRE, 1997).

The crop and livestock sectors in Ethiopia are closely interlinked (Fig. 3.1). The crop sector, which is dominated by cereals (mainly wheat, barley, teff and sorghum), provides fodder and cash for investment to the livestock sector in favourable years, whereas the livestock sector provides animal traction, manure and profit to be used for purchase of inputs in the crop sector.

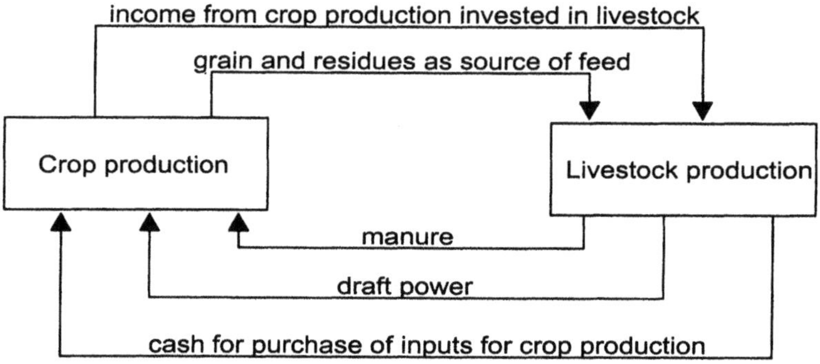

Figure 3.1: Major Crop Livestock Interactions in the Ethiopian Highlands.

The objectives of this study are to analyse the economic contribution to the farm household of the crop and livestock sectors, respectively and to examine through a modelling approach, different options for technology development. This chapter discusses different models that might potentially be used and their weaknesses. It describes the model selected, presents the input data used to run the model and its output, and discusses different development paths for the crop-livestock system.

3.2 Choice of Agro-ecological Models

Many models could potentially be used to model resource use in Ethiopian agriculture. Examples of such models are DSSAT (Tsuji *et al.*, 1994), Century (Parton *et al.*, 1987), and EPIC (Williams *et al.*, 1990). However, application of such process-based simulation models in a resource-constrained environment presents numerous problems. One bottleneck is their very high data requirements. Data are required, for instance, on chemical and physical properties of the soil in different layers, genetic coefficients of the crops, chemical composition of plant material, weather data, *etc.* Obtaining such data is both very costly and time consuming in tropical rural environments. Another criticism of these models is that many constraints that may influence yield are not accounted for (Mutsaers & Wang, 1999). In the models, yields are mainly limited by nitrogen, phosphorus and water availability whereas in real farming situations they are often constrained by factors such as weeds, pest, birds, potassium, sulphur, micro-nutrients, flooding, hailstorms, *etc*. Moreover, interactions between the factors are often not properly accounted for. These models have, therefore, been more successful in crop management studies in industrialised countries (Mutsaers & Wang, 1999), where yield-limiting and yield-reducing factors (micronutrients, *etc.*) are better controlled. A third criticism is their limited transparency, which makes it difficult to change the model when obvious problems are identified. Despite these criticisms, development of process-based simulation models is important in order to gain a better understanding of soil/plant dynamics. However, these models should be applied with caution for predicting yield under subsistence agriculture in the tropics.

Crop models may be classified in two categories: those that aspire to improve our understanding of physiological and environmental interactions of crops (mechanistic models) and those that are developed to provide sound management advice to farmers or predictions for policy makers (Passioura, 1996). The latter models are generally empirical and probably the best adapted to subsistence agriculture in the tropics, where many factors may influence yield and we have limited understanding of the interactions between the yield-determining and yield-limiting factors. The other advantage of empirical models is the possibility of making use of farmers' experiences in their construction. Limitations of the empirical models are, however, that they cannot be used outside the area where they were developed and cannot easily deal with effects of new technologies.

The type of model presented here belongs to the empirical category. The model was developed to study resource use in the crop-livestock system and to examine different options for its intensification in Ethiopia. It can be used to assess the effects of land availability, input use and herd composition on economic output, and nutrient flows. The soil-crop component is based on earlier models described by Aune (1995) and Aune & Massawe (1998). It is flexible and can be fairly easily incorporated into bio-economic models, as illustrated by Shiferaw *et al.* (this volume) and Glomsrød (this volume).

3.3 The Model

The model used in this chapter describes the use of crop residues (and natural vegetation) as sources of feed in animal production, and can be used to examine the profitability of alternative technological options. Other crop-livestock interactions shown in Fig. 1 are not included in the model.

3.3.1 Collection of Household Data and Model Construction

Household data were collected through Participatory Rural Appraisal (Mikkelsen, 1995) methods and household surveys in three villages in Lay Armacheo Woreda in the North Gondar Zone of Amhara National State (about 785 km north of Addis

Ababa) in 1998. The households in each village were first classified in three wealth classes, based on ownership of animals. Eight farmers were then randomly selected from each wealth class.

3.3.2 The Soil-Crop Component

The soil-crop component calculates crop production, in terms of grain and straw, as a function of nitrogen supply.

The release of mineral nitrogen from soil organic matter is described by first order kinetics with a mineralisation constant of 2% per year (Nye & Greenland, 1960):

$$N_{SON(t)}=N_{s(t-1)}\,(1-e^{-r}) \tag{3.1}$$

$N_{SON(t)}$ is the amount of nitrogen mineralised from soil organic matter, $N_{s(t-1)}$ the total amount of organic nitrogen in kg/ha and r the mineralisation constant (= 0.02).

Changes in soil organic nitrogen are calculated based on reductions due to decomposition of soil organic matter and additions through the root system:

$$N_{s(t)}=N_{s(t-1)}\,e^{-rt} + \beta\, N_{r(t-1)} \tag{3.2}$$

where β (= 0.3) is the proportion of nitrogen in roots entering the soil organic matter pool (Ladd & Amato, 1985). Additions through residues are not included, because in Ethiopia crop residues are used as fodder or as fuel.

Barley yield is calculated as a function of nitrogen supply. The response to nitrogen is estimated from the farm survey data:

$$Y_b = 907 + [fertiliser\ N + N_{s(t)}\,(1-e^{-r}) - N_{s(t=o)}\,(1-e^{-r})] * 6.67 \tag{3.3}$$

The expression between square brackets expresses N supply in year t relative to that in year 0, for which Eqn. (3.3) was calibrated.

Straw yield from barley is calculated assuming a harvest index of 30%.

3.3.3 The Livestock Component

The livestock component calculates energy requirements of oxen, cows and goats, as well as grass production from rangelands.

Annual maintenance requirements of ox and cow, expressed in feed dry matter are calculated as:

$$Maintenance\ requirement = Live\ weight\ (kg) * 2.5\% * 365 \tag{3.4}$$

The energy requirements of goats are calculated from requirements for maintenance, milk production and weaning of kids (Peacock, 1996):

- Maintenance energy for a goat of 30 kg live weight = 6 MJ/d.
- Energy requirement per kg of milk = 5 MJ.
- Energy requirement per weaned kid = 4 MJ/d.

Total energy requirement of goats for maintenance, weaned kid and 1 kg milk production = 15 MJ/d.

Metabolisable energy per kg dry matter is set to 7 MJ and 10 MJ for straw and urea-treated straw, respectively. Fodder production in rangeland was estimated at 6000 kg/ha, based on data from Mwendra *et al.* (1997).

3.3.4 Socio-economic Data

The socio-economic data (see Table 3.1) that were used as input to the base run of the model were obtained from the household survey. They typically present a household with low livestock holding.

Table 3.1: Socio-economic data (low livestock holding category) used in base run of the model.

Assets		Prices (birr/kg)	
Barley land	1.2 ha	Fertilisers	
Pasture land	0.2 ha	diammonium phosphate	2.3
Oxen	1	urea	1.5
Cows	1	Barley	2
Goats	0	Milk	1
		Meat	3

These socio-economic data are used to calculate profitability of different technological options. Different investment alternatives can be examined through the use of Net Present Value (NPV). The Solver in Excel is used to optimise NPV thereby identifying the best investment alternative.

3.4 Resource Efficiency of Different Crop-Livestock Systems in Ethiopia: Model Results

The results for the model's base run show that farmers are resource-constrained with respect to access to fodder. The availability of straw varied from 2825 to 4200 kg per household. As households with low livestock holdings only have access to 0.2 ha of grassland, dry matter production from grassland will be 1200 kg. Total availability of dry matter will be about 4000 kg without fertiliser use, but can be increased to 5400 kg if 66 kg N/ha (recommended rate) is used on barley. One ox and one cow (2 TLU) will consume about 5000 kg dry matter/year which shows that access to fodder is critical, even for households with low livestock holdings.

Nevertheless, farmers were on average able to sell animals and animal products for 475 birr/year, which constitutes about 50% of their total income (sale of crops accounts for only 13%). This amount is equal to the value of one heifer or 4 goats and shows the importance of livestock in the rural economy.

Nitrogen export in harvested product is based on grain and straw yield and their nitrogen content (Stoorvogel & Smaling, 1990) and will be about 36 kg N for a farm size of 1.2 ha, without fertiliser use. Manure that can be collected from a cow during 8 months coralling, the length of the dry season, amounts to 300 kg dry matter (Fernàndez-Rivera *et al.*, 1995) which corresponds to 9 kg N for 2 TLUs (Newcombe, 1989). In addition, losses of nitrogen occur when farmyard manure is used as fertiliser. This indicates negative soil nitrogen balances in the area.

The model was used to examine different options for changes in the management of the crop-livestock system. One option is to replace the ox with other animals, as it is of limited use, except for ploughing and threshing. Such a change implies a change in tillage system or the use of cows for animal traction. An alternative is zero tillage which has been successfully introduced in other parts of the tropics (Steiner, 1998).

Exclusion of oxen from the farming system leaves more fodder for other animals, for example dairy goats that have been introduced with success in Eastern and Southern Ethiopia (Peacock, 1996). This could be an interesting option for farmers not owning oxen and/or for farmers with limited availability of land. Energy requirements for maintenance of the goat, growth of the kid and one litre of milk production will be about 15 MJ per day, equivalent to consuming 2.2 kg straw. However, the goat and the kid cannot ingest more than 3% of their body weight (50 kg) daily, *i.e.* 1.5 kg. This implies no milk production, unless the fodder resources are up-graded qualitatively by growing fodder crops or by treating the straw with urea. The energy requirements of the animal can be met by feeding 1.5 kg of urea treated straw per day. This system will require about 550 kg of straw/year. The milk can either be consumed or sold at the local market. The cost per ton of urea treated straw is about 125 birr. One ton of urea treated straw will produce 540 litres of milk (from 2 dairy goats) for marketing, after weaning of the kids, with a value of about 500 birr and two kids, with a value of about 120 birr. The value-cost ratio (VCR) of this investment will be about 5, *i.e.* each birr invested in urea will give a return of 5 birr. An investment in fertiliser will, on the other hand, only give a VCR of 1.8. However, the VCRs of the two systems are not directly comparable, as they have different labour requirements.

3.5 Development Paths for the Crop-Livestock System in Ethiopia

Different development paths can be suggested for the crop-livestock system in the highlands of Ethiopia. Possible technological options are increased fertiliser use, zero tillage, cow tillage, oxen sharing, increased fodder production, stall feeding, increase in fodder quality through urea treatment of straw and change to milk production based on goats instead of reliance on the ox-cow system. The choice among these alternatives depends on such factors as their economic profitability, labour demand and supply, access to inputs, credit and markets, access to and price for oxen rental, access to fodder, need for erosion control and cultural acceptability

The livestock sector in the area has undergone major changes in the last 10 years. The number of animals has declined, with considerable difference between different types of animals. Currently, the average household owns 1.69 oxen, 1.40 cows, 1.17 sheep, 1.05 goats and 0.25 donkeys. In the last ten years, there has been a decline of 24, 61, 63, 66 and 82% for oxen, cows, sheep, goats, and donkeys, respectively. Hence, an increasing part of the fodder resources is consumed by oxen.

It is likely that the crop-livestock system in the Ethiopian highlands will continue to evolve in the coming decades. Change will most likely take place first in areas with small land holdings and where many farmers do not own oxen, *i.e.* the southern areas where hoe culture meets oxen tillage. These areas are also the most favourable for zero tillage, with high rainfall and maize dominance. Animal traction in areas with very small land holdings is to some extent a contradiction, as animal traction in other parts of Africa has been associated with an expansion of the cropping area and cash crop production (Speirs & Olsen, 1992). The area of cropping land per household in Ethiopia is decreasing rather than increasing. Oxen tillage would have been more justified in the past when land holdings were larger and fodder was more abundantly available.

The area of cultivated land in the region is probably too large for the control of weeds by hoeing, as 1 ha with the hoe will require 44 man-days, according to the household survey data. However, an interesting option could be to use cows as draft animals (Jabbar, 1993). Even though cows are weaker than oxen, the feasibility of using cows (cross-bred) for animal traction in Ethiopia has been shown (Belete *et al.*, 1994). Because of decreasing land holdings, cow traction has become more attractive as less land implies less work for cows, thereby reducing the risk of reduction in their reproductive capacity. In Wolaita and Tigray where land holdings are below 0.5 ha, hand hoeing might be an option (Wondimagegn, 1996). Hoe tillage might be an interesting option for those not owning oxen, comprising about 38% of the households in Tigray. These households have to pay up to 75% of their yield if the oxen owner undertakes tillage and threshing and provides seeds. A shift in tillage system might be anticipated when the land holding falls below a threshold level and where the feed resource is a critical factor.

The model results show that fertiliser is not necessarily the most profitable way to use scarce capital resources, but that it may be more remunerative to use funds to improve the quality of the fodder than to use fertiliser. However, this requires a change in the crop-livestock system to animals that are able to utilise high quality fodder resources efficiently for instance for milk production. The current livestock system dominated by oxen cannot pay the cost of up-grading the fodder resources. For such a change to take place, public investment in research, extension, infrastructure and rural credit is necessary. The pathways to intensification will vary across Ethiopia depending on local conditions.

3.6 References

Aune J.B. (1995): Predicting soil degradation in Tanzania - a system analysis approach. Norwegian Journal of Agricultural Sciences Supplement 21, 27-60

Aune, J.B. & A. Massawe (1998): Effects on soil management on economic return and indicators of soil degradation in Tanzania - a modelling approach. Advances in GeoEcology 31, 37-44

Belete, A., J.L. Dillon & F.N. Anderson (1994): The impact of using cross-bred cows and single ox on incomes of small scale farmers in the Ada and Baso-Worana areas of the central highlands of Ethiopia. Ethiopian Journal of Agricultural Sciences 14, 112-120

FDRE (1997): Agricultural Sample Survey Report on Livestock, Poultry and Beehives Population. 1996/97, Federal Democratic Republic of Ethiopia Central Statistical Authority, Private Peasant Holding, Statistical Bulletin No. 171, Vol. 2, Addis Ababa, Ethiopia

Fernàndez-Rivera, S., T.O. Williams, P. Hiernaux & J.M. Powell (1995): Faecal extraction by ruminants and manure availability for crop production in semi-arid West-Africa. In: J.M. Powell, S. Fernàndez-Rivera, T.O. Williams & C. Renard (Eds.) Livestock and sustainable nutrient cycling in mixed farming systems of sub-Sahara Africa. ILRI, Addis Ababa

Glomsrød S. (2000): Soil degradation in CGE models. This volume

Jabbar, M.A. (1993): Research on cow traction in Africa: some lessons to be learned. World Animal Review 74/75, 58-63

Ladd, J.N. & M. Amato (1985): Nitrogen cycling in legume cereal rotations. In: B.T. Kang & J.van der Heide (Eds.) Nitrogen management in farming systems in humid and subhumid tropics. Institute of Soil Fertility, Haren, The Netherlands, 105-128

McCann, J.C. (1995): People of the plow. The University of Wisconsin Press, 298 pp.

Mikkelsen, B. (1995): Methods for development work and research. Sage Publications, New Delhi

Mutsaers, H.J.W. & Z. Wang (1999): Are simulation models ready for agricultural research in developing countries? Agronomy Journal 91, 1-4

Mwendra, E.J., M.A. Saleem & Z. Woldu (1997): Vegetation response to cattle grazing in the Ethiopian highlands. Agriculture, Ecosystems and Environment 64, 43-51

Newcombe, K.J. (1989): An economic justification for rural afforestation: The case of Ethiopia. In: G. Schramm & J. Warford (Eds.) Environmental management and economic development. John Hopkins University Press, Baltimore, 117-137

Nye, P.H. & D.J. Greenland (1960): The soils under shifting cultivation. Technical Communication 51, Commonwealth Bureau of Soils, Harpenden UK, 144 pp.

Parton, W.J., D.S. Schimel, C.V. Cole & D.S. Ojima (1987): Analysis of factors controlling soil organic matter levels in the Great Plains grasslands. Soil Science Society of America Journal 51, 1173-1179

Passioura, J.B. (1996): Simulation models: Science, snake oil, education, or engineering. Agronomy Journal 88, 690-694

Peacock, C. (1996): Improving goat production in the tropics. An Oxfam/Farm-Africa Publication, 386 pp.

Shiferaw, B., S. Holden, & J. Aune (2000): Population pressure and land degradation in the Ethiopian highlands: a bio-economic model with endogenous soil degradation. This volume

Speirs, M. & O. Olsen (1992): Indigenous integrated farming systems in the Sahel. World Bank Technical Paper no 179

Steiner, K.G. (1998): Conserving natural resources and enhancing food security by adopting no-tillage. GTZ, Eschborn, 53 pp.

Stoorvogel, J.J. & E.M.A. Smaling (1990): Assessment of soil nutrient depletion in sub-Sahara Africa 1983-2000. Volume 1, Main report, Report 28, The Winand Staring Centre, Wageningen, the Netherlands

Tsuji, G.Y., G. Uehara & S. Balas (1994): DSSAT v3. University of Hawaii, Honolulu, Hawaii

Williams J.R., P.T. Dyke, W.W. Fuchs, V.W. Benson, D.W. Rice & E.D. Taylor (1990): EPIC: erosion/productivity impact calculator: 2. User manual. Technical Bulletin No. 1768, US Department of Agriculture, Washington, DC

Wondimagegn, F. (1996): Crop production in the central zone of Tigray. In: A.O. Øyhus & G. Gebru (Eds.). Rural exploratory studies in the central Tigray, Northern Ethiopia, Noragric, NLH, 45-54

4 Land Degradation as a Transformation Process in an Intertemporal Welfare Optimisation Framework

Peter J. Albersen[1)] **& Laixiang Sun**[2) *]

[1)] Centre for World Food Studies of the Vrije Universiteit (SOW-VU), Amsterdam, the Netherlands

[2)] International Institute for Applied Systems Analysis (IIASA), Laxenburg, Austria

Abstract

An intertemporal social welfare optimisation framework proposed by Fischer et al. (1996) describes desirable scenarios of land use and land cover change in the future for China. This chapter intends to demonstrate that it is feasible to integrate a land degradation dimension into this framework, if the required quantitative data are available. We show how to incorporate the land degradation dimension into the profit maximisation process of a representative farm as part of resource management. Land units are categorised according to their soil and climate characteristics to represent the spatial heterogeneity. They can shift from higher quality classes to lower ones as a result of land degradation, while it is also possible to reverse this process by land conservation measures. Land degradation is treated as an integral element of an economic transformation function. The basic assumption is that users of the land, farmers in particular, notice the negative effect of land degradation on their production potentials. They are supposed to have opportunities to cancel these negative effects by investing in land and/or by increasing the volume of their input packages.

---

[*] Both authors participate in the implementation of the land-use and land-cover change model for China, a joint effort of SOW-VU and IIASA. For recent works of SOW-VU and IIASA on land-use modelling, see Fischer *et al.* (1996), Keyzer (1998), Albersen *et al.* (in prep.).

4.1 Introduction

Land as a basic asset has been of fundamental importance for most rural households in developing countries. Good soil in combination with good climate is a scarce resource (Voortman *et al.*, 2000). Conservation of this asset is therefore essential for the rural community. Land degradation causes loss of farmland productivity with or without time lags. To a certain extent, the land degradation process is similar to the depreciation process of physical capital in industry. However, while physical capital in industry can be fully restored through renovation investment, a full recovery from land degradation may not be so straightforward due to the constraints inherent in the natural processes within the soil. Besides, land is by nature location-specific and rehabilitation and/or conservation options are not necessarily transferable to other locations.

This chapter presents an approach to incorporate the spatial and temporal aspects of land degradation in the technology of land users. Our starting set-up is the intertemporal welfare maximising policy analysis model described in Fischer *et al.* (1996). This welfare maximisation model is capable of integrating various socio-economic and biophysical factors that together drive land use and land cover change. Human-induced land degradation is added to this set-up as an extension of the land transformation. The basic assumption is that land users, farmers in particular, notice the effect of land degradation in their production possibilities and potentials. The effect might be a limited choice in crops and/or varieties, but it might also present a reduction in the production potential[1]. Users are given opportunities in the model to alleviate these negative effects by investing in land and by enhancing their input packages.

Land degradation is a general denominator for several, sometimes very different processes, such as nutrient mining, deforestation, loss of top soil, desertification, pollution, increase in water and wind erosion hazard, and loss of bio-diversity. It is questioned (Perman *et al.*, 1996; Bie, 1990) whether rehabilitation and restoration of the land is possible, even on slightly or moderately degraded land. Conservation as the alternative is mentioned by others (Sanders *et al.*, 1995; Van Kooten, 1993). A distinction can be made between rehabilitation and conservation in physical and in economic terms. In this chapter we focus on the economic perspective, *i.e.* restoring or conserving the productive capacity of land. Whether this is economically viable, depends on the circumstances and the behaviour of the actors in the model. The choice of the measure to be taken for conservation or rehabilitation is not part of the model.

---

[1] Potential production defined from the genetic characteristics of the plant is sometimes distinguished from maximum attainable production. The attainable production takes the soil and climate conditions into account, *i.e.* the site class definition. In this paper we use the latter definition of the production possibilities, which are time and location specific and we regard that as potential production.

The major constraint to the integrated modelling with a focus on land degradation is probably not the incorporation in economic theory, but the applied work, since quantitative information on land degradation and its effect on the production capacity of land, is available only for scattered plots and experimental sites. Usually, it is very difficult to generalise information from specific problem areas to characterise the situation in a much larger region (Wischmeier, 1976; Pierce & Lal, 1994). The impact on productivity of the same type of land degradation is often quite different across different soils. The interweaving of land degradation, various forms of soil conservation methods, and adequate land management practices, further makes the assessment of land degradation controversial, if not irrelevant (van Lynden & Oldeman, 1997; Heilig, 1999).

Recent examples of incorporation of land degradation in macro-economic modelling are *e.g.* Alfsen *et al.* (1996) and Aune *et al.* (1997). Both applications use general equilibrium modelling, but do not incorporate the spatial dimension to link the macro level of analysis with the disaggregate level at which land degradation occurs. Miller & Plantinga (1999) employ a behavioural model of profit maximisation to recover a parametric model of county-level land use share via maximum entropy. In their land use model, land use shares are functions of pre-determined variables such as output price, input costs, and land conditions. Soil erosion is treated as a passive consequence of land use only, with a technically predetermined soil erosion rate linked to each land use type. In comparison, we intend to explicitly model land transformation processes in terms of both active land conservation and passive consequences of land use (*e.g.* erosion), and furthermore, to integrate land transformation into the framework of social welfare optimisation.

4.2 The Transformation Process

4.2.1 Overall Framework

The welfare approach is the point of departure in Fischer *et al.* (1996). The T-period optimisation model aims at identifying and simulating socially desirable and economically efficient trajectories of investment and resource uses for the future. The social welfare maximisation problem is subject to technological and resource constraints. Let us briefly summarise the main features of the model (*ibid*, p. 40-41):

- Every consumer maximises the T-period sum of his own utilities, subject to an intertemporal budget constraint and with a constant discounting rate over time.
- Every producer maximises his own profit for each time period, subject to various technological and natural condition constraints. The transformation

functions describe the set of net supplies in period t and resources in period t+1 that are feasible at the given level of resources in period t.

- The commodity markets clear. For each commodity the account balances: the net demand by the consumers cannot exceed the net supply that is feasible under the transformation constraints (and from trading).
- Resource utilisation matches resource availability. The stock of each resource in the current period should not exceed the level carried over from the previous period.

A detailed description of this type of policy analysis tools is presented in *e.g.* Gunning & Keyzer (1995) and Ginsburgh & Keyzer (1997). For the purpose of this chapter, the profit-maximising producer is most relevant. The agent-specific transformation function G^t describes the set of technologically efficient production plans Y^t:

$$G^t(y^t, v^t, r^t, r^{t+1}, x^t) \leq 0 \tag{4.1}$$

where

y^t = vector of outputs (*e.g.* crop produce),
v^t = vector of inputs (materials, water, energy, …)
r^t = stock vector of resources (land, capital, labour, …); $r^t \geq 0$
r^{t+1} = stock vector of resources in the next period; $t = 1, ..., T$
x^t = vector of soil qualities (soil characteristics, climate, ...).

The vector $y^t \in Y^t$ denotes the supply of produce to the market, while v^t denotes the vector of demanded goods and services. Vector r^{t+1} is the stock of resources that has to be made available in the next period $t+1$. The stock may change through natural causes and/or as a result of decisions of the resource manager, in particular the farmer. Stock may increase through natural growth or can be generated by a production process. Decrease can *e.g.* occur through degradation or depreciation. The importance of the vector x^t lies in the fact that it describes the natural conditions that determine the production capacity of the land. For the finite horizon, model conditions for the stock in period $T+1$ have to be added to the program.

By assuming separability between outputs and inputs, the transformation function G^t of (4.1) can be written as:

$$Q^t(y^t) - F^t(v^t, r^t, r^{t+1}, x^t) \leq 0 \tag{4.2}$$

where Q^t is the output function (linear homogeneous, convex, and non-decreasing) and F^t the input-response or production function (linear homogeneous, concave and non-decreasing in (v,r)). The conditions given in the parentheses are necessary for the model to be well behaved. The scalar output production function is a special case of the general output function Q^t. Crop-specific inputs are required if (4.2) is to be estimated crop-wise. Certainly at an aggregate level, like village, county or district, often this information is not crop-specific and can only be related to an output package (Keyzer, 1998). Defining the multiple input-multiple output system (4.2), prevents an artificial distribution of inputs based on ex-ante description of the technology used.

Incorporating natural conditions x^l in the input-response function becomes essential if the area covered in the analysis has a large variability with respect to factors determining productive potential. Keyzer (*ibid.*) shows that *e.g.* a generalised version of the well-known Mitscherlich-Baule function in the agronomic literature is suitable to combine both the agronomic and economic aspects (Llewelyn & Featherstone, 1996). For China, Albersen *et al.* (in prep.) use a potential output component serving as an asymptote of actual output. This potential output is assessed with a China-specific implementation of the modified Agro-Ecological Zones (AEZ) methodology (Fischer *et al.*, 1999; FAO, 1978; FAO/IIASA, 1993). The AEZ assessment is a well-developed environmental approach, providing an explicit geographic dimension for establishing spatial inventories and databases of land resources and crop production potentials. The method is comprehensive in terms of coverage of factors affecting agricultural production, such as components of climate, soil and landform, considering basic conditions in supply of water, energy, nutrients and physical support to plants.

Two other recent studies on China by Carter and Zhang (1998) and Lindert (1999) incorporate, besides the conventional inputs, also climate and biophysical information. Carter and Zhang estimate a Cobb-Douglas model for grain productivity for the five major grain producing regions in China, with aridity indices using data over 1980-1990. Lindert estimates, for both North and South China and with a mixed translog and Cobb-Douglas specification, the agricultural and grain productivity, using soil chemical characteristics from soil profiles and input-output data at county level. These studies show that detailed information on land and aggregate information on inputs and outputs can be combined meaningfully.

4.2.2 Description of Land

The vector x^l represents the characteristics of the land, like location, soil characteristics, climate, *etc.* Land degradation represents a change in one or more characteristics of this resource, with consequences for long- and/or short-term production possibilities and potentials. Before incorporating specific features of land degradation into the model, we first describe the characteristics of a piece of land and their possible changes by means of attributes.

Two basic features with respect to land use and land cover change need to be distinguished in the transformation function given in (4.2):

- Inherent characteristics of the land, reflected in the natural attributes and/or conditions.

- Behaviour of the decision maker. For a farmer, given his economic setting, this refers to the crop-mix chosen in a particular period and the combination of inputs used. Agent-specific input and output functions have to reflect this. For a (local) government this refers *e.g.* to the development of industrial or residential areas or to a differential tax system for different types of land use.

All natural condition attributes are classified as the inherent characteristics of the given piece of land. Sites with the same inherent characteristics are categorised in the same *site class* s ($s = 1, ..., S$) and the corresponding area is denoted by A_s. For our purpose, only those characteristics, that are discriminatory with respect to land degradation by agricultural uses, need to be considered. Examples of such characteristics are the attributes of climate (temperature, radiation, precipitation, *etc.*), soil (type, erosion hazard, *etc.*), landform, altitude and slope.

The productive uses of land are denoted by the subscript h ($h = 1, ..., H$). Such *land use types* can be, for example, natural vegetation, arable land, grassland, forest, residence and infrastructure[2]. Within arable land, rainfed land can be distinguished from irrigated land. Further subdivision is possible by defining high-input- and low-input-level production within irrigated agriculture. The area of land use type h in the site class s is denoted by $A_{s,h}$. If the co-ordinates of the different site classes and land use types have been stored in a Geographical Information System (GIS), $A_{s,h}$ can be derived from an overlay of both layers. These classifications can be used *e.g.* in the estimation of the production function (4.2).

Total area does not change over time and the simple additive accounting rule applies. For each given time period t, the following equations hold:

$$A^t = \sum_s A_s^t \text{ : total area of land,} \tag{4.3}$$

$$A_h^t = \sum_s A_{s,h}^t \text{ : total area of land in land use type h.} \tag{4.4}$$

However, both h and s are subject to changes over time. The physical attributes are subject to autonomous changes and/or changes resulting from decisions of the user of the land. These changes may have consequences for potential use of the land. For example, changes in temperature or precipitation may create opportunities to cultivate new varieties or other crops, but may also impose constraints to maintain the current cropping patterns. Changes in the slope of the land by terracing may increase water-holding capacity of the soil and thus create scope for new crop selection. Soil erosion by water or wind may transform land from one site class to another. Other examples of changes in land use types include deforestation and industrial construction at the cost of agricultural land.

---

[2] Land reclamation can be also introduced in this set-up by defining a land use type 'others' as a potential source of land.

Whether a given land use change results in transfer of land between land use classes depends on the explicit definition of land use types. For example, if answering a research question requires to distinguish annual crops from perennial crops, it would not be wise to define a change in cropping pattern of annual crops as a change in land use type. However, if it is necessary to distinguish cropping patterns based on their impact on some kind of erosion hazard index, a change in cropping pattern may be associated with a change in land use type and may result in a change in site class. Growing crops with a low crop cover may result in wind and/or water erosion of topsoil and thus in reduced thickness of the productive soil layer, causing a change in the basic characteristics of the land (Kruseman & Van Keulen, this volume).

4.2.3 Costs of Productivity-Increasing Transformation

Land transformations that directly increase land productivity, and thus land value, are usually desirable for the decision maker, although some long-run environmental costs may need to be taken into account. Some of such transformations between land use classes (within a given site class s) may not be feasible due to physical constraints, regulations or significant economic costs.

Table 4.1: An example of land transformation possibilities for a given site class s.

	Destination				
Land use types Origin	Natural vegetation	Arable land, rainfed	Arable land, irrigated	Production forest	Residence, infrastructure
Natural vegetation	-	Yes	Yes	Yes	Yes
Arable land, rainfed	No	-	Yes	Yes	Yes
Arable land, irrigated	No	No	-	Yes	Yes
Production forest	No	Yes	Yes	-	Yes
Residence, infrastructure	No	No	No	No	-

Table 4.1 presents a hypothetical matrix of such productivity-increasing transformation possibilities as a simple example. Natural vegetation can be transformed into any other land use type and the process is irreversible. Although irrigated arable land can be used as rainfed, the production is supposed to

continuously use the water control characteristics of the irrigation scheme, because there is no reason to suppose that the farmer would voluntarily give up the significant cost-benefit advantage of irrigated crop production over rainfed crop production. If terracing is a prerequisite for irrigation, then it is considered a change in the slope of the land and thus a change in site class. Residence and infrastructure are regarded two permanent uses within the time horizon of the analysis.

For conversions between site classes, across land use types, and over time, some simple accounting rules apply. Total area of the site class s in the period t is:

$$A_s^t = A_s^{t-1} + \sum_{k \neq s} \Delta A_{ks}^t - \sum_{k \neq s} \Delta A_{sk}^t , \qquad k = 1,..,S \tag{4.5}$$

with

$$\Delta A_{ks}^t \geq 0 \text{ and } \Delta A_{sk}^t \geq 0 ,$$

where ΔA_{ks}^t is the area of site class k transformed into site class s. The transformation between site classes is assumed to take place between the productive use in periods t-1 and t. ΔA_{sk}^t represents the reverse process of transfer from s to k. A similar equation holds for land use type h on site class s in period t:

$$A_{s,h}^t = A_{s,h}^{t-1} + \sum_{m \neq h} \Delta A_{s,mh}^t - \sum_{m \neq h} \Delta A_{s,hm}^t , \qquad m = 1,..,H \tag{4.6}$$

with

$$\Delta A_{s,mh}^t \geq 0 \text{ and } \Delta A_{s,hm}^t \geq 0 ,$$

where $\Delta A_{s,mh}^t$ is the area of land in land use type m which is transformed into land use type h within a given site class s before the productive use in period t, and $\Delta A_{s,hm}^t$ is again the reverse process.

If the transformation from site class s to site class k is an activity of a farmer or (local) government, usually investments are involved, and this also holds for a transformation from land use type h to land use type m. Large transformation processes (projects) may take several years, but may be broken down into smaller subprojects with associated costs and benefits to the user(s) of the land. The investment costs per area unit are denoted by b_{sk}^t and $c_{s,hm}^t$, respectively. Such investment costs are non-negative:

$$b_{sk}^t \geq 0, \; b_{ss}^t = 0$$

$$c_{s,hm}^t \geq 0, \; c_{s,hh}^t = 0 .$$

Total costs of transformations among site classes are:

$$B^t = \sum_k \sum_s b_{ks}^t \Delta A_{ks}^t \tag{4.7}$$

and for the transformations among land use types in a given site class s:

$$C_s^t = \sum_m \sum_h c_{s,hm}^t \Delta A_{s,hm}^t \tag{4.8}$$

In Table 4.2 the entries of Table 4.1 have been replaced by the costs of transformation. Coefficient $c_{s,42}$, for example, indicates the costs of transforming production forest into rainfed arable land, such as the costs of removing the trunks and roots. Tree cutting and cleaning of the land necessary for replanting trees are part of the forestry production and should be accounted for as inputs in the forestry sector.

As mentioned before, human-desired land use changes are linked with investments in the targeted land, following the economic decision-making process. With a sufficiently long time horizon, the economic actor, typically the farmer, evaluates the possibilities, invests in and produces on the land (s)he owns or rents, aiming at maximum profitability, given the local conditions of each plot of land. The transformation possibilities described above, along with the accounting rules for maintaining total area, enter the welfare model as technical constraints.

Table 4.2: Land transformation costs per hectare $c_{s,hm}$ for a given site class s.

	Destination				
Land use types Origin	Natural vegetation	Arable land rainfed	Arable land irrigated	Production forest	Residence infrastructure
Natural vegetation	0	$c_{s,12}$	$c_{s,13}$	$c_{s,14}$	$c_{s,15}$
Arable land - rainfed	-	0	$c_{s,23}$	$c_{s,24}$	$c_{s,25}$
Arable land - irrigated	-	-	0	$c_{s,34}$	$c_{s,35}$
Production forest	-	$c_{s,42}$	$c_{s,43}$	0	$c_{s,45}$
Residence and infrastructure	-	-	-	-	0

If site classes and land use types are defined at plot level, a farm level model of profit maximisation of this type would result in specialisation and this is economically efficient in general. However, if the spatial level of analysis is less detailed, for example, at county level or above, such specialisation is unlikely and unacceptable. At an aggregate level, considerable variability in the characteristics of the land is likely to occur within land use classes and site classes. Increasing the number of classes may not be possible due to lack of information, or unattractive for computational reasons. To take this into account, the increasing marginal costs of land transformation should be introduced. The fixed coefficients in Table 4.2 are replaced by increasing functions of $A_{hm/mh}$. In practice, the plots within a site class can be ordered according to the actual costs of transformation. The last plot is more expensive in transformation than the preceding plot. This plot is probably also more productive than the last one in terms of yield.

4.3 The Degradation-Induced Transformation Process

The effects of land degradation are assumed to appear to the user of the land as a loss of current possibilities and/or loss of production potential[3]. This refers to the definition of site classes. Similar to the land use transformation described in the preceding section, land degradation may be considered as a reverse transformation of productivity-increasing ones indicated in Tables 4.1 and 4.2. Analogous to those tables, a matrix of opportunity costs for degradation-induced land transformations can be constructed (Table 4.3). Two different approaches could be followed.

First, let us focus on transformation within a given site class. A further subdivision within each of the land use types can be applied. As an example, let us classify 5 degrees of land degradation based on their impact on productivity for a given land use class and site class: *negligible*, *light*, *moderate*, *strong* and *extreme* (van Lynden & Oldeman, 1997). The costs in Table 4.3 now represent the opportunity costs (per area unit) of such productivity-reducing transformations. In other words, they represent the minimum investments per area unit required for reversing the degradation-induced land transformation.

The lower-left part of the Table is empty, since no investments are incurred in transformation into a more severe, or in maintaining the same, degree of degradation. The grey shaded area in the upper right triangle represents transformations spanning two or more classes.

[3] Degradation processes may, in practice, affect land qualities only gradually, so that in the short run the user will not notice the effects (Kruseman & Van Keulen, this volume).

Table 4.3: Opportunity costs (per area unit) of degradation-induced transformation $b_{h,sk}$ for a given land use class h and site class s.

Impact of degradation / Origin	Destination: Negligible	Light	Moderate	Strong	Extremely strong
Negligible	-	b_{h12}	b_{h13}	b_{h14}	b_{h15}
Light	-	-	b_{h23}	b_{h24}	b_{h25}
Moderate	-	-	-	b_{h34}	b_{h35}
Strong	-	-	-	-	b_{h45}
Extremely strong	-	-	-	-	-

A clear limitation of the opportunity cost approach is, that in case of irreversible damage caused by degradation, we have difficulty to define meaningful opportunity cost. An alternative approach would be an integrated definition of site class, *i.e.* to treat a shift in land degradation class as a change between site classes. Such an integrated representation is more flexible, since it accounts for explicit transformation costs from one site class to another at each level of the degree of degradation. Both approaches, the nested and the integrated one work out more or less the same in the remainder.

Suppose the nested (or subdivided) definition is used. We have an ordered set, starting with the 'undisturbed' (best) available land and ending with completely degraded land per given site class. Production potential associated with the first class would be maximal, and in the last class, yield potential is reduced to an economically unattractive level. Given a land use type h, the plot changes from sub-class ℓ in period t to sub-class ℓ+1 in the next period. As a result of soil degradation, the probability that $\Delta A^t_{\ell,\ell+1}$ and $\Delta A^t_{L,L}$ happen is 1 *i.e.*

$$A^t_h = TA^{t-1}_h \tag{4.9}$$

with A^t_h a L-dimensional vector and T an $L \times L$ transition matrix defined as:

$$T_{\ell,\ell+1} = 1, \text{ for } \ell = 1, \ldots, L-1,$$
$$T_{L,L} = 1,$$

and all other elements equal to zero.

The farmer may reverse this process by investing an amount of $b^{t}_{\ell+1,\ell}$ per hectare in this plot, so that the land returns to (or remains in) sub-class ℓ before the next cropping period and the yield potential does not change. Although some of the conservation measures need to be executed during the growing period, they are not defined as inputs for the current crop activity. The purpose of these measures is to prevent soil degradation and thus loss of yield potential for the next period. Whether or not the farmer engages in a land conservation activity depends on several groups of factors, which include prices, solvency and other financial constraints, land use type changes (*e.g.* crop mix), site class changes, and the mix of input levels.

4.4 Micro and Macro Level

Land degradation and land transformation can be incorporated in the decision making process of the economic agent. With the inclusion of land qualities and their intertemporal changes as presented above, the transformation function (1) of the profit-maximising producer incorporates this type of management. This transformation function is further embedded in an intertemporal welfare optimisation framework as indicated in Section 4.2. Other (partial) intertemporal frameworks that use the transformation function in combination with resource management are possible. However, since land is such a vital resource, its productive use and management cannot be isolated from the social and economic environment. This is particularly significant in developing countries. Therefore, a comprehensive tool such as welfare analysis is highly desirable. The welfare approach provides the opportunity for researchers to incorporate both the objectives of the farmer and the policy maker. Farmers and policy makers are likely to differ in their perceptions of land degradation, in their objectives and in their time horizon for decision-making. National impact of land degradation is not the responsibility of the individual farmer *per se*, and even if they share the same concern, say, the notion of 'sustainable' agriculture, their valuation of current events is not necessarily the same. Both groups of economic agents can be modelled with instruments at the appropriate level. Farmers decide on input levels and investments, given prevailing prices, while the government can use policy measures such as producer taxes and/or subsidies to influence farmers' decisions. The government also has the opportunity to invest in infrastructural projects for which the set-up costs are too high for an individual (or groups of) farmer(s).

Welfare models are usually implemented at national level, in some cases with a regional subdivision. If there are sufficient observations for making an econometric estimation of the transformation function, an analysis at a more disaggregated spatial level would be possible. If statistical data are sufficiently detailed, geographical distribution of actors can be achieved by classifying them on the basis of location and main income-generating characteristics: for example,

'Farmers in the North', 'Rural Non-farmers in the North', 'Urban Skilled in the North', *etc.* The natural features of land resources are often available in much more detail in GIS databases, though a precise match between the concrete pattern of land use and these features is not easy. In estimating the production relations, this information can be used to determine the production potential (Albersen *et al.*, in prep.; Fischer *et al.*, 1999). Transformation possibilities with respect to land use classes, site classes and degradation classes (Tables 4.2 and 4.3) can be incorporated in the model as technological and natural condition constraints at a more disaggregate level than the regional or national one, at which the economic analysis takes place. Consistent aggregation of the area and the increasing marginal costs of transformation can be preserved. To a certain degree of approximation, it is possible to map the simulation results of the model back to the GIS as well (Albersen *et al.*, in prep.; Keyzer, 1998).

4.5 References

Albersen, P.J., G. Fischer, M.A. Keyzer & L. Sun (in prep.): Estimation of agricultural production relations in the LUC Model for China, Amsterdam: SOW-VU, and Laxenburg, Austria: IIASA

Alfsen, K.H., M.A. De Franco, S. Glomsrød & T. Johnson (1996): The cost of soil erosion in Nicaragua. Ecological Economics 16, 147-163

Aune, J.B., S. Glomsrød, V. Iversen & H.Wiig (1997): Structural adjustment and soil degradation in Tanzania: A CGE model approach with endogenous soil productivity. Statistics Norway discussion paper no. 189. Statistics Norway, Oslo

Bie, S.W. (1990): Dryland degradation measurement techniques. Environment Working Paper No. 26. The World Bank, Washington, D.C.

Carter, C.A. & Bin Zhang (1998): The weather factor and variability in China's grain supply. Journal of Comparative Economics 26, 529-543

FAO (1978): Report on the agro-ecological zones project, Vol 3. Methodology and results for South and Central America. World Soil Resources Report 48/3. FAO, Rome

FAO/IIASA (1993): Agro-ecological assessments for national planning: the example of Kenya, FAO Soils Bulletin 67, FAO, Rome.

Fischer, G., Y. Ermoliev, M.A. Keyzer & C. Rosenzweig (1996): Simulating the socio-economic and biogeophysical driving forces of land-use and land-cover change: The IIASA Land-Use Change model. Working Paper WP-96-010, IIASA, Laxenburg

Fischer, G., H.T. van Velthuizen & F.O. Nachtergaele (1999): Global agro-ecological zones assessment: Methodology and results. Interim Report IR-99-053, IIASA, Laxenburg

Ginsburgh, V. & M.A. Keyzer (1997): The structure of applied general equilibrium Models. MIT Press, Cambridge, MA

Gunning, J.W. & M.A. Keyzer (1995): Applied general equilibrium models for policy analysis. In: J. Behrman & T.N. Srinivasan (Eds.) Handbook of development economics, Vol. IIIA. North Holland, Amsterdam

Heilig, G.K. (1999): ChinaFood. Can China Feed Itself? CD-ROM version 1.1. IIASA, Laxenburg

Keyzer, M.A. (1998): Formulation and spatial aggregation of agricultural production relationships within the Land Use Change (LUC): model, Interim Report IR-98-092/October. IIASA, Laxenburg

Kruseman, G. & H. van Keulen (2000): Soil degradation and agricultural production: economic and biophysical approaches. This volume

Lindert, P.H. (1999): The bad earth? China's soils and agricultural development since the 1930s. Economic Development and Cultural Change 47, 701-736

Llewelyn, R.V. & A.M. Featherstone (1996): A comparison of crop production functions, using simulated data for irrigated corn in Western Kansas, Kansas State University, mimeo.

Lynden, G.W.J. van & L.R. Oldeman (1997): The assessment of the status of human-induced soil degradation in South and Southeast Asia. UNEP, FAO, ISRIC, Wageningen

Miller, D.J. & A.J. Plantinga (1999): Modelling land use decisions with aggregate data, American Journal of Agricultural Economics 81, 180-194

Perman, R., Y. Ma & J. McGilvray (1996): Natural resource and environmental economics. Longman, London

Pierce, F.J. & R. Lal (1994): Monitoring soil erosion's impact on crop productivity. In: R. Lal (Ed.), Soil erosion research methods. SWCS, Ankeny, Iowa

Sanders, J.H., D.D. Southgate & J.G. Lee (1995): The economics of soil degradation: technological change and policy alternatives. Technical monograph soil management support services no. 22. Purdue Univeristy, West Lafayette

Van Kooten, G.C. (1993): Land resource economics and sustainable development: Economic policies and the common good. UBC Press, Vancouver, BC

Voortman, R.L., B.G.J.S. Sonneveld & M.A. Keyzer (2000): African land ecology: opportunities and constraints for agricultural development, CID Working Paper No. 37, Center for International Development, Harvard

Wischmeier, W.H. (1976): Use and misuse of the universal soil loss equation. Journal of Soil and Water Conservation 31, 5-9

5 Population Pressure and Land Degradation in the Ethiopian Highlands: A Bio-Economic Model with Endogenous Soil Degradation

Bekele Shiferaw[1], Stein Holden[1] & Jens Aune[2] *

[1] Department of Economics and Social Sciences, Agricultural University of Norway, Ås, Norway

[2] Centre for International Environment and Development Studies, Noragric, Agricultural University of Norway, Ås, Norway

Abstract

We apply a bio-economic modelling approach to trace important relationships between population pressure, poverty and management of land resources for a crop-livestock economy characterised by serious land degradation in the Ethiopian highlands. Our objective is to examine the interlinkages between population pressure and poverty, their impacts on household welfare and land management, and the consequent pathways of development in a low potential rural economy. Farm households are assumed to maximise their discounted utility over the planning horizon in a multi-period model, where the management of the resource base has feedback effects on the stock and quality of the resource base. Market imperfections lead to non-separability of production and consumption decisions of farm households. The models trace the dynamic interactions between crop and livestock production, the resource base, consumption preferences, and partial integration of the household economy into markets. Simulation results indicate that under high population pressure, land becomes more expensive, relative to labour. This induces labour-intensive conservation investments when off-farm employment is limited and labour is not in scarce supply. Availability of credit and fertiliser, however, seem to discourage labour-intensive conservation efforts. When markets are imperfect, poverty in vital assets (e.g. oxen and labour) limits the ability or the willingness to invest in conservation.

---

* We would like to thank the Norwegian Research Council for financial support and the International Livestock Research Institute for invaluable logistical support and collaboration during the fieldwork in Ethiopia.

5.1 Introduction

The East African Highlands are the largest densely populated areas on the African continent. The favourable agro-ecological conditions facilitated the development of ancient agriculture and thus the creation of large organised societies. In recent times, the expansion of health services and a further decline in death rates has led to growth in population and increased scarcity of land. Despite the rapid increase in population, as compared to the rest of the world, technological change has largely stagnated in much of sub-Saharan Africa. Moreover, policy failures, social conflicts and wars have also contributed to market failures, reinforcing the effect of increased population pressure. The widespread poverty and rapid degradation of natural resources in the region today is likely to be a cumulative outcome of all these interrelated factors.

The relationships among population growth or pressure, poverty and environmental degradation has been subject of debate and controversies over several hundred years with an upswing over the last 30 years (Malthus, 1798; Boserup, 1965; Meadows *et al.*, 1972; Dasgupta, 1995). Empirical studies in East Africa (Lele & Stone, 1989; Turner *et al.*, 1993; Cleaver & Schreiber, 1994; Tiffen *et al.*, 1994; Grepperud, 1996; Shiferaw & Holden, 1998) have yielded contradictory results. The Machakos case study (Tiffen *et al.*, 1994) provides hope that a market-friendly policy approach might lift some of these subsistence-oriented natural resource based economies out of poverty and provide sufficient incentives for investment in conservation.

The concern about sustainability of production systems in fragile areas, and the need to quantify the impact of resource degradation on the future productivity of resources, has led to the development of an interdisciplinary modelling approach that links biophysical and socio-economic factors. Policy analyses for sustainable land use require functional integration of agro-ecological and economic parameters. This study applies a bio-economic modelling approach to identify the most important relationships between natural resources and human behaviour and well-being in a case study area in the Ethiopian highlands, which represents a low potential crop-livestock system. An econometric study (Shiferaw & Holden, 1998), based on data collected in 1994, revealed that low or negative initial returns to conservation technologies may undermine households' incentives to invest in their implementation. High population pressure and the need to use land occupied by conservation structures even seemed to encourage removal of conservation structures introduced in the past through food-for-work incentives.

The objective of this study is to assess the interlinkages among population pressure, poverty and land management and the pathways of development in a low potential crop-livestock zone in the highlands of Ethiopia. In part two, we provide the conceptual basis for the bio-economic model of the population-poverty-environment nexus. Part three offers a description of the case study area. The basic structure of the bio-economic model is presented in part four. The simulation results are presented and discussed in part five. In the final part, we conclude by highlighting the major findings, policy implications and the strengths and weaknesses of the modelling approach employed in this study.

5.2 The Conceptual Framework

Our analysis and modelling approach is based on the theories of transaction costs and imperfect information, which, when applied to the type of setting we are studying, is called the economics of rural organisation (Hoff *et al.,* 1993). This theoretical framework draws on neo-classical and institutional economics. The basic underlying assumptions are far from those in the neo-classical Arrow & Debreu (1954) model. Instead, we assume that pervasive transaction costs and information asymmetries exist, leading to severe market imperfections in typical poor rural economies (Greenwald & Stiglitz, 1986). Binswanger and Rosenzweig (1986) draw on the same theoretical framework, in combination with the specific characteristics of tropical agro-ecosystems, and use this to develop more specific hypotheses about the production relations, and especially market characteristics, in tropical agriculture. We follow the same approach, combining it with empirical testing, when identifying the appropriate underlying institutional characteristics of the economy under study.

We assume that farm households maximise their welfare given their available biophysical, human, capital and information resources and institutional constraints (including policies, markets, cultural norms, *etc.*). They are assumed to be impatient and therefore to discount future welfare. Poverty may force them to discount the future more heavily (Holden *et al.,* 1998; Holden & Shiferaw, 1998). Missing or imperfect credit markets, causing particularly poor households to be rationed out of credit markets, contribute to the high discount rates of the poor. Poverty and subsistence requirements may thus constrain investments in conservation of the natural resource base. Under such conditions, rapid population growth may lead to a neo-Malthusian scenario, unless new technologies, improved market access, new policies, and/or collective action contribute towards a more sustainable pathway that could break the nexus. Our study area appears to be in such a trap, and the bio-economic model may help us better understand the nature of the poverty-environment nexus for different household types and their ability to adjust to increasing pressure.

A number of factors influences farm households' production, consumption and investment decisions. Figure 5.1 illustrates the role of household resource endowments and the conditioning influence of population pressure, markets, and the policy and institutional framework on the intertemporal resource use decisions. The trade-offs between household consumption, production and investment decisions, as conditioned by asset endowments and external factors, determine the quality of the resource base in every period. Non-separability results from imperfections in labour and credit markets, while the intertemporal nature of the model relates to the choices between consumption and investment, the process of land degradation, and the impact of population growth on consumption needs and labour availability. When adverse external conditioning factors hamper households to make suitable adjustments in land use and technology choice, poverty and high population pressure may lead to degradation of natural resources (Shiferaw & Holden, 1997; Templeton & Scherr, 1999).

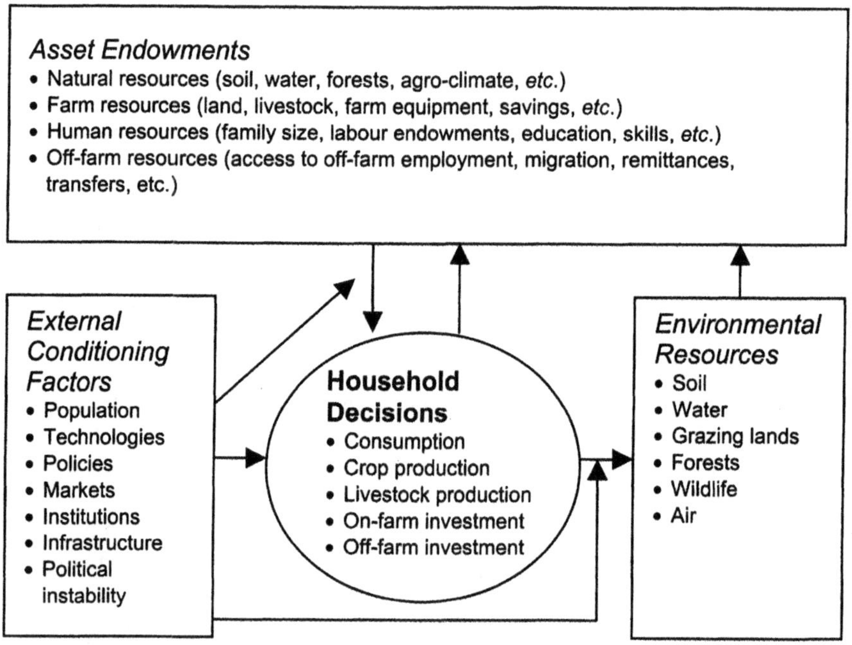

Figure 5.1: A conceptual farm-household model with poverty-environment links (Shiferaw & Holden, 1997).

Understanding farm households' incentives and constraints to intensification of land use, technology choice and investment behaviour, and analyses of the resulting pathways of development, requires integration of biophysical and economic modelling approaches at the household level (Ruben *et al.*, 1998). Integrated bio-economic modelling is therefore a logical approach for analyses of policies for sustainable land management and rural development. It provides a suitable analytical tool to investigate possible impacts of proposed new technologies (both production and conservation methods) and policies (*e.g.*

property rights, access to credit, access to input and output markets, taxes/subsidies, relative prices, *etc.*) on household welfare and the environment.

The intertemporal bio-economic model simultaneously determines the optimal use of household resources (land, labour, oxen, *etc.*) and optimal levels of consumption, production and land-improving investments, given their resource availability and market constraints. Although everything is endogenous in the long run, for a shorter planning horizon of 10-15 years, some variables may be considered exogenous. Prices of tradables and population growth are in this category. For other types of variables their initial conditions (stocks) are given, while their changes during the period for which the model is run, are endogenously determined. Examples of these are land quality, livestock holdings, and subsistence requirements. A third class of variables are those endogenously determined in each period, like *e.g.* input use, output, and consumption.

5.3 The Biophysical and Economic System

The study site (Andit Tid) is located some 180 km North East of Addis Ababa along the national highway in the northern Shewa region. The Soil Conservation Research Project (SCRP) opened a field station at the site in 1982 and gathered extensive biophysical data on soil erosion, run-off, sediment loss, conservation practices and crop yields. The SCRP climatic data show that Andit Tid has a bimodal rainfall pattern averaging 315.4 mm in the *Belg* (Jan.-May), and 1056.8 mm in the *Meher* (June-Dec.) season, with an annual average (1986/92) of 1372.2 mm. Topographically it is situated roughly between altitudes of 3000 and 3500 m.a.s.l., and is characterised by highly dissected and very steep terrain with over three quarters of the land area having a slope of more than 25%, which makes it highly vulnerable to erosion (Gebremichael, 1989). The area represents one of the low potential crop-livestock zones in the country facing a serious problem of soil degradation. Due to the bimodal nature of the rains, *Belg* and *Meher* crops are grown often without the use of mineral fertilisers. The *Meher* season crops include barley, wheat, faba beans, field peas, lentils and linseed. In the *Belg* season, only barley, field peas and lentils are grown.

There are strong crop-livestock interactions in the farming system. The common types of livestock in the area include cattle, sheep, goats, donkeys, horses and chicken. Crop residues are typically used as animal fodder. Oxen provide traction power. Animal manure is used for enhancing soil fertility (mainly on plots close to homesteads) and as a major source of household energy. Land may be used for crops, grazing, or tree (mainly eucalyptus) production. Use of fertiliser is limited but contributes towards increasing both crop and livestock production. High population pressure and scarcity of land seem to increase competition between crop and livestock production, especially for larger stock. Sale of the small stock complements both household consumption and crop production activities.

The soil types in the area include Andosols (66%), Regosols (32.4%), Cambisols (1%) and Fluvisols (0.6%) (Gebremichael, 1989). Andosols, locally know as *Tikur Afer*, relatively rich in organic matter, are often found above 3300 m.a.s.l., the frost prone zone mainly suited for growing of barley in the short rainy season (*Belg*). Fallowing and burning (locally known as *Ga'ay*) is commonly practised on these soils. Regosols, locally known as *Merere*, are found at lower altitudes around the settlements where frost is not a major problem. They are relatively poor in N, but richer in P. Regosols are the most important and intensively cultivated soils in the area. Soil erosion and nutrient depletion are the most important problems of soil degradation in the area.

In Ethiopia, the state maintains ownership to all land and farm households only retain uncertain use rights that may be revoked through periodic land re-distributions. Since the 1975 land reform, land has been distributed frequently to peasants, often on the basis of family size and land quality. The latest such redistribution in the study region occurred in early 1997, though its fairness and economic justification is questionable (Ege, 1997). Since the user does not own land and use rights are not sufficiently protected by law, farm size was not considered a suitable wealth indicator to classify households. Households were, therefore, classified on the basis of their oxen ownership, a vital resource and only means of traction power for smallholders in much of highland Ethiopia. In 1994/95, 26.5%, 15.3%, and 56.5% of the smallholders in the study area owned 0, 1 and 2 oxen, respectively. Only 1.7% owned more than a pair of oxen. Basic characteristics of the three farm household categories (no oxen, one ox, and two or more oxen) are presented in Table 5.1. Household size and farm size are positively correlated with oxen ownership, as is ownership of other livestock assets. Households without oxen obtain a relatively larger share of their income from renting out land and off-farm employment (see Table 5.1). For simulation of the biophysical system and variations in land quality, we classified land into different soil depth and slope classes by soil type. As indicated above, the soil types also represent the altitudinal variation in the study area (Regosols and Andosols below and above 3300 m.a.s.l., respectively). Distribution of land by soil type and soil depth class is presented in Table 5.2. Land quality classification by soil type, soil depth, and slope classes is presented in Table 5.3.

Due to its location near a major highway, Andit Tid has good access to relatively well-developed output markets in the vicinity. Most of the production is for home consumption and only wealthier households produce a significant surplus for sale. But there are significant transportation costs to some of the accessible markets in the area. Whilst households may trade labour, land, trees, and livestock and borrow credit within the village, other major transactions often occur in one of the nearby markets. Households also enter into short-term informal land rental contracts (share or fixed rent tenancy). The rental value of land depends on its quality, but it is not necessarily higher for land treated with conservation practices. Informal credit markets are generally underdeveloped, but may charge up to 60%

interest per annum. Formal credit facilities are scarce, but some NGOs provide limited credit at low interest rates (10-12%). Households may also get some credit for fertiliser (12% interest) from the peasants' co-operative located at the site. There is seasonality in product prices, indicating a high negative covariance between supply and prices. Thin local markets and relatively high transaction costs contribute to the demand-driven nature of output prices. The reservation wage varies seasonally: high during the peaks and low during the slack season. Labour may be hired in for cash, in fixed share of the crop output, or in exchange for traction power.

Table 5.1: Basic farm household characteristics in Andit Tid, 1994.

Variable	Household Type			All
	No ox	ONE OX	Two+ oxen	
Family size	2.80	5.80	7.20	6.10
Male work force	0.80	1.37	2.12	1.68
Female work force	0.70	1.53	1.73	1.53
Consumer units	2.60	5.17	6.58	5.55
Labour units	1.53	2.78	3.98	3.23
Own farm size (*Timad*)[a]	5.55	7.68	10.96	9.05
Tropical Livestock Units (TLU)	1.45	3.52	7.10	5.10
Oxen	0	1.00	2.30	1.53
Total cultivated land	3.30	5.08	8.84	6.73
Own cultivated area	3.00	4.79	7.80	6.07
Rented in land	0.30	0.28	1.04	0.66
Rented out land	1.55	0.18	0.10	0.31
Grazing land	0.25	0.32	0.80	0.55
Fallow land	0.75	2.38	2.26	2.12
Total household income (Birr)	1215	1489	2297	1859
Off-farm income	297	256	251	258
Livestock income	81	248	267	236
Rental income (land and oxen)	123	25	23	36
Crop income	714	961	1755	1327
Cash input expenditure	172	263	395	318

[a]Land areas are measured in *Timad*, approximately a quarter hectare. The sample size was 10, 30 and 40 households for households with 0, 1, and ≥2 oxen, respectively.

Table 5.2: Land areas of surveyed households (in *Timad*) by soil type and soil depth.

Soil depth (cm)	Soil type		All
	Regosols	Andosols	
0-30	339	157	496
30-60	148	102	250
> 60	50	51	101
Total area	537	310	847

Table 5.3: Land area (in *Timad*) by farm household category, soil type, soil depth and slope classes.

Soil type	Codes	Soil depth class (cm)	Slope Class (%)	Household category		
				No ox	One ox	Two+ oxen
Andosols		All	All	2.03	2.82	4.02
"	AD1S12	0-30	0-20	0.91	1.26	1.8
"	AD2S12	30-60	0-20	0.57	0.78	1.12
"	AD3S34	>60	0-20	0.32	0.44	0.63
"	ADSS34	0-30	>20	0.24	0.33	0.48
Regosols		All	All	3.52	4.88	6.98
"	RD1S12	0-30	0-20	1.62	2.25	3.21
"	RD2S12	30-60	0-20	0.86	1.19	1.69
"	RD3S34	>60	0-20	0.31	0.44	0.62
"	RDSS34	0-30	>20	0.73	1.01	1.44

5.4 The Bio-economic Model

The bio-economic model uses a non-separable farm household model as a basis whereby production, consumption and investment decisions are determined interdependently in a dynamic setting. The household's choice is constrained by market imperfections (*e.g.* credit, labour, livestock feed, *etc.*) emanating from seasonality of prices and transaction costs. This may encourage the household to attach a mark-up price for tradables that ease the credit constraint and to be self-sufficient in non-tradables thereby making the model non-separable (Singh *et al.*,

1986; de Janvry *et al.*, 1991). Household characteristics and asset endowments therefore play an important role in decisions. The on-site costs of soil erosion and nutrient depletion are endogenous, and their impacts on land productivity in consecutive years influence the household's choice of land management practices. If the shadow value of soil degradation is high (*e.g.* due to high relative returns to farming, low discount rates and/or long planning horizons) the household will implement conservation methods until marginal returns equal marginal costs of such investments.

The farm household maximises a discounted utility function defined as the sum of the household's net income and the value of home time (leisure)[1]:

$$U = \sum_{t=1}^{T}\left(\sum_{g=1}^{G}\sum_{c=1}^{C} A_{cgt}\left\{\rho_{ct}\, y_{cgt}(x_t) - \sum_{i=1}^{n} e_{icgt}\, x_{icgt}\right\} + \sum_{V=1}^{V} L_{vt}\right.$$
$$\left.\left\{\rho_{vt}\, y_{cgt}(x_t) - \sum_{i=1}^{n} e_{icgt}\, x_{ivt}\right\} + \sum_{s}^{S} w_{st}\,(h_{st} + z_{st})\right)(1+r)^{-(t-1)} \qquad (5.1)$$

subject to resource supply, market access and subsistence constraints[2]. A is the area of crop c produced on land type g in period t. L is the production of units of livestock v in each period. $\boldsymbol{x}_t$ is a vector of inputs $(x_{1t}, \ldots, x_{nt})$ used in crop and livestock production. ρ is the per unit price of crop or livestock output and e is the per unit input cost for input x_i. y is the yield function for the production of crop c or livestock v. Family labour 'consumed' as home time (leisure) is h while w is the seasonal reservation wage (after transactions costs). The seasonal off-farm labour supply is z. The rate of discount is r.

The decision variables in the model include crop and area choice by land type; levels of fertiliser use; allocation of land, labour and traction power across different seasons; seasonal labour and oxen hiring; livestock production; selling and buying of crops and livestock; buying of farm inputs; choice of consumption goods to meet subsistence needs; savings and borrowing of credit; and choice of level of soil conservation investments[3]. Land use decisions are made seasonally among different crops as dictated by agro-ecological conditions. Family labour is used on-farm, employed elsewhere or consumed as leisure. The constraints include the household's resource endowments (land, labour, oxen and other livestock); subsistence consumption needs, livestock feed requirements, limits on

---

1 A non-linear formulation of the utility function may improve the model but adds complexity which makes it time-consuming to solve. The estimated production function with fertiliser response curves and endogenous soil degradation already makes our model non-linear in constraints.

2 The actual model is more complex than the highly aggregated presentation given here.

3 Seasonal activities and constraints were defined for ten periods within a year, reflecting the cropping calendar and other seasonal activities in a given year.

borrowing of credit and off-farm employment, and restrictions on crop rotations. A price band of 5-10% was included, depending on item and season, for buying activities over selling activities to account for transactions and supervision (for labour) costs.

The level of soil erosion and nutrient depletion on each land type is determined by crop type and technology choice. Changes in rooting depth (and hence fertility) of the soil depend on soil type, slope, crop and conservation technology choice. Soil depth (*sd*), and hence soil erosion, in each period depend on soil type (*s*), slope (*sl*), rainfall (*rf*), land management practice used (*k*) and type of crop grown (*c*).

$$sd_t = \varphi(s, sl, rf_t, k, c_t) \tag{5.2}$$

Soil erosion levels for each land type and land use under the two land management options (with or without conservation) were obtained from the Soil Conservation Research Project (SCRP) experiments at the site or have been estimated using regression methods based on plot-level survey data for some land use types where experimental data were unavailable. Depending on soil type, slope and crop grown on each type of land, choice of conservation technology (*k*) determines the level of soil erosion for the average level of rainfall in the area.

Under given growing conditions, crop yield in each period depends on soil type, rooting depth, and plant available nutrient stock in the soil. The level of use of organic and inorganic fertilisers, initial nutrient content of the soil, and choice of land management practices (conservation technologies) determine the plant available nutrient stock in each period. The rooting depth also depends on conservation technology choice. Hence, the yield of crop *c* on land type (*g*) in period *t* is given as a non-linear function of:

$$y_{cgt} = f(s, sd_t, N_t, P_t) \tag{5.3}$$

where *s* is soil type, *sd* is soil depth, and *N* and *P* are nitrogen and phosphorus available to plants either from the soil stock or from external additions. The soil depth-yield relationship is econometrically estimated based on a plot-level time series data collected by SCRP, while responses to N and P fertiliser are estimated from the FAO fertiliser response studies (Ho, 1992) and the soil productivity calculator (Aune & Lal, 1995). Depletion of plant available soil nutrients through soil erosion and exhausting production practices (for nitrogen) reduces crop response to fertiliser use.

The change in the N stock is given by:

$$N_{t+1} = N_t - \delta\,(N_t - \eta(se_t)) - \eta(se_t) \tag{5.4}$$

where *se* is the period *t* rate of soil erosion, δ is the decomposition (mineralisation) constant for soil organic matter in each period, and η is the N composition of the soil. An enrichment ratio of 2 is used for eroded soil together with an annual mineralisation rate of 1% for soil nitrogen. Nutrient depletion reduces the plant available N and P. The change in plant available nitrogen (ϖ) due to nutrient depletion is computed as:

$$\varpi = \delta(N_t - N_{t+1}) \tag{5.5}$$

where δ is as defined above. Hence, the cumulative depletion of soil N is included into the production function and feeds back into the model to influence future crop yields. Currently only the effect of depletion of rooting depth and N stock on crop yields are incorporated into the model. This is because incorporating the effect of depletion of the P stock on land productivity requires additional data on P-fixation, conversion of stabile P to labile P, and the total P-stock in the soils.

Because of their relative suitability for high rainfall areas, the model uses graded soil-stone structures as erosion-control methods. Construction of conservation structures is estimated to require 100-120 working days per hectare (ha) in the first year, while maintenance requires 15-20 days/ha per year, depending on the slope gradient of the land. This is in line with the norms developed by the Ministry of Agriculture for remuneration of conservation labour through food-for-work schemes. The model determines both the seasons for undertaking land-improving investments and the amount of labour used. Although conservation structures may occupy productive land and reduce yields in the initial period, the model assumes that the effect of conservation investments on crop yields is only through mitigating soil erosion. If area loss leads to lower production, the reduced relative profitability of conservation may lead to lower optimal investments than estimated by the model[4]. On the other hand, more than merely mitigating soil depletion, conservation may improve certain soil properties or may raise relative returns to intensification (*e.g.* fertiliser use). Therefore, the net effect is indeterminate.

The model is solved for the two household types: poor household (no ox), and better-off household (two oxen). Hence, four scenarios are developed: poor land-scarce, poor land-abundant, better off land-scarce, and better off land-abundant (See Table 5.4)[5]. This means that the effect of the starting population pressure (as defined by the per capita land availability for the different household types) on intertemporal resource use is incorporated through changing the land-labour ratio of the households at the initial period. The model is calibrated using real household and biophysical characteristics of the study area in 1994. We use a

---

[4] When structures occupy some portion of a unit of cropland, the yield per effective area cultivated may be the same, but the response for the unit as a whole (crop production) decreases on land treated with conservation.

[5] Our classification of 'better-off' and 'poor' is for lack of a better term. Indeed the term better-off is relative, since most (if not all) households in the area are poor. In this regard, the term 'less-poor' may be preferred to 'better-off'.

planning horizon of 10 years and a discount rate of 10%[6]. A 2% annual growth rate of the household is also included in the model. Coupled with depletion of land productivity due to soil degradation, population growth makes meeting subsistence more difficult during the terminal years of a longer planning horizon. However, growth of the family and degradation of resources increase the pressure on the land at different levels for the four stylised household types outlined above.

Table 5.4: Household sizes in the selected scenarios.

	Better off		Poor	
	large[a]	small	large	small[a]
Family size	7.2	4.2	5.0	2.8
Worker units	4.0	1.5	3.0	1.5
Consumers units	6.6	3.0	4.0	2.6

[a] These are average values from the study area (see Table 5.1).

5.5 Simulation Results

5.5.1 Impact of Population Pressure on Better-off Households

According to the Boserupian hypothesis (Boserup, 1965), intensification of land use and investments to enhance its productivity will be limited when land is more abundant relative to labour. This suggests that smaller families with large farms will have less incentives to increase intensity of labour and other inputs per unit of land to enhance productivity of their farm. The results from our simulations also indicate that when land is relatively more abundant, land users lack the impetus to make significant investments to mitigate soil erosion. For a better off household with abundant land, satisfying subsistence requirements of the small household is less demanding even when markets are imperfect and soil degradation is taking place. Hence, when the impact of soil degradation on household welfare is low, the incentive to mitigate it is likely to be less. Shortage of labour relative to land also means that the labour-scarce household will have to hire in labour to install labour-intensive soil conservation investments. When the shadow value of family

[6] Although a planning horizon of 10 years may be relatively long, especially when use rights to land are insecure, the 10% discount rate is on the low side for poor, subsistence-constrained households. Higher discount rates will discourage erosion-control investments on land (Holden *et al.*, 1998; Shiferaw & Holden, 1999). Endogenising the discount rate and making it specific to household types is likely to improve specification of the model.

labour is high, the household lacks the impetus to implement labour-intensive conservation investments, unless the user cost of soil degradation is very high. Results also show that the shadow value of land was much lower for the land-abundant household than for the land-scarce household. The marginal user cost of soil erosion was lower on farms with higher land-to-labour ratio. However, for most of the seasons, the marginal return to labour in agriculture (shadow value) was higher for the household with higher land-to-labour ratio. The cumulative effect of scarcity of labour and land abundance is lower soil conservation effort in all market scenarios (see Table 5.5).

Table 5.5: Percentage of land area treated with conservation practices at the terminal period (better off household under different market and resource constraints).

Household type	Household welfare ('000)	Regosols				Andosols			
		RD1S12	RD2S12	RD3S12	RDSS34	AD1S12	AD2S12	AD3S12	ADSS34
Land-abundant[a]	24.72	9.5	51.9	0	0	0	0	0	0
Land-scarce[a]	*37.67*	*11.2*	*56.4*	*40.9*	*0*	*16.2*	*0*	*0*	*0*
Land-abundant[b]	20.43	13.7	27.0	0	0	17.6	0	0	0
Land-scarce[b]	*33.49*	*31.6*	*63.2*	*61.0*	*3.8*	*35.8*	*45.2*	*37.0*	*0*
Land-abundant[c]	20.39	13.7	27.1	0	0	17.5	0	0	0
Land-scarce[c]	*32.48*	*36.6*	*59.2*	*82.2*	*0*	*50.1*	*67.3*	*90.6*	*0*

[a] Unlimited credit and fertiliser supply; [b] Limited credit and fertiliser supply; [c] Without credit but limited fertiliser supply.

In the base scenario of limited credit and fertiliser supply, scarcity of land relative to family labour due to population pressure induces more soil conservation (see Table 5.5). Therefore, when off-farm employment is limited, the land-scarce household, which is not constrained by shortage of family labour, tends to implement more erosion-mitigating practices than the land-abundant household. Moreover, access to credit reduces the incentive to conserve land. For the small household, shortage of labour and low shadow value of land seem to prevent increased soil conservation investments. Unlike the large household, access to credit has a limited effect on land management since the need for fertiliser credit is limited. In the presence of land rental markets, the small better-off household also finds it more profitable to rent out more land than to hire in labour for cultivation. On the other hand, the land-scarce household puts more effort into soil conservation when the credit market is missing and restitution of soil fertility through increased use of mineral fertilisers becomes more difficult. Hence, when family labour is abundant, land is scarce, and labour-intensive conservation technologies are available, capital market imperfections seem to encourage investment in soil fertility maintenance.

The small household will rent out surplus land when land markets exist and use fertiliser on the remaining land with some hired labour. The small better-off household uses more fertiliser per unit of cultivated land. The land-scarce household rents in additional high quality land mainly on Regosols, while it may rent out some poor quality land especially on Andosols to generate some cash. While the credit constraint is binding in the base scenario in almost all years for the large land-scarce household, the small land-abundant household does not require credit after a few initial years. If access to credit is limited, households rent out more land or shift it into grazing to specialise more in livestock production. The household shifts more land into grazing when lack of credit prevents investment in fertilisers, but increased production and selling of small stock generates the required capital for investment and purchasing some food to satisfy subsistence demand. When off-farm employment is possible, a boost in off-farm labour supply, made possible through a reduction in leisure time, also compensates shortage of capital credit.

If access to credit and fertiliser is unlimited, maintenance of soil productivity via increased use of mineral fertilisers per unit of land becomes a more preferred strategy[7]. This further discourages soil conservation efforts, even when family labour is abundant, since the yield gap between alternative management practices progressively disappears with increased use of fertilisers. The outcome may be different when a longer planning horizon is used or if soil conservation is a cheaper alternative than fertiliser use to improve soil fertility.

Therefore, improved access to credit and fertilisers increases household welfare and encourages intensification in terms of increased fertiliser use per unit of land, but discourages investment in labour-intensive soil conservation practices other than allowing mitigating nutrient depletion through investment in soil fertility maintenance. Although fertiliser use may temporarily offset soil degradation and improve yields in the short-term, sustained loss of organic matter, water holding capacity, and soil depth is very difficult to replace (Nandwa & Bekunda, 1998; Bationo *et al.*, 1998). Therefore, without proper soil fertility management, increased use of fertilisers could be a non-sustainable path in the long run, since mineral fertilisers are unlikely to substitute for the eroded soil stock. An important policy implication of this result is the need to interlink credit and fertiliser subsidies with conservation requirements to avoid substitution of fertiliser for soil conservation investments.

---

[7] In the short-run, this is an unrealistic scenario but access to credit and fertiliser is likely to improve in the long run. It is included here to examine the possible impact of an improved access to credit and fertiliser on land management.

5.5.2 Impact of Population Pressure on Poor Households

The impact of population pressure on management of resources and household welfare becomes more pronounced when the household is less-endowed in terms of animal traction. The effect of higher pressure on achieving the household's basic needs and countering soil degradation is even more severe when credit markets fail and off-farm employment opportunities are limited. Rural market imperfections, therefore, reduce the ability of poor households to adjust to high population pressure, the impact being stronger on poor households with more consumer units per unit of land. When rental markets for oxen and land exist, a small land-abundant household rents out the surplus land and cultivates the rest by hiring in some oxen. As for the small better off household, scarcity of labour relative to land discourages the household from allocating its scarce labour to soil conservation. The household instead intensifies land use on the remaining land by using capital inputs (fertiliser). Satisfying subsistence needs of the small household becomes much easier, even in the absence of a credit market, when rental markets exist for land and oxen and the household is able to purchase food shortfalls from local markets. However, a missing credit market forces the small household to reduce leisure time and increase off-farm labour supply, to reduce the area cultivated, to increase rented out land or land used for grazing. Hence, livestock sales or income from renting out land increases when credit markets fail. Therefore, in the integrated crop-livestock economy, livestock not only provides traction power and organic manure for use in cropping, but sale of animals, especially small stock, also provides cash for credit-constrained households. However, shortage of labour and its high shadow price still prevent the household from investing in mitigating soil degradation (see Table 5.6).

When the household is less-endowed in both oxen and land, population pressure (increased family size) puts high pressure on its ability to meet basic needs. When higher transaction costs exist and food markets are imperfect, the high shadow value of subsistence and scarcity of land limit the household's ability to rent out land and buy its subsistence consumption. The household rents in oxen to grow part of its subsistence needs, but meeting the consumption needs of a large household becomes difficult, unless the surplus labour finds some employment off-farm. If the labour market is missing, the household is unable to meet its basic needs unless external assistance is made available. In such situations, some members of the family might migrate in search of employment elsewhere. Therefore, for a poor, land-scarce, labour-rich household, a vital means of attaining livelihood requirements is to find off-farm employment. When off-farm jobs are available, the incentive to implement improved land management practices depends on the profitability of farming relative to off-farm employment. If off-farm employment opportunities are unlimited, the household allocates up to twice as much labour to off-farm than to on-farm. In this case, the strong dependence on off-farm income seems to discourage soil conservation efforts. When off-farm employment is limited, the household invests the surplus labour in soil conservation practices (see Table 5.6). Improving the relative profitability of soil conservation through cutting the costs or increasing yields on conserved land is therefore more likely to encourage households to invest in more sustainable land management practices.

Table 5.6: Percentage of land area treated with conservation practices at the terminal period (poor household under different market and resource constraints).

Household type	House-hold welfare ('000)	Regosols				Andosols			
		RD1S12	RD2S12	RD3S12	RDSS34	AD1S12	AD2S12	AD3S12	ADSS34
Land-abundant[a]	15.50	2.5	4.7	12.5	0	0	0	0	0
Land-scarce[a]	*22.87*	*2.5*	*26.8*	*12.5*	*0*	*0*	*0*	*0*	*0*
Land-abundant[b]	14.72	2.5	4.7	12.5	0	0	0	0	0
Land-scarce[b]	*21.07*	*8.4*	*43.9*	*15.5*	*23.1*	*0*	*22.7*	*0*	*0*
Land-abundant[c]	14.52	2.5	4.7	12.5	0	0	0	0	0
Land-scarce[c]	*20.25*	*8.4*	*53.3*	*17.5*	*43.1*	*0*	*32.7*	*0*	*0*

See Table 5.5 for explanation of superscripts.

The need for off-farm employment becomes more pronounced when credit markets are missing. Unless the household compensates lack of borrowing opportunities through increased deployment of its labour off-farm, it cannot meet its livelihood requirements, possibly leading to famine, unless out-migration or external assistance improves its situation. Under credit market imperfections, limited access to off-farm work may encourage reduced use of credit-constrained commodities (*e.g.* fertiliser) thereby slightly encouraging soil conservation effort (see Table 5.6). However, such effects under the credit constraint disappear when the household does not face problems for off-farm employment. A perfect credit and fertiliser market, which allows mitigation of the consequences of soil degradation through increased use of fertiliser, also discourages land-improving investments. While population pressure and scarcity of land seem to encourage land-improving investments, poverty in household assets (*e.g.* oxen and labour) generally limits the ability to invest in improved land management. Labour-scarce households without oxen find it difficult to invest in land, since lack of traction power forces them to rent out land. Construction of conservation structures may in this case be unattractive, unless the rental value of land increases through conservation and other land-improving investments. This is unlikely, unless farmers' land rights are more secure and leasing out land for a relatively longer period of time is possible.

5.6 Conclusions

One of the major challenges in degrading hilly and mountainous regions with high population pressure is to increase the productivity of agriculture while maintaining or improving the production potential of the resource base. Model simulations indicate that, when the land-to-labour ratio is high, better off households are unlikely to implement labour-intensive soil conservation investments to improve the productivity of the resource base. A growth in family size and increased scarcity of land encourage households to undertake some soil conservation practices, especially when off-farm employment is limited. Access to credit and fertiliser markets, which allows yield increases through addition of plant available soil nutrients, generally discourages such investments, since yields can be maintained despite accelerated degradation of the soil. This outcome may change if soil conservation can be made cheaper than fertiliser use to improve soil fertility or if conservation is positively correlated with returns to fertiliser use. This suggests the need for policies that link access to credit and fertiliser with soil conservation requirements to create short-term complementarity between fertiliser use and conservation. Policies that link access to program benefits to compliance with approved conservation standards are typical examples. Improving the response to fertiliser use on conserved land has a similar effect.

When the household is poor, both in labour and traction power, it has no option but to rent out land to other relatively better off households. Such short-term contracts, however, do not provide a sufficient incentive to those renting in land to undertake significant land-improving investments. On the other hand, access to credit encourages land-rich households to intensify land use by increasing the use of fertiliser without significant effort to abate soil degradation. This may also be the case when land is scarce, but the system of property rights constrains households from considering a longer planning horizon to internalise long-term effects of soil degradation. For the poor and large household, the incentive to carry out conservation measures very much depends on the relative returns to on-farm and off-farm employment. When off-farm employment is available, the poor household, lacking traction power, prefers to rent out more land and deploy much of its labour off-farm. Lack of credit for acquiring oxen reinforces this outcome, while rental markets for oxen have the opposite effect. Poverty in some asset categories (oxen and labour) reduces the incentive to use labour-intensive land management practices. Hence, poverty in oxen has decreased the household's ability to adjust to high population pressure. Scarcity of land and cash capital, however, seems to increase the incentive to conserve the soil.

Summarising, the effect of population pressure and the associated pathway of development depend on economic and institutional constraints, and on the asset position of the household. Household poverty in some asset categories, lack of profitable conservation technologies, lack of information thereof, or insecurity of land tenure may prevent land users from following a more sustainable intensification path. However, population pressure and scarcity of land may encourage increased land-improving investments when other enabling conditions are fulfilled.

Furthermore, the modelling approach applied in this study links degradation of the soil to household decision-making by endogenizing cumulative soil erosion and nutrient depletion into a model where consumption, production and conservation investment decisions are jointly determined. This approach is more appropriate given the pervasive market imperfections that small farmers face in rural economies of the developing world. The multi-period model also attempts to account for the lagged on-site effects of soil degradation. However, the model can be improved in several ways: (1) defining a non-linear objective function (although this may increase solving time and make policy analysis more difficult), (2) including the effect of depletion of phosphorus and other limiting nutrients, investment in trees, and risk choice. Our work in progress tries to address these issues.

5.7 References

Arrow, K.J. & G. Debreu (1954): Existence of an equilibrium for a competitive economy. Econometrica 22, 265-290

Aune, J. & R. Lal (1995): The tropical soil productivity calculator: A model for assessing effects of soil management on productivity. In: R. Lal & B. Stewart (Eds.) Soil Management Experimental Basis for Sustainability and Environmental Quality. Lewis Publishers, 499-520

Bationo, A., F. Lompo & S. Koala (1998): Research on nutrient flows and balances in West Africa: state of the art. Agriculture, Ecosystems and Environment 71, 19-36

Binswanger, H.P. & M.R. Rosenzweig (1986): Behavioral and material determinants of production relations in agriculture. The Journal of Development Studies 22, 503-539

Boserup, E. (1965): The conditions of agricultural growth. The economics of agrarian change under population pressure. Earthscan Publications Ltd, London

Cleaver, K.M. & G.A. Schreiber (1994): Reversing the spiral. The population, agriculture and environment nexus in sub-Saharan Africa. The World Bank: Washington, D.C.

Dasgupta, P. (1995): The population problem: Theory and evidence. Journal of Economic Literature 33, 1879-1902

de Janvry, A., M. Fafchamps & E. Sadoulet (1991): Peasant household behaviour with Missing Markets: Some Paradoxes Explained. Economic Journal 101, 1400-1417

Ege, S. (1997): The promised land: The Amhara land redistribution of 1997. Working Papers on Ethiopian Development No.12. Centre for Environment and Development, University of Trondheim

Gebremichael, Y. (1989): Land-use, agricultural production and soil conservation methods in the Andit Tid Area, Shewa Region. Research Report 17. Soil Conservation Research Project, Addis Ababa

Greenwald, B. & J. E. Stiglitz (1986): Externalities in economies with imperfections and incomplete markets. Quarterly Journal of Economics 101, 229-264

Grepperud, S. (1996): Population pressure and land degradation: The case of Ethiopia. Journal of Environmental Economics and Management 30, 18-33

Ho, C.T. (1992): Results of NPK fertiliser trials conducted on major cereal crops by ADD/NFIU (1988-1991). ADD/NFIU Joint Working Paper 43, Ministry of Agriculture, Addis Abeba

Hoff, K., A. Braverman & J. E. Stiglitz (Eds.) (1993): The economics of rural organization. Oxford University Press, Oxford

Holden, S.T. & B. Shiferaw (1998): Poverty and land degradation: Peasants' willingness to pay to sustain land productivity. Discussion Paper D-06/1998, Department of Economics and Social Sciences, Ås

Holden, S. T., B. Shiferaw & M. Wik (1998): Poverty, credit constraints, and time preferences: of relevance for environmental policy? Environment and Development Economics 3, 105-130

Lele, U. & S.W. Stone (1989): Population pressure, the environment and agricultural intensification. Variations on the Boserup hypothesis. MADIA Discussion Paper 4. The World Bank, Washington, D.C.

Malthus, T. (1798): An essay on the principles of population, as it affects the future improvement of society. Johnson, London

Meadows, D.H., D.L. Meadows & J. Randers (1972): The limits to growth. Potomac Associates, Washington, D.C.

Nandwa, S.M. & M.A. Bekunda (1998): Research on nutrient flows and balances in East and Southern Africa: state of the art. Agriculture, Ecosystems and Environment 71, 5-18

Ruben, R., H. Moll & A. Kuyvenhoven (1998): Integrating agricultural research and policy analysis: analytical framework and policy applications for bio-economic modeling. Agricultural Systems 58, 331-349

Shiferaw, B. & S.T. Holden (1997): Peasant agriculture and land degradation in Ethiopia: Reflections on constraints and incentives for soil conservation and food security. Forum for Development Studies 2, 277-306

Shiferaw, B. & S.T. Holden (1998): Resource degradation and adoption of land conservation technologies in the Ethiopian highlands: A case study in Andit Tid, North Shewa. Agricultural Economics 18, 233-248

Shiferaw, B. & S.T. Holden (1999): Soil erosion and smallholders' conservation decisions in the Ethiopian Highlands. World Development 24, 739-752

Singh, I., L. Squire & J. Strauss (Eds.) (1986): Agricultural Household Models: Extensions, Applications and Policy. Baltimore: The Johns Hopkins University Press.

Templeton, S.R. & S.J. Scherr (1999): Effects of demographic and related microeconomic change on land quality in hills and mountains of developing countries. World Development 27, 903-918

Tiffen, M., M. Mortimore & F. Gichuki (1994): More people - less erosion: environmental recovery in Kenya. Wiley and Sons, London

Turner, B.L., G. Hyden & R. Kates (1993): Population growth and agricultural change in Africa. University of Florida Press, Gainesville

6 Imperfect Food Markets and Household Adoption of Soil Conservation Practices in the Dominican Republic Highlands: Household Probit and Duration Models

Bénédicte de la Brière*

Development Economics Research Group, The World Bank, Washington D.C., U.S.A.

Abstract

This paper formalises the determinants of adoption and maintenance of soil conservation practices on peasant households' food plots in the Dominican Republic highlands, where a local rural development project provides training, technical assistance and subsidies in the form of food-for-work. We model these decisions as an intertemporal labour allocation choice in the context of household-specific food markets imperfections. From the theoretical model, we derive consistent empirical models (probit and duration models). Results show, that food subsidies help bring marginal households into soil conservation. Households facing a higher return to their labour on the labour market tend nevertheless to abandon conservation practices once the subsidy stops. Large landholdings are also associated with less soil conservation. The main adopters are thus households strongly vested in agriculture, especially if they depend on own production for their consumption.

6.1 Introduction

Soil erosion and resulting yield and welfare losses severely threaten the livelihoods of poor households in the fragile lands of watersheds[1]. Even directly

---

* An earlier version of this paper formed a chapter of my Ph.D dissertation in the Department of Agricultural and Resource Economics at UC Berkeley.

[1] We do not deal in this paper with the external costs of soil erosion, for instance sedimentation of reservoirs for downstream users. In addition to concerns with rural poverty, internalisation of these externalities will justify social investment in rural development programs such as the one we analyse.

faced with soil erosion detrimental effects, many households typically do not adopt soil conservation techniques. Subsidised adoption through rural development programs may end in subsequent abandonment, thus failing to achieve sustainability. Policies may affect both decisions differentially, thus calling for dynamic analyses of soil conservation programs.

This paper analyses the determinants of adoption and maintenance of soil conservation in two remote watersheds of the Dominican Republic, characterised by deforestation, swidden agriculture, soil erosion, casual labour markets, and widespread receipts of remittances. In 1979, to address pervasive malnutrition and improve poor peasants' food production, Plan Sierra, a rural development program, started promoting soil conservation techniques[2], with the provision of two years of food-for-work subsidies, tools, and training. Diffusion was extensive, as was however subsequent abandonment, thus calling for understanding behavioural determinants to improve Plan Sierra's methods.

We build an intertemporal behavioural model of households' labour allocation to agriculture, soil conservation, and labour market activities, in the context of household-specific market failures, which Plan Sierra partially mitigates. We focus on food market imperfections, as food security was the main goal of Plan Sierra's 'Food Self-Sufficiency Program'. Specifying this model and analysing its comparative static implications allow us to formulate a consistent empirical model of adoption and abandonment (Besley & Case, 1993). The following hypotheses (de Janvry *et al.*, 1994) guided the formulation of the model:

1. Autarkic households, which face an endogenous food price, will be more prone to use soil conservation techniques[3]. When separability does not hold (de Janvry *et al.,* 1991; Sadoulet & De Janvry, 1995), household characteristics in consumption affect time allocation decisions, and hence soil conservation. The converse is true, if separability holds (Burt, 1981; Mc Connell, 1983).
2. Food subsidies may either artificially boost adoption, if some households adopt only to receive the food, or they may help cover initial capitalisation costs and hence increase both adoption and maintenance.

---

[2] The conservation techniques promoted included: (1) erosion control methods: terraces, slope drainage ditches, canals (water storage, control and drainage), stone and branches barriers, low tillage; (2) crop yield enhancement methods: contour sowing, crop rotation, inter-cropping, hedgerows, live fences, green mulching, grass or nitrogen-fixing barriers, productive ditches.

[3] The fragmentation of markets for foods that can be produced locally is notable. Communities in the Sierra have no farmers' markets and local shopkeepers mainly sell dry foods such as noodles and rice that are not produced locally. For five staple crops, reported self-sufficiency ranged from 34% to 70% –beans (34%), corn (48%), cassava (51%), sweet potato (61%), and bananas (70%)–.

3. Smaller farms may face tighter land constraints in a swidden system and thus have greater incentives to use soil conservation techniques.

4. With off-farm earnings setting the opportunity cost for time allocation, higher labour wages deter time allocation to soil conservation and encourage abandonment.

5. Perceived security of tenure matters for soil conservation investments, because of sunken costs, both in physical infrastructure and in knowledge acquisition. The magnitude of the land-saving effect of soil conservation depends on the idiosyncratic managerial ability of households, particularly in the use of the more complex yield-enhancing techniques, such as multiple cropping and green manure.

Because access to the food market is household-specific, we develop the behavioural model and its comparative statics under both autarky and integration in markets. The empirical strategy then consists of two levels of analysis: a) probit functions for probabilities of adoption and continued use of soil conservation after food subsidies end, and b) duration equations for the sustainability of use among adopters.

6.2 A Household Model of Adoption and Maintenance of Soil Conservation Techniques

A utility-maximising household derives utility from food consumption c and monetary income y, a proxy for non-food consumption. In each period i, the household allocates its labour $\overline{L}$ to agricultural production L_{iA}, soil conservation L_{iC}, and off-farm work L_{iW}[4].

Food output q is a function of land area A, soil fertility F, and agricultural labour L_{iA}. Land area is fixed but its fertility varies. Natural fertility is declining at a constant rate ρ over time unless labour L_{iC} is allocated to soil conservation[5], with an efficiency factor η.

The household's planning horizon extends over three periods. In the first period[6], when subsidies are available, the household may decide to allocate labour to soil conservation. In the next period, as subsidies disappear, it decides whether to

[4] As households in the Sierra are poor, we assume they face no trade-off between time worked and leisure.

[5] We assume that the household has perfect information on the effect of time allocated to soil conservation practices on soil fertility.

[6] Period is related but not equivalent to years. In particular, in the setting of Plan Sierra, subsidies lasted for two years (period 1), but withdrawal can occur over many years (period 2).

continue with soil conservation. The third period closes the model in a finite horizon.

The household maximises an additively separable utility function u over the three periods and over food and income, given its discount rate δ, the production technologies for food and 'soil fertility', the wage rate w, land and household labour availability.

We assume increasing and strictly concave food production and utility functions:

$$q'_{iF} > 0, q'_{iL} > 0, q''_{iFF} < 0, q''_{iLL} < 0, q''_{iFL} > 0,$$
$$u'_i > 0, u''_i < 0, \quad i = 1,2,3.$$

with the optimal solution such that: $L_{iA}{}^* > 0$ and $L_{iW}{}^* > 0$.

6.2.1 Households Facing Food Market Imperfections

Market failure results from price bands in food transactions. The household, faced with a food market failure, maximises utility, under the additional requirement that food consumption equals production, at each period:

$$Max_{L_{iA}, L_{iC}, L_{iW}} \sum_{i=1}^{3} \delta^{i-1} \left[u\left(c_i; z^h\right) + y_i \right] \qquad \text{objective function}$$

subject to

$$y_i = wL_{iW} + T, \quad i = 1,2,3 \qquad \text{monetary income}$$

$$L_{jW} = \overline{L} - L_{jA} - L_{jC}, \quad j = 1,2$$
$$L_{3W} = \overline{L} - L_{3A} \qquad \text{period-specific time constraints}$$

$$F_{j+1} = \rho F_j + \frac{\eta L_{jC}}{A}, \quad j = 1,2 \qquad \text{soil fertility in period } j+1$$

$$c_1 = q_1\left(A\overline{F}_1, L_{1A}; z^q\right) + \alpha L_{1c}$$
$$c_k = q_k\left(AF_k, L_{kA}; z^q\right), \quad k = 2,3 \qquad \text{period-specific food consumption}$$

$$L_{iA} \geq 0, L_{iW} \geq 0, i = 1,2,3 \text{ and } L_{iC} \geq 0, \quad i = 1,2 \qquad \text{non-negativity constraints,}$$

where, in addition to the variables previously defined,

$\overline{F}_1$	natural soil fertility in period one
T	exogenous monetary transfers (remittances)
z^h, z^q	household characteristics in consumption and in production
α	food subsidy rate in period 1 if soil conservation is undertaken ($L_{IC} > 0$).

Solving the household's constrained maximisation for L_{iA} and L_{IC}, yields the following first-order conditions, with μ_i^C the Lagrange multiplier for the non-negativity constraint on soil conservation:

$$u_i' q_{iL_A}' - w = 0, \quad i = 1,2,3 \tag{6.1}$$

$$\alpha u_1' - w + \delta\eta u_2' q_{2F}' + \delta^2 \eta\rho u_3' q_{3F}' + \mu_1^C = 0 \tag{6.2}$$

$$-\delta w + \delta^2 \eta u_3' q_{3F}' + \mu_2^C = 0 \text{ if } L_{1C} > 0;\ L_{2C} = 0 \text{ if } L_{1C} = 0 \tag{6.3}$$

Condition (6.1) states that the shadow price of agricultural labour, as measured by the increased utility of increased food consumption, must equal the price of off-farm labour at each period. Conditions (6.2) and (6.3) reflect the fact that the discounted marginal increase in food production (and utility), due to soil conservation must also equal the wage. Condition (6.1) remains unchanged in the varying scenarios of soil conservation behaviour, which we now proceed to examine.

A No soil conservation adopted: $\mu_1^C > 0, L_{1C} = 0$

Condition (1) determines L_{iA} in each period, as a function of $A, \overline{F}_1, z^q, z^h$ and w. (6.2) and (6.3) yield:

$$\alpha u_1' < w - \delta\eta u_2' q_{2F}' - \delta^2 \eta\rho u_3' q_{3F}' \tag{6.2'}$$

$$L_{2C} = 0 \tag{6.3'}$$

From Equation (6.2'), we can derive $\alpha_{min}(L_{1A}, L_{2A}, L_{3A}, w)$, the minimum food subsidy for the household to engage in soil conservation. Econometrically, this suggests the formulation of a probit function where conservation occurs when $\alpha \geq \alpha_{min}(A, \overline{F}_1, z^q, z^h, w)$.

B Conservation only in the first period: $\mu_1^C = 0, L_{1C} > 0, \mu_2^C > 0, L_{2C} = 0$

The food subsidy is high enough to induce adoption of soil conservation techniques $(\alpha > \alpha_{min})$, but the household abandons in the second period. The first-order conditions then yield for (6.2) and (6.3):

$$\alpha u_1' - w + \delta\eta u_2' q_{2F}' + \delta^2 \eta\rho u_3' q_{3F}' = 0 \tag{6.2"}$$

$$w - \delta\eta u_3' q_{3F}' > 0 \tag{6.3"}$$

Equation (6.1) determines L_{1A}. Equations (6.2"), and (6.3") jointly determine L_{1C}, L_{2A} and L_{3A}. The household will not maintain soil conservation if the discounted increase in food production (and therefore consumption) due to soil conservation labour, is less than the wage (from (6.3")). The subsidy does not come into play in this decision. Econometrically, this suggests a probit analysis of maintenance if $u_3' q_{3F}' \delta\eta - w > 0$, and possibly, a Tobit analysis of the extent of soil conservation (*e.g.* the area) or a multiple-choice model according to the determinants of the profitability of soil conservation.

C Conservation in both periods: $\mu_1^C = 0, L_1^C > 0, \mu_2^C = 0, L_2^C > 0$

$$\alpha u_1' - w + \delta\eta u_2' q_{2F}' + \delta^2 \eta\rho u_3' q_{3F}' = 0 \tag{6.2'''}$$

$$w = \delta\eta u_3' q_{3F}' \tag{6.3'''}$$

Equation (6.1) determines L_{1A}, while (6.2'''), (6.3''') and (6.2") jointly determine L_{1C}, L_{2C}, L_{2A} and L_{3A}.

6.2.2 Households Facing Functioning Food Markets

In the same environment, there now exists a market where food is sold at price p (constant over time). The household maximisation problem becomes:

$$Max_{c_i, L_{iA}, L_{iC}, L_{iW}} \sum_{i-1}^{3} \delta^{i-1} \left[u\left(c_i; z^h\right) + y_i \right] \qquad \text{objective function}$$

subject to the same time constraints, soil fertility law of motion and non-negativity constraints as in Section 6.2.1 but with:

$$y_i = p\left(q_i + \delta^{i-1} \alpha L_{1C} - c_i\right) + wL_{iW} + T, \quad i = 1,2,3 \qquad \text{monetary income}$$

$$q_i = q_i\left(AF_i, L_{iA}; z^q\right), \quad i = 1,2,3 \qquad \text{food production.}$$

Assuming again that $L_{iA} > 0$ and $L_{iW} > 0$, the first-order conditions become:

$$pq_{iL}' - w = 0, \quad i = 1,2,3 \tag{6.1s}$$

$$p\alpha - w + \delta\eta p q_{2F}' + \delta^2 \eta\rho p q_{3F}' + \mu_1^C = 0 \tag{6.2s}$$

$$-\delta w + \delta^2 \eta p q_{3F}' + \mu_2^C = 0 \text{ if } L_{1C} > 0;\ L_{2C} = 0 \text{ if } L_{1C} = 0 \tag{6.3s}$$

$$u_i' = p. \tag{6.4s}$$

Conditions (6.1s) to (6.3s) are identical to the first-order conditions that would be derived from maximising total income:

$$Max_{L_{iA},L_{iC},L_{iW}} \sum_{i=1}^{3} \delta^{i-1}\left[p(q_i + \delta_{i1}\alpha L_{1c}) + wL_{iW} + T\right].$$

Hence, separability holds and the household behaves as if it was taking production and consumption decisions sequentially. The condition for the household to undertake soil conservation becomes:

$$\alpha \geq \alpha_{min} = \frac{w}{p} - \delta\eta q'_{2F} - \delta^2 \eta p q'_{3F} .$$

The subsidy must equal the relative wage minus the discounted marginal profitability of soil conservation, which is the 'net' opportunity cost of labour.

For the household to maintain soil conservation practices after the end of the subsidy (*i.e.* $L_{2C} > 0$), the marginal productivity of fertility q'_{3F} must be larger than $w/p\delta\eta$.

6.3 Some Comparative Statics Results

Details of the proofs are available from the author.

Proposition 1. *The minimum food subsidy, α_{min}, decreases with*

- *the discount factor δ,*
- *and the efficiency of labour in conservation η,*

and increases with

- *farm size A and initial fertility $\overline{F}_1$,*
- *some food utility-increasing household characteristics z^h,*
- *the wage w if there is a food market or if the one-period risk aversion ($-u''/u'$) is moderate.*

Proof: For the non-separable household, the differentiation of the minimum subsidy: $\alpha_{min} = \dfrac{w - \delta\eta u'_2 q'_{2F} - \delta^2 \eta p u'_3 q'_{3F}}{u'_1}$, yields the above results.

For the separable household, marginal utilities u'_i are replaced by price p, and second derivatives u''_i set to 0. $\partial\alpha_{min} / \partial w$ is also unambiguously positive.

Proposition 2. *Higher wages provide disincentives to undertaking soil conservation.*

Proposition 3. *Adoption in the first period takes place if its marginal net benefit (NB_1) is positive. NB_1 increases with*

- *the subsidy α,*
- *discount factor δ,*
- *efficiency of labour η,*

and decreases with:

- *farm size A, initial fertility $\overline{F}_1$, and*
- *the wage w if the food market functions or the one-period risk aversion ($-u''/u'$) is moderate.*

Proof: Condition (6.2') states that a household will adopt conservation if the marginal net profitability of conservation work is greater than the opportunity cost of labour:

$$NB_1 = \alpha u_1' + \delta \eta u_2' q_{2F}' + \delta^2 \eta \rho u_3' q_{3F}' - w > 0 .$$

Using proposition 1, differentiation of NB_1 yields the above mentioned signs.

Proposition 4. *Maintenance in the second period takes place if the marginal net benefit of maintenance (NB_2) is positive. NB_2 increases with*

- *the discount factor δ,*
- *and the efficiency of labour in conservation η.*

NB_2 decreases with

- *farm size A,*
- *initial fertility $\overline{F}_1$,*
- *and with the first-period subsidy α.*

Proof: Condition (6.3'') states that households on the food market will maintain conservation if its marginal net profitability is greater than the opportunity cost of labour:

$$NB_2 = \delta \eta u_3' q_{3F}' - w > 0 .$$

Differentiation of NB_2 yields proposition 4.

Since conservation and maintenance in the first two periods are substitutable, higher subsidies α may artificially induce high levels of adoption with subsequent abandonment. If the household allocates a large amount of labour in first-period adoption, it will need comparatively less maintenance labour in the second period to keep the optimal level of fertility, and may even decide not to undertake any maintenance. This might explain the sudden decrease in the use of some practices, such as terraces.

6.4 Profitability over Time and Decision to Withdraw

We now consider varying efficiency of soil conservation (η_t) over time[7]. Learning-by-doing improves efficiency of household practices. On the other hand, if practices are not optimal, fertility loss – albeit slower – might still occur, leading to decreasing land productivity, and in turn reduced efficiency of conservation. Most food plots are eventually abandoned, which indicates that fertility losses exceed conservation benefits.

Fig. 6.1 presents curves of net profitability $NB_2(\eta_t)$ for different household categories: 1) households, which never adopted, or 2) abandoned when the subsidies stopped, or 3) abandoned some years after the subsidies ended, or 4) had not abandoned by the time of the survey, showing truncation of the duration spell. Empirical analysis will support this representation.

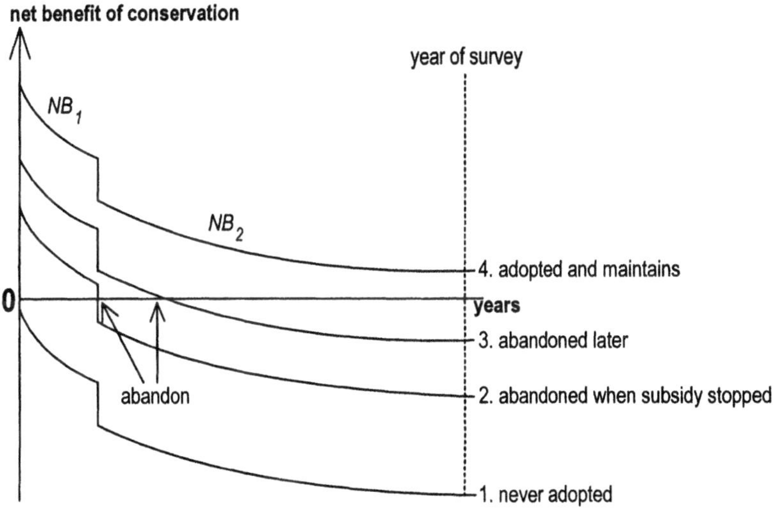

Figure 6.1: Decision to adopt and maintain soil conservation.

[7] In the household choice model, we considered instantaneous adoption decision-making with perfect information, assuming that Plan Sierra intervention occurred once in each village. We therefore exclude strategic behaviour in the timing of adoption, whereby some farmers not wanting to incur the costs of learning-by-doing would acquire information from their neighbours' results, as in Foster & Rosenzweig (1995). Hence, their decision to adopt is based on the net benefit $NB(\eta_1)$ computed with the efficiency of land conservation labour in the first year.

6.5 Descriptive Statistics

We collected household survey data in 1994 with Plan Sierra (de Janvry *et al.*, 1994).

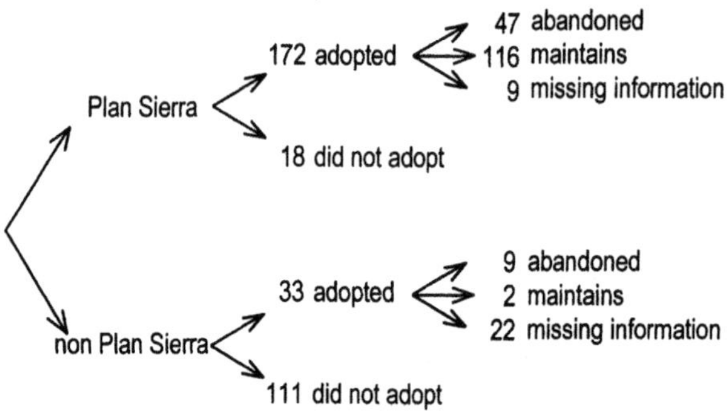

Figure 6.2: Distribution of the sample by membership in Plan Sierra and adoption.

As shown in Fig. 6.2, the 334 households with complete interview forms comprise 190 members of Plan Sierra and 144 non-members[8]. Among Plan Sierra members, 90.5% adopted[9] and of these, 27.3% subsequently abandoned all practices. Fewer non-members adopted (22.9%) and most abandoned[10] (81.8%).

Table 6.1: Correspondence between the model concepts and survey variables.

Theoretical concept	Observed variable	Expected sign in models	
		non-separable	separable
$A\rho$	Agricultural land	-	-
δ	Land 'owned' [11], age of head	+	+
η	Highest level of education	+	+
T	Amount of remittances	+	0
z^h	Share of uneducated adults	-/+	0
z^q	Number of children 12 to 15 years old	+/-	0
w	Community-level wage	-	-
α	Food subsidies from Plan Sierra	+	+

[8] This is based on the sample design. Of the 385 households that were interviewed, 51 either did not engage in agriculture or had very incomplete data on that section.

[9] Adopters are households, which answered "yes" to the question "have you ever done any soil conservation?"

[10] Only for households with complete questionnaire sections about maintenance.

[11] This includes land in ownership with and without formal titles.

Table 6.1 shows the proxies we use in establishing a correspondence between the household model concepts and the survey variables. We also indicate the expected signs on the probability of adoption according to the two versions of the model.

Descriptive statistics in Table 6.2 contrast adopters and non-adopters. Contrary to diffusion studies where adopters typically are wealthier and larger farmers (Feder *et al.*, 1985), here, adopters earn 34% less per capita income, own 36% less land and receive 37% less remittances than non-adopters. They also access less total land, which limits their ability to use fallow for soil fertility restoration. They do not engage more in off-farm activities, which would compensate for small land and low remittances. Hence, they need to use more intensively their land for food and cash crops (coffee).

Adopters have larger families to feed, as measured in adult equivalents (Goungetas & Johnson, 1992), although their food plots are smaller. They thus seem trapped in agriculture, a social group that depends heavily on meagre land resources for own consumption and cash income[12]. By default, this suggests two categories of households that might not adopt soil conservation practices: those with a high commitment to off-farm income (labour market or migratory networks); and those with large farms on which slash-and-burn rotation remains the cheapest option.

Consistent with these observations, in the second panel of Table 6.2, we then divide households into two groups according to the size of their food plot per adult equivalent. We expect households with smaller and larger food plots to participate in the market as buyers or sellers, respectively, while the middle group should be self-sufficient[13].

[12]Comparing Plan Sierra adopters to those who adopted on their own, again shows that Plan Sierra tends to cater to a clientele of limited resource households. Plan Sierra adopters grow smaller food plots (10.2 vs. 13.2 tareas), their household heads are older (50.5 vs. 44.5 years), and more uneducated adults (26 vs. 12%) are part of the family.

[13]We tried to estimate an equation of participation in the food markets but the predictive power was very low, which led us to the present classification.

Table 6.2: Summary statistics, all households by adoption or separability class.

	All households, N=334			Separable and non-separable, N=296		
	Adopters Mean (CV)	Non-adopters Mean (CV)	Test of difference of means	Non-separable	Separable	Test of difference of means
Percentage of households	38.6	61.4	na	32.1	67.9	
Percentage of adopters				70.5	57.7	
Agricultural assets						
Owned area (tareas)	69 (1.75)	108 (1.94)	– –	83.0 (1.84)	54.3 (1.83)	+
Family jointly-owned area (tareas)	18 (3.94)	48 (3.98)	– –	30.2 (3.93)	13.3 (2.93)	
Total land area (tareas)	120 (1.43)	191 (1.51)	– – –	132 (1.37)	98.9 (1.54)	
Human capital assets						
Family size	5.7 (0.39)	5.3 (0.46)	+	5.0 (0.37)	4.9 (0.39)	
Age of the household head	49.5 (0.24)	52.2 (0.31)	–	47.6 (0.26)	50.3 (0.26)	
Number of adults	4.0 (0.47)	3.7 (0.49)				
Family size (number adult equivalents°)	3.1 (0.65)	2.61 (0.67)	++			
Number of children 12-15 years old	0.52 (1.41)	0.47 (1.48)		0.43 (1.41)	0.5 (1.48)	
Share of uneducated adults	16.6 (1.35)	17.30 (1.33)		24.9 (0.01)	0.3 (1.08)	
Highest level of education in the hh°°)	2.96 (0.31)	2.76 (0.32)	+	2.61 (0.35)	2.71(0.30)	
Prevailing wage in the village (RD$)	46 (0.15)	47 (0.15)		45.2 (0.15)	46.6 (0.14)	
Household income strategies						
Plan Sierra participation (%)	84	14	na	89.6	78.4	na
Area in food plot (tareas)	10.6 (0.88)	12.7 (0.96)	–	9.4 (0.41)	11.5 (1.03)	
Percentage of households with coffee	82 (0.48)	74 (0.60)	+	91 (0.32)	76.7 (0.55)	++
Number of off-farm workers	1.2 (1.07)	1.2 (1.23)				
Remittances received (RD$)	3179 (2.57)	5055 (1.85)	–	3946 (2.46)	3057 (2.65)	
Monetary income (RD$)	24104 (0.93)	30380 (0.93)	– –			
Income per capita	4570 (0.98)	6901 (1.37)	– – –			

Legend of Table 6.2:
'Non-separable' household: 1.4 to 2.36 tareas in food plot per adult-equivalent;
'Separable' household: less than 1.25 or more than 2.75 tareas in food plot per adult-equivalent;
+++ or - - - significant at 99% level; Exchange rate 1 US$ = 12.85 RD$ in the Sierra, Summer 94;
++ or - - significant at 95% level; 1 ha = 16 tareas;
+ or - - significant at 90% level;
° Male adult = 1, female adult = 0.8, child = 0.6;
°° Illiterate = 1, primary education = 2 or 3, secondary education = 4, higher education = 5.

6.6 Adoption Behaviour: Probit Analysis

From the results of the comparative static analysis, we estimate adoption as a probit, as in Lutz *et al.* (1994), so that:

Prob (adoption by household i) = $\phi(\gamma X_i / \sigma_{vi})$,

where:

X_i is the set of previously described exogenous variables;
γ set of coefficients;
ϕ standard normal p.d.f.;
v_i i.i.d. farm specific *ex ante* random shock, normally distributed as $\aleph(0,\sigma^2)$.

To address the potential self-selection bias in Plan Sierra membership, we first estimate a probability of membership and use the Inverse Mills Ratio (IMR) computed from that equation in the adoption equation[14].

We run the analysis separately for the two categories of households, as described in Table 6.2. The test for separability is provided by the significance of the coefficients for the share of uneducated adults in the household and for the number of children aged 12 to 15, who may provide labour on-farm[15] but not on the market. The test reported in Table 6.3a confirms the division of households, according to their participation in the market.

[14]The Hausman type test is generally consistent with exogeneity of Plan Sierra membership. The results from these probits are available upon request. They show that Plan Sierra enrolls households with larger family size, and with a higher frequency of poorly maintained houses, both of which are symptoms of poverty.
[15]No detailed information is available about children below 12.

Table 6.3a: Probit analysis of adoption of soil conservation practices by separability class (217 separable and 117 non-separable households).

	Separable households			Non-separable households		
	Coefficient	z-statistic	Significance	Coefficient	z-statistic	Significance
Institutional assets						
Plan Sierra membership	4.76	1.78	*	4.46	0.73	
Inverse Mills Ratio of PS membership†	-1.65	-1.02		-0.37	-0.11	
Agricultural assets						
Agricultural land used (100 tareas)	-0.19	-2.25	**	-0.04	-0.19	
Land owned (100 tareas)	0.01	0.13		-0.09	-0.39	
Human capital assets						
Number of children 12 to 15 years old	0.04	0.27		1.23	2.32	**
Share of uneducated adults	0.41	1.00		-0.97	-1.34	
Age of the household head	-0.02	-2.43	**	-0.03	-1.72	*
Highest level of education in the hh	0.35	1.36		0.92	1.64	*
Prevailing wage in the village (RD$)	-0.01	-0.31				
Remittances (10^4 RD$)	0.11	0.79		0.11	0.46	
Regional characteristics						
Jicome pole dummy	0.17	0.38		-0.62	-0.70	
Janico pole dummy	0.10	0.20		1.94	1.92	**
Las Placetas pole dummy	0.17	0.40		0.10	0.12	
Juncalito pole dummy	-0.12	-0.29		-0/07	-0.08	
Constant term	-1.42	-0.69		2.25	0.60	

Table 6.3a: (continued).

Predictive power	Predicted			Predicted		
Observed	Does not adopt	Adopts	Total	Does not adopt	Adopts	Total
Does not adopt	73	28	101	29	9	38
Adopts	18	108	126	1	78	79
Total	91	136	217	30	87	117
% correct	80.2	85.7	83.4	96.7	89.7	91.5
Goodness-of-fit: Pseudo-R2	0.41			0.67		
Joint-test of significance for children 12 to 15 and share of uneducated adults	Chi2 (2) = 1.00 Prob > chi2 = 0.61			Chi2 (2) = 6.42 Prob > chi2 = 0.04		

† From the first probit equation (results available from the author)

Table 6.3b: Probit analysis of adoption without or beyond subsidies by separability class (195 separable and 103 non-separable households; the same set of variables is included in the regression).

	Separable households			Non-separable households		
	Coefficient	z-statistic	Significance	Coefficient	z-statistic	Significance
Institutional assets						
Plan Sierra membership	4.43	2.02	**	3.22	0.92	
Inverse Mills Ratio of PS membership†	-1.56	-1.16		-0.01	-0.01	

Step 1. Probit analysis of adoption across all households in the sample (Table 6.3a)

The predictive power of the Probit is satisfactory (83%-91% correct predictions). Plan Sierra membership, with its associated technical assistance and subsidies, is, as expected, a determinant of adoption among households participating in the market. That decision seems exogenous to the adoption decision as shown by the non-significant negative[16] IMR.

As higher total land area ($A\rho$) might allow the rotation of slash-and-burn plots, smaller holders are most receptive for adoption, especially among market participants. The age of the household head is always significantly negative, reflecting lower discount rates (δ) and longer planning horizons. Younger households may also be more likely to have higher consumer-worker ratios and more child labour.

Step 2. Probit analysis of adoption without subsidy (non-Plan Sierra adopters) or continued use after subsidy (Plan Sierra adopters after the first two years) (Table 6.3b).

The results are similar to those in Step 1, but allow a slightly different interpretation of Plan Sierra's role. For all adopters (Step 1), Plan Sierra subsidies could have artificially boosted adoption if some households adopted primarily to receive the food. Now, results in Table 6.3b show that continued use of conservation is linked with Plan Sierra membership. The subsidies capitalised limited resource households, especially among those on the market, so that they were subsequently able to continue using their soil conservation practices. This suggests the existence of a threshold that the poor need to overcome to start conserving and that subsidies can be effective for this purpose. A welfare system then acts as a start-up instrument, to help the poor enter into profitable activities, which they could not have engaged in without the subsidy.

6.7 Maintenance Behaviour: Duration Analysis

Some households adopt conservation and subsequently abandon it, as was foreseen through Proposition 4. Sustainability of soil productivity cannot be achieved unless practices are continued. We model the behaviour, illustrated in Fig. 6.1, as a duration phenomenon, where duration is the length of time from adoption until drop-out or until the survey year (in which case truncation occurs).

One particularly flexible functional form of the hazard function, that captures the non-linearity in the profile of abandonment, is the Weibull form (Greene, 1993):

[16] A negative significant IMR would indicate that the Plan is slightly biased in its choice of members toward the households less likely to adopt (mainly because they are poor).

$$h(t,x(t),\theta) = \lambda^{p-1}$$

where $\lambda = e^{-\beta x(t)}$ and $\theta = (\beta', p)$.

The hazard rate is decomposed into two effects: λ is a function of the covariates $x(t)$ while p expresses the role of spell duration on the hazard rate.

In Table 6.4a, the first duration model contains only time-invariant covariates. Plan Sierra membership is effective in extending the duration of soil conservation. Total land owned, which captures the discount factor, increases the duration spell. By contrast, total land used accelerates abandonment for self-sufficient households. Low educational levels also accelerate abandonment.

Finally, the Weibull p equals 1.47 for non-separable households, suggesting a slight drift toward abandonment as time passes, which corroborates Fig. 6.1. These households, whose livelihood depends more on land, might especially feel the limited efficiency of their mix of soil conservation practices.

To capture the role of the termination of subsidies with Plan Sierra, we introduce a shock of ending the subsidies, shown in Table 6.4b. This second year shock does result in abandonment, especially for households on the market. It does not affect significantly the positive effect of Plan Sierra on sustainability, though: Plan Sierra members continue to abandon conservation practices at a much lower rate than other adopters (Fig. 6.3).

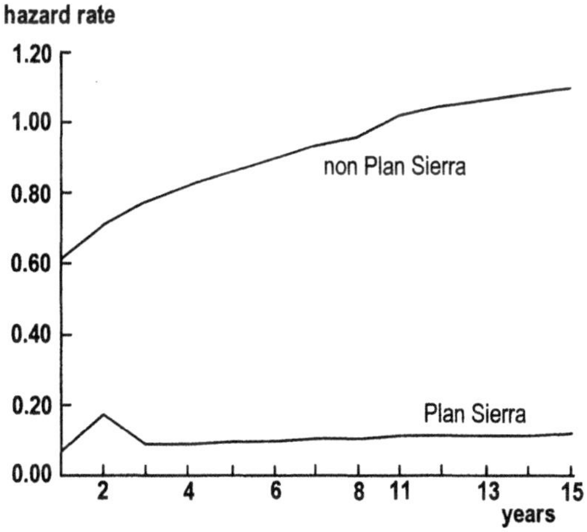

Figure 6.3: Hazard rates for Plan Sierra and non-Plan Sierra adopters.

Table 6.4a: Soil conservation duration model by separability class (constant covariates).

	Separable households			Non-separable households		
	Coefficient	z-statistic	Significance	Coefficient	z-statistic	Significance
Number of households	105			67		
Institutional assets						
Plan Sierra membership	1.71	3.51	***	2.64	3.80	***
Agricultural assets						
Total land used (100 tareas)	-0.12	-0.88		-0.33	-2.39	***
Land owned (100 tareas)	0.29	1.17		0.48	2.14	***
Human capital assets						
Number of children 12 to 15 years old	-0.18	-0.83		0.20	0.66	
Share of uneducated adults	-1.70	-0.80	***	-1.19	-1.51	*
Age of the household head	0.03	2.07	**	0.02	1.05	
Highest level of education in the hh:						
- primary schooling	-0.24	-0.42		-0.82	-1.24	
- secondary schooling	-0.74	-1.36		-0.75	-1.00	
Prevailing wage in the village (RD$)	-0.01	-0.44				
Remittances (10^4 RD$)	0.22	1.05		0.30	0.68	
Constant term	1.33	0.87		0.76	0.54	
Parameter of the Weibull function		Std. Err.			Std. Err.	
Parameter p at data mean	1.02	0.14		1.47	0.66	
Log-likelihood	-98.3			-47.7		
Joint-test of significance for children	F (2,85) =	6.30		F (2,49) =	2.83	
12 to 15 and share of uneducated adults	Prob > F =	0.04		Prob > F =	0.24	

Table 6.4b: Soil conservation duration model by separability class (one time-varying covariate).

	Separable households			Non-separable households		
	Coefficient	z-statistic	Significance	Coefficient	z-statistic	Significance
Institutional assets						
Plan Sierra membership	1.28	1.93	*	2.69	3.36	**
Year-two shock: end of subsidies °	-1.07	-2.84	**	-0.56	-0.61	
Agricultural assets						
Total land used (100 tareas)	-0.09	-0.63		-0.22	-1.71	*
Land owned (100 tareas)	0.24	1.32		0.33	2.11	**
Human capital assets						
(Same signs as in Table 6.4a)						
Parameter of the Weibull function		Std. Err			Std. Err.	
Parameter p at data mean	1.66			2.14		
Log-likelihood	-98.3			-47.7		

° Time-varying covariate

6.8 Conclusions

The results about adoption and maintenance of soil conservation measures suggest the following policy implications:

1. *Welfare measures, including food subsidies, are effective in promoting conservation*, as shown by the positive effect of Plan Sierra limited-resource household membership on both adoption and longer maintenance, enhancing sustainability.

2. *Food autarky, due to market failure, induces soil conservation*, as reflected by the higher proportion of adopters among self-sufficient households. The non-significant coefficients for Plan Sierra and the IMR indicate that they would do so, even without the program and therefore benefit relatively less than other types of households from Plan Sierra's help.

3. *Actual land ownership*[17] *favours maintenance of soil conservation*. This effect is not a main determinant of adoption, a finding also reported by White & Runge (1994) in Haiti. Increasing security of access to land will favour the maintenance of practices with a large sunk cost dimension.

4. *Large land holdings are associated with lower soil conservation adoption and maintenance*. Large landholders will thus remain a continued source of soil erosion (either because of labour constraints to implement labour-intensive soil conservation or because slash-and-burn rotation is still a viable option to them).

5. *Lack of minimal education may lead to earlier abandonment of soil conservation practices*. Training for adopters may need to go beyond information for the adoption of the new practices, to include integrated farm management. This confirms previous findings about Plan Sierra extension classes (de la Brière & Jímenez, 1994).

6. *Soil erosion on farmers' plots is not totally halted*. Soil conservation practices help to extend the productive life of food plots, from an average three years in the traditional slash-and-burn cycle to six years, a significant achievement. However, new plots will still be cleared thereafter, which implies that Plan Sierra efforts to diffuse the adoption of soil conservation techniques need long-term support for these, in particular through food subsidies.

[17] Owned here is contrasted with occupied, jointly-owned and rented land.

General Conclusions on the Targeting of Technology Diffusion Efforts

Households strongly vested in agriculture for their livelihood, with large families to feed, and with sufficient educational levels to make profitable use of new technologies, are the prime candidates for assistance by rural development programs. They will adopt and continue using the techniques acquired through such programs' technical assistance and temporary subsidies. Targeting this class will yield the most effective results. Improving security of access to land, education beyond primary school and training, and possibly access to credit to finance conservation investments will enlarge this class.

Households heavily committed on the labour market, or with precarious access to land are not interested in soil conservation due to poverty or life plans beyond agriculture. Households with access to large areas of land, for which soil conservation is expensive, will also remain an important source of soil erosion and externalities.

6.9 References

Besley, T. & A. Case (1993): Modeling technology adoption in developing countries. American Economic Review 83, 396-402

Brière, B. de la & A. Jímenez. (1994): Adopción de prácticas agroforestales para una agricultura sostenible en la Sierra: el Programa de Autosufficiencia Alimentaria. Department of Agricultural and Resource Economics, UC Berkeley, mimeo

Burt, O. (1981): Farm level economics of soil conservation in the Palouse area of the Northwest. American Journal of Agricultural Economics 63, 83-92

Feder, G., R.E. Just & D. Zilberman (1985): Adoption of agricultural innovations in developing countries: a survey. Economic Development and Cultural Change 33, 255-298

Foster, A. & M. Rosenzweig (1995): Learning by doing and learning from others: human capital and technical change in agriculture. Journal of Political Economy 103, 1176-1209

Goungetas, B. & S. Johnson (1992): The impact of household size and composition on food consumption. Iowa State University: Center for Agricultural and Rural Development.

Greene, W.H. (1993): Econometric analysis. Englewood Cliffs, Prentice Hall, NJ

Janvry, A. de, M. Fafchamps & E. Sadoulet (1991): Peasant household behaviour with missing markets: some paradoxes explained. Economic Journal 101, 1400-1417

Janvry, A. de, B. de la Brière & E. Sadoulet (1994): Adoption and maintenance of soil conservation techniques in the Dominican Sierra. A report prepared for the OECD Watershed Project and IDB Environment Division, mimeo

Lutz, E., S. Pagiola & C. Reiche (1994): The costs and benefits of soil conservation: the farmer's viewpoint. The World Bank Research Observer 9, 273-295

Mc Connell, K.E. (1983): An economic model of soil conservation. American Journal of Agricultural Economics 65, 83-89

Sadoulet, E. & A. de Janvry (1995): Quantitative development policy. John Hopkins University Press, Baltimore, MD

White, T. & C. Runge. (1994): Common property and collective action: lessons from cooperative watershed management in Haiti. Economic Development and Cultural Change 43, 1-41

7 Soil Conservation and Imperfect Labour Markets in El Salvador: an Empirical Application of a Dynamic Control Model of Farm Production

Claudia B. Romano*

Department of Agricultural and Resource Economics, University of Maryland at College Park, U.S.A.

Abstract

An important issue in the study of soil conservation is the impact that imperfect markets can have on farmers' decisions regarding resource management. Rural areas in developing countries are characterised by high transaction costs and imperfect markets. Since soil conservation is labour-intensive, the functioning of agricultural labour markets and the availability of off-farm work opportunities are likely to affect soil conservation effort. In this paper I develop a dynamic control model of agricultural production in the context of lack of access to labour markets in order to analyse how soil management decisions are affected. An empirical application of the model is developed using data from a 1996 farm household survey in El Salvador. The results of the model are confirmed by the econometric estimation, showing that soil conservation is negatively affected by off-farm employment and smaller family labour force only for farmers who face a missing agricultural labour market. Soil degradation seems to be closely associated to conditions that prevent farmers from fully participating in rural markets. Moreover, rural policies that promote the non-farm rural sector may have an unwanted negative impact on soil conservation, and soil conservation programs would be more effective if they promoted labour-saving practices.

---

* This paper is part of my Ph.D dissertation work in the Department of Agricultural and Resource Economics at the University of Maryland. I thank Ramón López for his guidance, and Sara Scherr and Stefano Pagiola for important comments early on in this study. I am also grateful to Bekele Shiferaw and other participants at the workshop Economic Policy Reforms and Sustainable Land Use in LDCs, Wageningen, the Netherlands.

7.1 Introduction

Soil degradation is a major environmental problem world-wide, with negative on-site effects for agricultural productivity, as well as off-site effects such as siltation of rivers and reservoirs. Despite substantial effort by various governmental and non-governmental organisations to promote the adoption of soil conservation practices among farmers, the results of such programs have not been very encouraging, particularly in developing countries. Understanding the incentives that farmers perceive regarding soil management, can help policy makers in understanding the effects that sectoral and macro-policies may have on resource quality and in the design of more effective conservation policies.

An important question regarding the behaviour of agricultural producers with respect to conservation is whether and how their decisions are conditioned by imperfect rural markets. Rural markets in developing countries are characterised by high transaction costs, which cause the same factors of production and commodities to have different purchase and sale prices, driving certain households to be self-sufficient in these markets. When a farmer faces a missing market, his rational responses to policies and market incentives can be different and, sometimes, even opposite to what would be expected in conditions of perfectly functioning markets (Ali & Byerlee, 1991; de Janvry *et al.*, 1991; Feder, 1985). Various studies have examined the link between imperfect markets and resource conservation, including the food market (de la Brière, this volume), credit and insurance markets (Holden *et al.*, 1998; Barbier & Lopez, 1998), land markets (Ruben *et al.*, this volume) and the interplay between labour and credit markets (Pender & Kerr, 1998).

Since soil conservation is a labour-intensive activity (FAO, 1994), the functioning of labour markets and access to off-farm work opportunities are likely to affect conservation investments (Pender & Kerr, 1998). Rural labour markets in Latin America are affected by poor infrastructure and other transaction costs and are typically characterised by inelastic labour supply, while many farm households have labour surpluses (Edwards, 2000). Labour allocation in a farm household depends on various factors, including market wages, transaction costs, on-farm shadow wages and behaviour towards risk (Barret, 1999). Market conditions can affect all of these factors and, consequently, labour allocation and soil management.

In this respect, one important question is how off-farm employment interacts with rural market characteristics to affect farmers' decisions on soil conservation. This question is particularly relevant given the significant increase in non-agricultural activities in rural areas of developing countries in the past two decades. Participation in off-farm activities results in reduced family labour on the farm. Keeping all else constant, when this reduction is not compensated by hired labour there can be a reduction in the amount of work on the farm, with negative consequences for soil conservation as well as agricultural production. If, however, labour is hired to compensate for lost family labour, or if there are changes in production technology and crop composition that improve soil management, the result may be an improvement in soil quality. Income from off-farm employment can also provide the cash necessary to finance conservation and other technology improvements.

A recent review by Templeton & Scherr (1997) summarises the available evidence on the effect of off-farm employment (among other factors) on soil conservation. Examples from Nepal, Kenya and Rwanda indicate that improving non-farm employment opportunities promoted environmental quality through land improvement with more tree planting or the use of more capital intensive soil conservation techniques, financed by off-farm income. Other evidence, however, indicates the opposite effect. Examples from the Bolivian Andes, Ecuador, Kenya, Philippines and Oaxaca in Mexico indicate that the use of soil erosion control practices decreased as more farmers were employed off-farm, resulting in reduced soil quality.

In the present study I develop a dynamic control model of a farm household in the context of missing agricultural labour markets and examine whether this situation can affect the household decisions regarding soil conservation investment, with particular attention to the impact of increased opportunities for off-farm employment. In this way I also attempt to elucidate the opposing evidence on the relationship between off-farm employment and soil conservation, which could be partially the result of differential market access. An empirical application of the model is carried out based on data from a 1996 farm household survey in El Salvador. The country is well-suited for this analysis. It is the most densely populated in Latin America, much of its area is mountainous and land degradation is widespread. Moreover, El Salvador has experienced a strong increase in non-farm rural employment in the period between 1964 and 1994, when the proportion of the economically active rural population employed in the non-agricultural sector doubled to 36% (World Bank, 1998). The results of this empirical analysis can indicate whether the current trend of increase in non-farm rural activities affects soil conservation and whether labour market imperfections condition this effect. Rural policies that promote the non-farm rural sector may have an unwanted negative impact on soil conservation and would thus require additional compensation measures.

7.2 Labour Allocation in a Model of Farm Production with Missing Agricultural Labour Markets

Rural areas in developing countries are characterised by poor infrastructure, shallow local markets, high search and recruitment costs, imperfect information and high transportation costs. Although a market may exist in such circumstances, it may not function perfectly for all households. According to Sadoulet & de Janvry (1995, 149):

> '...a market may fail for a particular household when it faces wide price margins between the low price at which it could sell a commodity or factor and the high price at which it could buy that product or factor. Faced with this wide price band, the household may be better off choosing self-sufficiency in that good or factor if its subjective price (defined as the price which equates its supply and demand) falls inside the band.'

The higher the transaction costs, the wider the price band and, therefore, the more likely are households to fall in the self-sufficiency category. This means that many households in rural areas do not buy or sell certain commodities and factors of production. If we consider the agricultural labour market, those households whose on-farm subjective wage falls within the wage band will not sell or hire labour. Those falling outside the band may hire or sell, depending on the relative position of their on-farm shadow wage.

Other considerations, however, may lead even those households whose subjective wage falls inside the price band to engage in off-farm work, at the expense of farm production. They may choose to work off-farm to diversify household income sources in order to reduce the risk associated with agricultural production (Reardon & Barret, 2000). Another reason may be lack of liquid assets to meet consumption and production needs. Income earned off-farm can provide the needed cash (Alwang & Siegel, 1999).

Access to higher rural wages may also attract farm workers. Non-farm activities usually pay higher wages[1], and certain large-scale agricultural operations, such as large plantations of cash crops, may also pay higher wages due to efficiency considerations. For all of these reasons, some households whose on-farm shadow wage falls within the wage band may actually have some family members working off-farm. Yet, these households likely do not hire in any labour as the subjective wage for those family members that remain at the farm is below the market agricultural wage in the peasant sector[2].

Since these households can engage in off-farm employment but do not hire-in farm labour, it is expected that, keeping everything else constant, those households where more family members work off-farm will have less family labour time

---

[1] Sadoulet *et al.* (1996) explain farm households' differential labour regimes based on a model where households have family members with differentiated skills and opportunity costs. High-skilled members always choose to work off-farm when non-farm jobs are available, because the wages are higher.

[2] One exception to this behaviour is possible when the need for labour is accentuated in peak seasons. In such circumstances, farmers may hire-in during the peak season only, despite finding themselves in a position of self-sufficiency the rest of the year. Implications of this for the econometric estimation are considered latter on. These situations could be seen as a mixture of the two extreme stylised cases discussed here.

available to work on the farm and, thus, practice less soil conservation. It is also expected that households with larger families use more labour on conservation and on productive farm work than smaller families.

For households that are not self-sufficient in farm labour, either because they have higher demand for labour or because they are located in areas where the agricultural labour market works better (the wage band is small), these relationships should not be observed. So, there should be no significant link between more family members employed off-farm or a larger family, and conservation or other productive farm work for this group of farmers.

Figure 7.1 illustrates the on-farm labour allocation decision framework with imperfect labour markets. Since decisions about off-farm employment depend either on criteria related to risk and financial liquidity, or on the higher wages paid for certain types of jobs, we assume that the amount of labour employed off-farm is determined exogenously to this system. Therefore, the decisions on labour allocation for farm work can be represented separately and based only on market wages and transaction costs in the peasant sector. The household demand curve for labour, or value of marginal product of labour, is represented by D. The curves S^1 and S^2 represent the supply of family labour for on-farm work of two different households. Each is assumed fixed. S^2 is the supply curve of a household that either has a larger family or less people employed off-farm (thus more family available for farm work) than the household represented by S^1. Since neither household hires-in labour, because their shadow wage is lower than the market wage paid, the household represented by S^1 will apply less effort on conservation and productive farm work.

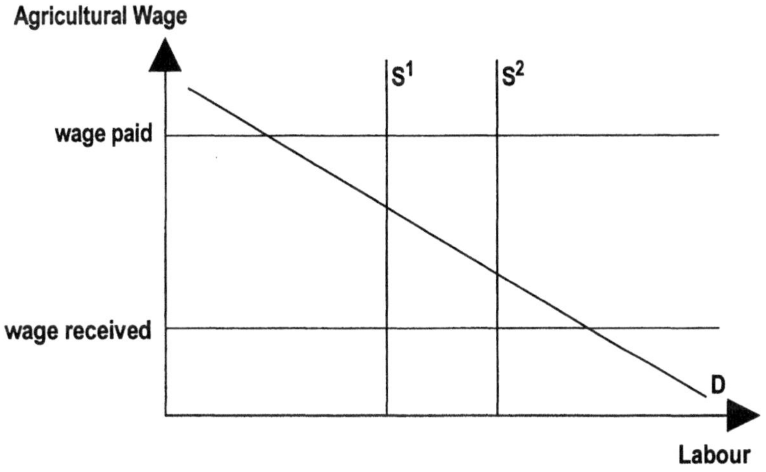

Figure 7.1: Farm household labour allocation to on-farm work, with wage band.

Even when labour on-farm decreases, it does not necessarily mean that soil quality deteriorates. The effect of less on-farm work on soil quality is ambiguous, because reduced soil conservation work may be compensated by less work effort on directly productive activities that increase soil degradation. For instance, there could be a decrease in the area cultivated throughout the year. Other possible effects of reduced labour are changes in crop composition and technology, which can affect soil quality. For example, tree planting, which is less labour-intensive, may substitute crops in areas where trees are valuable for wood, fruits, or for fuel wood and fodder. Tree plantations improve soil erosion control. On the other hand, there can be labour-saving choices that are not beneficial to soil quality, such as more use of machinery causing more soil compaction. Off-farm labour could also have a positive effect on soil conservation, as the extra earnings it provides can be an important source of cash for investments. Particularly when credit is scarce, family savings are the main source for land investment, including soil conservation. Therefore, the final effect of off-farm employment on soil quality depends on the relative importance of each of these effects and can only be determined empirically.

A dynamic farm profit maximisation model is developed in order to analyse the importance of different factors on the use of soil conservation practices, in the context of missing agricultural labour markets. For simplification, farm production is determined only by family labour and soil quality. The only household characteristic that affects production is family work force. Soil quality is affected negatively by labour used in directly productive activities and positively by labour in soil conservation activities. Soil quality can be interpreted as an index of those soil characteristics that affect the agricultural productivity of soil, such as topsoil depth, fertility or nutrient content and water retention capacity (Pagiola, 1996).

The amount of family labour employed off-farm is determined exogenously by the availability of and accessibility to off-farm work. Total family labour available is assumed fixed, thus we disregard the possible substitution between work and leisure[3]. Later on in the econometric specification, non-labour income (a factor related to consumption of leisure) is included in the regressions in order to examine whether this assumption is valid.

The farmer maximises the present value of production:

$$\int_{t=0}^{\infty} F(S, L_p) e^{-\delta t} dt \tag{7.1}$$

subject to:

[3] This assumption is consistent with the development literature on household models that deal with low income rural contexts. It means that the substitution between leisure and work activities is not important in our analysis, and leisure and non-leisure time are considered fixed.

$$\dot{S} = a + g(L_C) - bL_P,$$

$$\overline{L} - L_O = L_P + L_C,$$

$$S(0) = \overline{S}_0,$$

where $F(S, L_P)$ is a concave increasing farm production function in both its arguments. The output price is normalised to one. S is the soil quality index, L_P is family labour used in directly productive farm activities and δ is the discount rate. The time evolution of the soil index describes how the soil quality index changes; a is the natural regeneration of soil, labour used in soil conservation activities, L_C, increases soil quality according to $g(L_C)$, where $g(.)' > 0$ and $g(.)'' < 0$[4]. The last term in the equation represents the negative effect of cultivation on soil quality, where a fixed rate represented by the coefficient b transforms labour in directly productive activities into soil degradation[5]. Total family labour, $\overline{L}$, is fixed and equal to the sum of labour on the farm plus L_O, the exogenously determined level of labour employed off-farm. $S(0)$ is the initial level of soil quality.

The first-order conditions are:

$$F_2 - \lambda(b + g') = 0, \quad (7.2)$$

$$\dot{S} = a + g(\overline{L} - L_P - L_O) - bL_P, \quad (7.3)$$

$$\dot{\lambda} = \delta\lambda - F_1, \quad (7.4)$$

$$S(0) = \overline{S}_0; \ \lim_{t\to\infty} e^{-\delta t}\lambda(t)S(t) = 0, \quad (7.5)$$

where λ is the shadow value of soil quality and the subscripts of the F(.) function denote partial derivatives with respect to the corresponding function argument. Condition (7.2) indicates the maximising condition that applies for all t. At steady-

[4] A soil evolution equation as a function of a natural rate of regeneration and of inputs in production is commonly used in models of optimal soil depletion (McConnell,1983; Barret, 1991). Explicit inclusion of the positive effects of conservation activities has also been used (see Barbier, 1990).

[5] This assumption facilitates the algebra but does not affect results if instead of a linear relation one assumes an increasing and convex function where labour in production degrades soil at an increasing rate. Other studies have assumed a convex function for the effect of cultivation on degradation of natural resources (López & Niklitschek, 1991; Barbier, 1990).

state, conditions (7.3) and (7.4) equal zero. From (7.2) we can get the implicit solution to L_P and its partial derivatives with respect to S, L_O and λ:

$$\frac{\partial L_P}{\partial S} = \frac{-F_{21}}{\lambda g''+F_{22}} > 0, \tag{7.6a}$$

$$\frac{\partial L_P}{\partial \lambda} = \frac{b+g'}{\lambda g''+F_{22}} < 0, \tag{7.6b}$$

$$\frac{\partial L_P}{\partial L_O} = \frac{-\lambda g''}{\lambda g''+F_{22}} < 0 \tag{7.6c}$$

Given the partial derivatives of L_P, it is possible to obtain the slope of the evolution schedules. The slope of the $\dot{S} = 0$ schedule is:

$$\left.\frac{\partial \lambda}{\partial S}\right|_{\dot{S}=0} = -\frac{-(g'+b)\partial L_P/\partial S}{-(g'+b)\partial L_P/\partial \lambda} > 0, \tag{7.7}$$

while the slope of $\dot{\lambda} = 0$ is given by:

$$\left.\frac{\partial \lambda}{\partial S}\right|_{\dot{\lambda}=0} = -\frac{-F_{11}-F_{12}(\partial L_P/\partial S)}{\delta - F_{12}(\partial L_P/\partial \lambda)}, \tag{7.8}$$

where the denominator is obviously positive. In order to get the sign of the numerator we need to substitute the solution to $\partial L_P/\partial S$ and after some manipulation it becomes:

$$-\frac{F_{11}F_{22} - F_{12}^2 + F_{11}\lambda g''}{\lambda g''+F_{22}}. \tag{7.9}$$

Given the concavity of the production function, the term above is always positive. Therefore, the total expression and the slope are negative. Figures 7.2(a) and 7.2(b) illustrate the dynamics of the system, where the equilibrium point E is a saddle point, as can be seen by the streamlines in 7.2(b) depicting the direction of movements from any point in the S/λ space[6].

---

[6] The characteristic roots test confirms that E is locally a saddle point, since the root of the Jacobian at the equilibrium point is negative (Chiang, 1992).

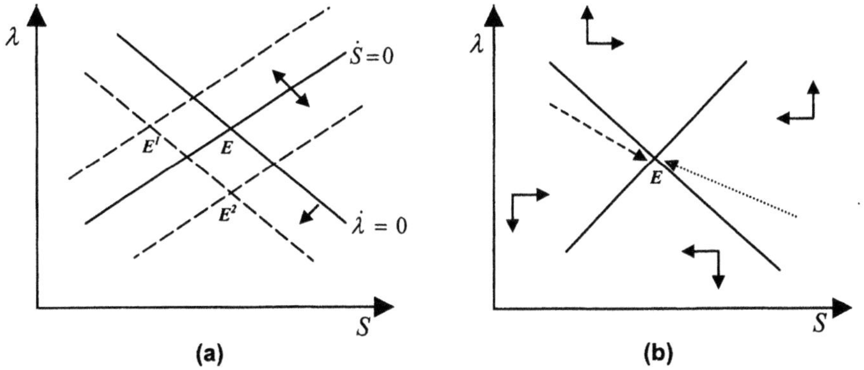

Figure 7.2: The impact of an increase in off-farm employment on soil quality.

With an exogenous increase in off-farm employment, the schedule $\dot{\lambda} = 0$ shifts downwards:

$$\left.\frac{\partial \lambda}{\partial L_O}\right|_{\dot{\lambda}=0} = \frac{F_{12}\dfrac{\partial L_P}{\partial L_O}}{\delta - F_{12}\dfrac{\partial L_P}{\partial \lambda}} \langle 0. \tag{7.10}$$

The direction of the shift in the $\dot{S} = 0$ curve is:

$$\left.\frac{\partial S}{\partial L_O}\right|_{\dot{S}=0} = \frac{g'(1+\dfrac{\partial L_P}{\partial L_O}) - b\dfrac{\partial L_P}{\partial L_O}}{-(g'+b)\dfrac{\partial L_P}{\partial S}}, \tag{7.11}$$

which is ambiguous and depends on the relative magnitudes of the terms in the numerator. Substituting the solution for $\partial L_P / \partial L_O$ and after some manipulation of the numerator, it turns out that the shift is towards the right or left as:

$$\left|\frac{b}{b+g'}\frac{L_P}{L_C}\frac{g''L_C}{g'}\right| > or < \left|\frac{F_{22}L_P}{F_2}\right|.$$

If the $\dot{S} = 0$ schedule moves to the left, the new short-run equilibrium level, E', is certainly lower than the initial level. Thus, according to the condition above, if the function $g(L_C)$ is less steep than the production function, the effect of an increase in off-farm labour will be a decrease in the equilibrium level of soil stock. In other words, if the farm production technology is such that a unit decrease in L_P implies a larger decrease in output (more concave production function) than the decrease in soil quality due to a unit decrease in L_C, farmers will tend to allocate the reduction in labour more to L_C than to L_P, thus negatively affecting the equilibrium soil quality level. This condition is simply a re-statement of the classic problem of balancing current gains with future losses, since losing soil quality now means a loss of productivity in the future.

If the $\dot{S}=0$ curve shifts to the right, the new soil equilibrium level, E^2, can be either higher or lower than the initial level, depending on the relative impacts, or shifts, in the steady-state schedules. A large shift in the $\dot{S}=0$ schedule and a small shift in the $\dot{\lambda}=0$ schedule make it more likely that the new soil quality level is higher. Thus, again, the relative concavity of the production and the *g(.)* functions affect the new soil equilibrium level in the same way as discussed above.

The ambiguity in the final effect on soil quality results from the opposite effects of farm labour on soil quality and is determined by the relative changes in L_P and L_C. At steady-state, the first two conditions of the maximisation problem (7.1) or, equivalently, Equation (7.3), determine the optimal allocation of labour to productive and conservation work, thus we can obtain the total effect of off-farm employment on allocation of farm labour:

$$\frac{dL_P}{dL_O}=\frac{-g'}{g'+b}\langle 0 \qquad (7.12a)$$

$$\frac{dL_C}{dL_O}=\frac{-b}{g'+b}\langle 0 \qquad (7.12b)$$

As can be seen from (7.12), labour in farm production and in conservation decrease by less than a unit as off-farm work increases by one unit. Once L_P and L_C have been determined, the equilibrium values of soil quality and the shadow value, λ, at steady-state can be obtained recursively using these values in conditions (7.2) and (7.4). The final impact on soil quality is ambiguous, because the reductions in L_P and L_C have opposite effects: less labour on productive activities implies less soil degradation[7], but less L_C results in lower soil conservation. The outcome is determined by the relative magnitude of the changes and these are, in turn, functions of the relative marginal effects of labour on soil quality, represented in this model by g' and b[8].

---

7 As pointed out earlier, decreases in L_P could in some cases imply an increase in soil degradation. In that case the final effect on soil quality would be unambiguously negative.

8 The results do not change if variable inputs are included as a factor in farm production, even if they also impact negatively on soil quality. As long as there are perfect markets

For farmers that do not face a missing agricultural labour market, the maximising problem changes in the sense that farm labour is not limited to available family labour. The household model is separable, so consumption and production decisions are taken sequentially. Therefore, neither family labour force nor off-farm labour should affect the level of total labour effort on the farm, or the use of soil conservation practices. Instead, labour allocation should be a function of the market wages only.

7.3 An Econometric Application of the Model

7.3.1 Data

The data come from a survey of rural households conducted in early 1996, administered jointly by the El Salvador Foundation for Economic and Social Development (FUSADES) and the World Bank. The survey incorporated 302 farm households from the 14 administrative regional Departments of El Salvador and is representative of the rural farm population at the 10% level of significance. The proportion of farmers in each Department was chosen according to the labour force census of 1992, which contained information on labour force per Department and their municipalities (World Bank, 1998).

The survey contains detailed information for each household, including household composition, education, a detailed account of on-farm and off-farm labour by all family members, and credit accessibility. Characteristics of agricultural production, including land area cultivated, output, use of intermediate inputs and use of conservation practices per plot, as well as hired labour, and capital are included. It also contains the farmer's evaluations about soil quality, slope and erosion problems in each plot. Unfortunately, these evaluations are subjective and qualitative in nature, thus limiting their usefulness in quantitative analysis of soil quality[9]. In the sample, approximately 35% of farmers indicated that erosion already caused or would soon cause a decline in agricultural production in their farms, while 54% use some type of soil conservation in their main plot. About half of them hired in at least one temporary farm worker in the year surveyed.

7.3.2 Estimation Procedures

The empirical model is derived from the first-order conditions at steady-state. As implied by these conditions, the equilibrium values of the choice variables,

---

for inputs and prices are given, the optimal solution to inputs would be given by equality between the value of marginal product of inputs and the sum of its price and its marginal impact on soil quality.

[9] A general description of sample characteristics is not included here for briefness, but is available from the author.

including labour allocated to soil conservation, are functions of the total available family labour, $\overline{L}$, family labour employed off-farm, L_O, initial soil quality, $\overline{S}_0$, discount rate, δ, coefficient of soil regeneration, a, and the coefficient that determines the impact of agricultural production on soil quality, b, and of exogenous variables in the farm production function.

For tractability purposes, the production function in the theoretical model was reduced to include only two variables, labour and soil quality index. However, at the estimation stage, other relevant variables are included. We do not postulate any particular format for the production function:

$$F_i = f(A_i, K_i, T_i, L_{Pi}, S_i)$$

where F_i is total farm output for farmer i, K_i is farm capital owned, T_i is land size operated, L_{Pi} is labour used in directly productive activities on the farm and S_i is soil quality index. A_i, an index of total factor productivity, is a function of characteristics of the farmer and family, such as education and age (as a proxy for experience), as well as farm characteristics that may affect technical productivity, such as land tenure situation, location and distance to infrastructure, such as roads. Therefore, the optimal value of labour used for soil conservation activities is a function of all exogenous variables (prices are normalised to one),

$$L_C^*(\overline{L}, L_0, \overline{S}, \delta, K, T, A, a, b).$$

Non-labour income is also included as an explanatory variable. As pointed out, if the assumption that there is no trade-off between leisure and work for poor households is correct, non-labour income should not have a significant impact on farm labour decisions. If, however, the trade-off is important, labour theory suggests that non-labour income should increase demand for leisure (because it increases the reservation wage) and consequently decrease farm work and conservation effort. The other possible, opposite, effect of non-labour income is as a source of cash for productive investments, including investment in soil conservation work[10]. The final effect is, therefore, ambiguous. As in El Salvador, remittances from family members that migrated are the most important source of non-labour income (World Bank, 1998), they are included in the regressions as a proxy.

The available data do not permit direct separation of farmers into those facing or not facing missing agricultural labour markets. One possible way to separate the sample is to use the information on hired farm labour. As discussed, households in

[10] For the same reason, the availability of credit for investment can also significantly affect the use of soil conservation, but it is usually scarce and expensive in rural areas. In regressions where access to credit was included, it did not significantly affect adoption of soil conservation practices nor did it change the other parameters (results available from the author).

a situation of missing agricultural labour markets do not hire in any labour because their on-farm subjective wage is lower than the effectively paid wage. Thus, one could expect that households that hire-in farm work are in a situation of perfect labour markets. However, farmers that do not hire labour can be either facing a missing market or not, since the decision to use only family labour can be just their optimal choice given household assets.

By dividing the sample in farmers that do and do not hire in labour we can say that the first group contains mostly farmers in a situation of well-functioning markets[11], while the latter includes households for which the market fails, but possibly also others for which the optimal choice is self-sufficiency. Other farm characteristics that could be related to market access, such as distance to closest paved road and connection to electricity were examined for these two groups of farmers separately. It was found that farmers that hire are closer to roads and more frequently connected to electricity than farmers that do not hire (difference in means are statistically significant at 5% and 10%, respectively). In the remainder of this paper the two groups will be referred to as the 'hire' and 'do not hire' groups of farmers.

Since this division of the sample does not completely correlate with the groups defined in the theoretical part, it is important to examine what can be the biases for the main estimated parameters. According to the theory results, the use of conservation practices by the group of farmers that face a missing labour market should be negatively affected by smaller family size and more family members employed off-farm, while in the other group these factors should not be significant. Thus, if the sample of farmers that do not hire includes households that actually do not face a missing agricultural labour market, the parameters to the relevant variables could be biased toward non-significance. This means that if these parameters are found to be significant and negative, the 'unbiased' effect can only be stronger.

The available data include information on type of soil conservation practices used, but not on their costs or labour required. The types of conservation activities include, among others, minimum tillage, use of crop residue as mulch, cover crops, hedges, contour barriers and terraces. Since we only have information on the use or not of conservation practices, the estimation must be of a discrete nature. The probability of adoption of soil conservation practices depends on the optimal level of labour effort in conservation activities, and the discrete nature of

[11] As mentioned earlier about seasonality considerations, strong peaks of demand can cause even households that face missing labour markets to hire-in temporarily. Therefore, with this sample separation some farmers that actually face a missing market would be included in the group representing households in a perfect labour market. According to the theoretical model, this fact implies that estimated parameters of family labour force and off-farm work could be biased toward significance. Thus, if the estimated parameters turn out not significant, the unbiased results should only be less significant.

the estimation captures this probability. A binary choice model is estimated to measure the probability of adoption (see also Lee & Stewart, 1983 and Rahm & Huffman, 1984).

Some of the explanatory variables are not measured directly in the survey. Certain proxies are used to represent these variables in the estimation. The index of soil quality, S, is represented by the qualitative answer of the farmer about soil fertility. The impact of cultivation activities on soil erosion, b, increases with the natural erosion potential of soil, which, in turn, is represented in the Universal Soil Loss Equation (USLE) by factors such as slope steepness and length and weather conditions. Slope is used in the regressions as a binary variable describing whether the plot inclination is steep or not; the steeper the slope, the higher is the 'potential to erode'. Farmers' perception of soil quality is included also as a binary variable, equal to one if it is considered of good quality. Climatic characteristics are not available.

Farmer and family characteristics are represented by the average education of all family members older than 12 years of age and the age of the head of household. Other characteristics that can influence total factor productivity include land tenure situation, access to technical assistance and location variables, represented by 14 regional dummies. Total family work-force, land size and farm capital, the main household assets, are also included.

In order to measure off-farm labour by family members, three different variables are used. They are: 1) a dummy that equals 1 if any family member works off-farm during at least some period; 2) proportion of adult family members that work off-farm at least part of the year; 3) annual number of hours worked off-farm averaged for all adult family members.

The binary choice model is estimated in per capita terms, using a logit specification model. The sample includes only the main plot for each farm, since this is the only one for which specific characteristics are available.

7.4 Results and Discussion

In Table 7.1 the results of the two sets of logit regressions are presented. Different specifications are used, including three different measures of off-farm employment. The effect of off-farm employment on the use of soil conservation is consistently negative and significant (at 1% in all three specifications) for the group of households that do not hire, while it is not significant in any of the specifications for farmers that hire, confirming the theoretical results.

Table 7.1: Logit estimates of adoption of soil conservation practices [1] (Dependent variable = 1 if household adopts; 0 otherwise).

	Regression 1		Regression 2		Regression 3	
	'do not hire'	'hire'	'do not hire'	'hire'	'do not hire'	'hire'
Intercept	7.73^{**}(3.32)	4.44 (3.32)	6.32^{*}(3.77)	4.12(3.33)	6.89^{*}(3.93)	4.35(3.35)
Family work force	0.99^{*}(0.59)	-0.52(0.56)	0.95(0.59)	-0.52(0.56)	1.00^{*}(0.61)	-0.54(0.56)
Dummy=1 if any family member works off-farm	-1.83^{***}(0.56)	-0.48(0.50)	--	--	--	--
Average hours worked off-farm per family	--	--	-0.15^{***}(0.04)	-0.03(0.04)	--	--
Proportion of persons above 15 yrs old working off-farm	--	--	--	--	-5.79^{***}(1.56)	0.21(1.22)
Log of average education of persons over 12 yrs	-0.21^{*}(0.11)	0.02(0.11)	-0.21^{*}(0.11)	0.02(0.11)	-0.27^{**}(0.12)	0.01(0.11)
Log of age of head of household	-2.07^{**}(0.91)	-0.80(0.86)	-1.99^{**}(0.91)	-0.81(0.87)	-1.82^{**}(0.93)	-0.91(0.88)
Log of remittances	-0.03(0.04)	0.03(0.03)	-0.03(0.04)	0.03(0.03)	-0.03(0.04)	0.04(0.03)
Log of land size per capita	-0.24(0.32)	-.010(0.23)	-0.26(0.32)	-0.10(0.23)	-0.10(0.33)	-0.07(0.22)
Log of per capita capital attached to land	0.05^{*}(0.03)	0.04(0.03)	0.05^{*}(0.03)	0.03(0.03)	0.05(0.03)	0.03(0.03)
Log of per capita capital non-attached to land	-0.02(0.05)	0.02(0.05)	-0.02(0.05)	0.02(0.05)	-0.01(0.05)	0.03(0.05)
Have title to land	0.94(0.64)	0.45(0.54)	0.99(0.63)	0.48(0.54)	1.198*(0.67)	0.54(0.53)
Steep slope	0.21(0.48)	0.46(0.49)	0.17(0.48)	0.45(0.49)	0.35(0.49)	0.50(0.49)
Land of good quality	-0.85^{*}(0.50)	-0.43(0.49)	-0.81^{*}(0.50)	-0.43(0.49)	-0.67(0.51)	-0.43(0.49)
Received technical assistance	1.73^{***}(0.71)	2.26^{***}(0.73)	1.67^{**}(0.71)	2.25^{***}(0.73)	1.47^{**}(0.72)	2.28^{***}(0.73)
Distance to closest paved road	-0.02(0.03)	-0.01(0.03)	-0.02(0.03)	-0.01(0.03)	-0.01(0.03)	0.001(0.03)
Concordant predicted probabilities	82.7%	78.9%	82.6%	78.8%	84.0%	78.0%
Discordant predicted probabilities	17.2%	21.0%	17.2%	21.0%	15.9%	21.7%
log-likelihood (multiplied by (-2))	149	152	149	153	141	153
Percentage of households that adopt	49%	58%	49%	58%	49%	58%
Sample size	150	143	150	143	150	143

[1]Standard errors in parentheses; * p<10%, ** p<5%, ***p<1%; thirteen regional dummies were included, but are not reported.

As discussed in the previous section, if we consider that the 'do not hire' group can include farmers that actually do not face missing agricultural labour markets, the possible bias in the family labour parameters is towards non-significance. Therefore, finding significant parameters suggests an even stronger real effect. Moreover, the empirical results show that if there are significant positive effects of off-farm employment on conservation due to increased access to cash for investment, this effect is dominated by the negative labour effect.

Family work force has a positive effect on conservation only in the group 'do not hire', while being insignificant in the other group. Family characteristics, such as education and age of head, also have significant impacts only in the 'do not hire' group. Thus, separability does not hold for this group, confirming that the sample division has been appropriate in discriminating between households operating under imperfect and perfect agricultural labour markets.

Contrary to expectations, education has a negative and significant effect on the use of soil conservation for the 'do not hire' group of farmers, while it has no significant effect for other farmers. Most studies in developed countries have found a positive relationship between education and conservation investment (Ervin & Ervin, 1982; Norris & Beatie,1987; Rahm & Huffman, 1984; Wu & Babcock, 1998), although some studies in developing countries have more commonly shown neutral or negative relationships. Education is expected to increase farmers' understanding of the benefits of soil conservation, as well as their ability to use conservation techniques.

The reason for the negative effect of education found in this study should be further explored, but there are at least two possibilities. When more children go to school, they cannot work on the farm. Even at young ages, children usually help in less demanding jobs such as house-keeping, thus freeing-up work hours by the older family members. Therefore, either directly, or indirectly, as more children go to school, less labour is available for soil conservation. Additionally, the trade-off between leisure and work is stronger at higher wages, and more educated workers likely earn better salaries. In better educated families, therefore, more leisure may be selected over farm and conservation work.

Age of the farm operator is also negative only in the constrained group. Shorter planning horizons and, consequently, higher discount rates, reduce the benefits of long-term investments, such as conservation, at an older age. However, if conservation investments are reflected in land values, the negative effect of age should not be significant (McConnell, 1981; Norris & Beatie, 1987). The fact that for farmers not facing missing markets, age is not significant may indicate that these households may also have access to better functioning land markets.

The effect of remittances is not significant in either group, suggesting that the trade-off between leisure and work is not affected by non-labour income for most farmers.

Technical assistance is strongly positive and significant for both groups. In El Salvador there have been various programs for soil conservation, public and private, that included both conservation and productivity technologies. But many such programs have not been very effective after their completion. The results here indicate, however, that access to technical assistance has unequivocally improved adoption by all farmers.

Soil quality affects conservation in the same way for both groups, although only significantly in two regressions for the 'do not hire' group. The better the soil, the lower the probability that a farmer uses soil conservation. The effect of land tenure is positive but only significant in regression 3 for the group of farmers facing missing labour markets. This indicates that for this group, tenure security has a positive impact on investments in conservation, as it affects the technical productivity of the farm.

Extreme poverty and subsistence considerations may considerably affect farmers' behaviour with respect to conservation. Since most conservation investments have returns in the long-run, while the costs are felt in the present, high rates of discount would decrease the 'profitability' of such investments. Since farmers at subsistence have higher discount rates, they would be less likely to adopt conservation. The estimations in Table 7.1 already partially take into account household wealth with land and farm capital. An additional dummy variable was included in other regressions[12] to account specifically for annual income. The dummy indicates whether the per capita household income falls below a 'subsistence' poverty line estimated for El Salvador in previous studies (World Bank, 1998). The results are very similar to those presented here but, interestingly, extreme poverty seems to impact positively on conservation investments. Pagiola (1995) argues that households facing subsistence constraints are usually more likely to adopt sustainable land use practices, because of their heavy dependence on agricultural production, no access to fertilisers and few alternative income sources.

7.5 Conclusions

The results confirm the hypothesis that rural market conditions affect the behavioural responses of farm households with respect to soil conservation. Additionally, off-farm employment decreases use of conservation practices only in farm households facing a missing agricultural labour market.

Other studies that econometrically estimated the effect of off-farm employment on conservation, have not considered differences in behavioural responses of households with different access to labour markets (Gould *et al.*, 1989; Norris & Beatie, 1987). Some have found positive and some negative effects. The results of

[12] Available from the author.

the present study suggest that differential access to markets partially determines households' behavioural responses and, therefore, conclusions about the link between off-farm employment and conservation should consider the conditioning effect of markets.

Although this paper focuses only on the labour market, other market imperfections may impact on use of conservation practices directly, or indirectly through labour allocation. As mentioned before, financial constraints, lack of access to credit and insurance markets for agricultural production, can increase the importance of off-farm employment as a source of income and to reduce risk. As such, households may choose to work off-farm even when the loss in agricultural productivity on the farm is too high. The functioning of these other markets should be taken into consideration in the design of sectoral and conservation policies for rural areas.

The econometric analysis did not include an evaluation of the final effects on soil quality. The data requirements for such an analysis are much higher than available here. Bio-economic models are much more appropriate for analysis of soil quality changes, since they can include more specific interactions and evolution equations of soil quality. Additionally, in order to evaluate soil quality in this study, it would be necessary to consider what happens to labour used in directly productive farm activities and whether these activities themselves change when increased off-farm employment takes place. As discussed before, labour saving changes in productive agricultural activities could be either soil depleting or conserving. This is a topic for future analysis.

The results indicate that soil degradation and labour market conditions are closely associated. They also indicate that the opposing evidence on the relationship between off-farm employment and soil conservation could be partially explained by the existence of labour market failures. For households that can hire-in farm labour and invest in conservation, more off-farm job opportunities can even improve soil quality (although the results obtained here do not show positive significant effects of off-farm employment in this case). On the other hand, farmers in a situation of missing labour markets are not able to adapt to the new situation of increased off-farm employment and tend to abandon or not to adopt conservation practices on their farm.

Rural policies that promote non-farm rural economic activities may have an unwanted negative impact on soil quality, wherever labour markets do not function properly. Policy design should take into account the variability in household characteristics that can affect their access to markets and adoption of conservation practices. Not doing so would render the design of conservation policies and targeting strategies inefficient. Additionally, policies would be more effective if they promoted labour-saving conservation practices given that labour is a limiting factor. On a broader level, investment in infrastructure that decreases transaction costs in rural markets should contribute to reducing the negative effect of off-farm employment on soil quality.

7.6 References

Ali, M. & D. Byerlee (1991): Economic efficiency of small farmers in a changing world: a survey of recent evidence. Journal of International Development 3, 1-27

Alwang, J. & P.B Siegel (1999): Labor shortages on small landholdings in Malawi: implications for policy reforms. World Development 27, 1461-1475

Barbier, E.B (1990): The farm level economics of soil conservation: the uplands of Java. Land Economics 66, 199-211

Barbier, E.B. & R. Lopez (1998): Debt, poverty and resource management in a rural small-holder economy. Paper presented at the World Congress of Environmental and Resource Economics, Venice, 25-27 June, 1998

Barret, S. (1991): Optimal soil conservation and reform of agricultural pricing policies. Journal of Development Economics 36, 167-187

Barret, C.B. (1999): Stochastic food prices and slash-and-burn agriculture. Environment and Development Economics 4, 161-176

Chiang, A.C. (1992): Elements of dynamic optimization. McGraw-Hill, Inc. New York

de Janvry, A., M. Fafchamps & E. Sadoulet (1991): Peasant household behaviour with missing markets: some paradoxes explained. The Economic Journal 101, 1400-1417

de la Brière, B. (2000): Imperfect food markets and household adoption of soil conservation practices in the Dominican Republic highlands: household probit and duration models. This volume

Edwards, A.C. (2000): The labor market in rural Latin America. In: R. López & A. Valdés (Eds.) Rural poverty in Latin America: analytics, new empirical evidence and policy. St. Martin Press, New York, MacMillan, London

Ervin, C.A. & D.E. Ervin (1982): Factors affecting the use of soil conservation practices: hypotheses, evidence, and policy implications. Land Economics 58, 277-92

FAO (1994): Marco normativo de la politica de desarrollo agricola y rural sostenible de El Salvador. Informe Tecnico, Project FAO/TCP/ELS/2251(A), San Salvador

Feder, G. (1985): The relationship between farm size and farm productivity: the role of family labor, supervision and credit constraints. Journal of Development Economics 18, 297-313

Gould, B.W., W.E. Saupe & R.M. Klemme (1989): Conservation tillage: the role of farm and operator characteristics and the perception of soil erosion. Land Economics 65, 167-182

Holden, S.T., B. Shiferaw & M. Wik (1998): Poverty, market imperfections and time preferences: of relevance for environmental policy? Environment and Development Economics 3, 105-130

Lapar, M., A. Lucila & S. Pandey (1999): Adoption of soil conservation: the case of the Philippine uplands. Agricultural Economics 21, 241-256

Lee, L.K. & W.H. Stewart (1983): Land ownership and the adoption of minimum tillage. American Journal of Agricultural Economics 65, 256-64

López, R. & M. Niklitschek (1991): Dual economic growth in poor tropical areas. Journal of Development Economics 36, 189-210

McConnell, Kenneth E (1983): An economic model of soil conservation. American Journal of Agricultural Economics 65, 83-89

Norris, P.E. & S.S. Beatie (1987): Virginia farmers' soil conservation decisions: an application of tobit analysis. Southern Journal of Agricultural Economics 19, 79-90

Pagiola, S. (1995): The effect of subsistence requirements on sustainable land use practices. Paper presented at the Annual Meetings of the AAEA, Indianapolis

Pagiola, S. (1996): Price policy and returns to soil conservation in semi-arid Kenya. Environmental and Resource Economics 8, 255-271

Pender, J. & J. Kerr (1998): Determinants of farmer's indigenous soil and water conservation investments in semi-arid Inidia. Agricultural Economics 19, 113-125

Rahm, M.R. & W.E. Huffman (1984): The adoption of reduced tillage: the role of human capital and other variables. American Journal of Agricultural Economics 66, 405-13

Reardon, T. & C.B. Barret (1999): The ambiguous effects of policy reform on sustainable agricultural intensification in Africa: renewed threats to fragile margins? Mimeo.

Ruben, R., M. v.d. Berg & S. Tan (2000): Land rights, farmers' investment, and sustainable land use: modelling approaches and empirical evidence. This volume

Sadoulet, E. & A. de Janvry (1995): Quantitative development policy analysis. The Johns Hopkins University Press, Baltimore

Sadoulet, E., A. de Janvry & C. Benjamin (1996): Household behavior with imperfect labor markets. Working Paper Series, Dept. of Agricultural and Resource Economics, University of California, Berkeley

Templeton, S.R. & S.J. Scherr (1997): Effects of demographic change on hilly-mountainous land in developing countries: a microeconomic analysis of empirical evidence. International Food Policy Research Institute, Washington D.C.

World Bank 1998. El Salvador: Rural development study. The World Bank, Washington D.C.

Wu, J. & B. Babcock (1998): The choice of tillage, rotation, and soil testing practices: Economic and environmental implications. American Journal of Agricultural Economics 80, 494-511

8 Assessing the Effects of Policy Measures on Household Welfare and Agro-Ecological Sustainability: an Overview of Farm Household Modelling Approaches

Gideon Kruseman

Development Economics Group, Department of Economics and Management, Wageningen University and Research centre, Wageningen, the Netherlands

Abstract

Farm household modelling is a powerful tool to aid policy makers with the analysis of sustainable land use. This paper gives the methodological foundations for a farm household simulation approach encompassing the results from quantified agronomic models, based on parametric optimisation. The paper discusses shortcomings and advantages of different approaches underlying the methodology, and demonstrates how these can be integrated into a pragmatic framework. Starting with a basic farm household model extensions are made and methods described to make those extensions operational. While econometric analysis alone is an insufficient tool to fully explain all relevant household behaviour, it is an important element in the approach. The results of the optimisation model are used in metamodelling exercises that summarise the relationship between exogenous parameters and indicator variables, to account for endogenous prices at the regional level.

8.1 Introduction

Soil degradation is considered a threat to development in many parts of the developing world (Pinstrup-Andersen, 1994). Soil degradation occurs when the level of exploitation continuously remains above the soil's capacity to regenerate. In many parts of Africa fallow cycles have become increasingly shorter without alternative strategies being in place to maintain soil fertility. Income levels have not yet fallen dramatically, as existing stocks of soil nutrients are being exploited (van der Pol, 1993). This process is commonly referred to as soil mining. Soil capital is being converted into income with no, or insufficient, replacement investments taking place.

As long as readily available employment outside agriculture remains limited, large portions of the growing population will depend on farming for their livelihood. With this in mind, policy makers will have to address the issue of sustainable rural development. This implies inducing changes in markets, institutions and available technologies in such a way that agriculture can remain productive, generate employment and guarantee food security either through food production or high value cash crops that can be traded for food.

Policy measures influence development processes in an indirect way through their effect on socio-economic circumstances. Farmers react to these circumstances, adapting their farming systems to the new challenges. For policy makers trying to enhance sustainable development in rural areas, it is therefore important to understand the impact of policy change on household decisions and the subsequent effect on welfare and sustainability indicators.

Farm household modelling is potentially a powerful tool to aid policy makers. It permits analysis of decisions at the level of land users to assess the potential impact of technological change and the implementation of policy measures. Economic models do not usually take into account the agro-ecological processes underlying agricultural production. Instead those processes are treated as a black box and *simplified* relationships between relevant input parameters and output variables are used. Moreover, specifications of production functions are based on observed allocations of scarce resources and hence cannot be extrapolated to deal with new or different technologies.

Agronomic models of plant growth and crop-soil interactions take into account biophysical processes only, disregarding behavioural aspects of farming. As a result, biophysical studies commonly indicate scope for improvement of farming systems (Wolf *et al.*, 1991; Penning de Vries *et al.*, 1995). Biophysical models give very useful insights into the relationship between land use, organic and inorganic fertiliser applications, and output in terms of production and change in soil quality.

This chapter presents a methodological framework for bio-economic farm household models to assess the effectiveness of policy measures on household welfare and agro-ecological sustainability indicators. In Section 8.2 the theoretical underpinnings of farm household modelling are briefly reviewed with an emphasis on the underlying assumptions that permit certain commonly used simplifications. Section 8.3 briefly discusses where traditional econometric analysis fails when estimating bio-economic farm household models. Section 8.4 introduces mathematical programming as a technique to deal with some of the problems that have arisen. The general structure of a bio-economic household model is described, that combines an econometrically specified non-linear expenditure part with a linear production structure. Mathematical programming has its own shortcomings, some of which can be dealt with through metamodelling techniques, discussed in Section 8.5. Finally in Section 8.6 the key points of the methodology are summarised and critically reviewed.

8.2 Theoretical Underpinnings of Farm Household Modelling

A model that analyses farm household production is the natural starting point in dealing with the issue of sustainable land use. Traditionally, economists have used a profit function to explain household behaviour. When there are no missing or incomplete markets and low transaction costs, consumption and production decisions are separable, and a profit function might be used to represent production decisions by family farms. This function is also useful in analysing behaviour of plantation agriculture. But since consumption and production decisions are usually not separable, the proposition that profit is the principal objective is not tenable. Consumption as an objective itself may also pose problems, because a household may have more and sometimes conflicting goals and aspirations. The notion of a household model that links consumption and production dates back to early twentieth century Russian economists (Chayanov, 1923) and has been revived in the past twenty years (Barnum & Squire, 1979; Singh *et al.*, 1986; De Janvry *et al.*, 1991). The basic model, omitting all complexities, consists of an objective function that households are assumed to maximise:

$$u = u(\mathbf{c}, l) \tag{8.1}$$

where $\mathbf{c}$ is a vector of consumption goods and l is leisure. Utility is maximised, subject to a cash income constraint:

$$\mathbf{p}^m \mathbf{c}^m = \mathbf{p}^a(\mathbf{q} - \mathbf{c}^a) - w(L - F) \tag{8.2}$$

where $\mathbf{c}^m$ and $\mathbf{c}^a$ are market purchased and household produced commodities, $\mathbf{p}^m$ and $\mathbf{p}^a$ are vectors of market and farm-gate prices, $\mathbf{q}$ is a vector of the household's production, w is the wage rate, L is total labour input and F is family labour input. This implies that if $(L\text{-}F)$ is positive, labour is hired, conversely it represents off-farm labour supply.

The household faces a time constraint, specifying that the household cannot allocate more time to labour input F or leisure l than available:

$$l + F = T \tag{8.3}$$

where T is the total stock of household's time. The household also faces a production constraint or technology function that depicts the relationship between inputs and outputs (in this simple case labour and commodities):

$$\mathbf{q} = q(L, A, \tau) \tag{8.4}$$

where A is the household's fixed quantity of land and τ is the fixed technology level.

The basic model does not account for risk, it assumes hired and family labour to be perfect substitutes and the labour market to be free of imperfections. If, in addition, transaction costs are assumed to be negligible the choice between subsistence and cash crop production is one of mere semantics. The result of these assumptions is a separable model, in which profit is maximised, followed by solving an indirect utility function. All the constraints can be collapsed into a single constraint that is identical to the concept of full income (Becker, 1965). Subsequently, first-order conditions are uncomplicated and it is easy to derive reduced-form equations relating decision variables to exogenous parameters (prices and wages, quasi-fixed inputs such as land, technology level).

The basic model fails to describe the reality of farmers in developing countries for four reasons. Household decisions are assumed to be separable, while both prices for tradable goods and technology are exogenous parameters, and the dynamic character of sustainability is ignored. The use of separable household models is only permitted when there are no market imperfections. This is seldom the case in developing countries and hence, consumption and production decisions must be considered simultaneously. Market imperfections exist in several markets. Farm households face severe weather risk against which there is no insurance. Transportation and other transaction costs are high, especially in many parts of Africa where physical and social infrastructure is poor, and the maximum carrying capacity of the land is reached at relatively low population densities due to the poor land quality.

In addition to the existence of market imperfections, modelling should also consider the degree to which households are heterogeneous (Holden *et al.*, 1998). If there is a fair amount of heterogeneity the interactions of households in local markets must be considered. If these interactions are not integrated in outside markets, local price formation occurs. Heterogeneity is the most common circumstance in developing countries and any number of commodities is at least poorly tradable at the aggregate level. For some commodities, markets are intrinsically shallow (perishable crops such as vegetables). For staple food crops that are grown by most farmers in all regions aggregate supply tends to fluctuate with weather conditions. The implication of local price formation is the need to endogenise prices.

The basic model includes technology as an exogenous parameter. In reality technologies are often adapted to changing circumstances. An additional issue which is solved simultaneously with inclusion of endogenous technology, is the fact that the original basic model considers a single agricultural output and does not distinguish between crops. A further refinement of the biophysical aspects in the farm household model concerns crop-livestock interactions and the effects of these interactions on technology choice and agro-ecological sustainability indicators[1].

Non-separability is introduced into the model as high transaction costs exist on factor and commodity markets, and there is severe weather risk. Transaction costs imply market imperfections. They can be illustrated as the existence of a price difference between effective purchasing and selling prices in the case of commodities, and between returns on factor use for own production and rented or hired production factors. If the shadow price of a production factor or commodity lies between the buying and selling price, *i.e.* within the price band (De Janvry & Sadoulet, 1994), a non-separable household model has to be applied. Following Omamo (1995), concentrating on missing markets for commodities, the budget constraint becomes:

$$\mathbf{pc} + \mathbf{d}^c\mathbf{c}^m = \mathbf{pq} - \mathbf{w}(\mathbf{x} - \mathbf{R}) - \mathbf{d}^q\mathbf{q}^m + y^* \qquad (8.5)$$

where $\mathbf{p}$ is a vector of commodity prices, $\mathbf{d}^c$ and $\mathbf{d}^q$ are vectors of marketing costs, $\mathbf{c}$ and $\mathbf{c}^m$ are vectors of all and market purchased consumption goods, respectively, $\mathbf{q}$ and $\mathbf{q}^m$ are vectors of all and sold commodities, respectively, $\mathbf{w}$ is a vector of rewards for production factors, $\mathbf{x}$ a vector of production factors used, $\mathbf{R}$ a vector of resource endowments, and y^* is exogenous income. The underlying assumption is that marketing costs are constant with respect to scale. Notice that $\mathbf{p}$ denotes a vector of market prices, whereas the equilibrium of a household not participating in the market, is determined by its shadow price.

Weather risk affects the household in two ways; (i) directly due to production risk related to weather risk; and (ii) indirectly through the covariance of weather and price risk. Since weather is unknown at the outset of the growing season, decisions are made on basis of expectations about the probability of weather (in the case of semi-arid regions this is the probability of precipitation levels and distribution). The impact is at three levels. Prices become expected prices conditional on weather, the production function contains weather as a parameter, and thirdly the utility function is adapted to take into account risk:

$$u^e = \int u(\mathbf{c})dF(\mathbf{c}) \qquad (8.6)$$

where u^e is expected utility and F is the cumulative distribution function $F:\mathbb{R}\rightarrow[1,0]$ associated with uncertainty of outcomes due to price and weather risk.

The last issue that needs to be resolved is the objective function. To address sustainability issues soil quality dynamics have to be included into the model. This

---

[1] See Kruseman & van Keulen (2000) for further details on incorporating biophysical aspects.

implies extending the basic model to take into account intertemporal aspects through discounted utility:

$$u^{npv} = \int u(\mathbf{c})e^{-rt}dt \tag{8.7}$$

where r is the discount rate and t is time, ignoring time subscripts. This corresponds to the general formulation of optimal control models (Bulte & Van Soest, 1999).

Combing Equations (8.6) and (8.7) yields:

$$u = \int\int u(\mathbf{c})e^{-rt}dF(\mathbf{c})dt \tag{8.8}$$

This objective function in combination with Equations 8.1 and 8.2 renders the model non-separable, because production depends critically on consumption decisions in earlier periods.

8.3 Empirical Estimation of Bio-economic Farm Household Models

The farm household models developed in the 1980s (Singh *et al.*, 1986) were estimated econometrically. Full estimation of a complete farm household model, even without the extensions suggested in Section 8.2, is tedious and requires a large consistent data set. Very often a model is not estimated completely, but for a specific purpose only. In that case, the model can be simplified to a great extent, and reduced forms can be estimated (Goetz, 1992; Benjamin, 1992; Skoufias, 1994; Sadoulet *et al.*, 1998). This section briefly reviews the main characteristics of the model developed in Section 8.2 and determines the possibilities and data requirements for its econometric estimation.

Maximising under the assumption of an interior solution, the necessary conditions depend on where the household is situated with respect to the price bands. The existence of price bands implies that household supply and demand curves cannot be differentiated in the border cases where market prices corrected for transaction costs equal the household shadow price. This leads to first order conditions that are conditional on whether the households participate in the market, and on whether they are net buyers or sellers. This leads to very complex first-order conditions, and requires switching regression models to discern between the types of households, which makes the estimation of the model more tedious.

All equations are linked due to the non-separability of production and consumption decisions. These relationships are complex, as can be deduced from deriving the first-order conditions. Furthermore, endogenous technology choice also has consequences for the functional form. Taking into account the synergistic effects of input combinations, both the vector of production functions and the soil quality function are likely to be highly non-linear in both variables and parameters (Hengsdijk *et al.*, 1996; Kruseman & van Keulen, this volume). Hence the functional form of the reduced form equation nor the signs of its coefficients can be predetermined.

The general format of the reduced form equation is that the relevant endogenous variable is described in terms of the relevant exogenous parameters. In the case of the sustainability indicator this is (see also Kruseman &Van Keulen, this volume):

$$\dot{s} = f(\mathbf{p}, \mathbf{d}, dF(\mathbf{c}), r, \mathbf{w}, y^*, \mathbf{R}, s_0, \xi, \zeta) \tag{8.9}$$

where $\dot{s}$ denotes change in soil quality, s_0 denotes initial soil quality, the vector of prices **p** denotes expected prices, ξ and ζ denote household and production characteristics, respectively. Following the line of thought of Nerlovian supply response models (Nerlove, 1979; Askari & Cummings, 1976) expected prices are a function of past prices. This implies that **p** is now a matrix of prices. Subjective expected prices are the easiest problem to solve. Transaction costs **d** and the subjective cumulative distribution function of weather risk are difficult to obtain, as is the subjective discount rate. The data requirements for full econometric estimation of this reduced-form equation deserve attention. Household level data on *e.g.* household characteristics, prices, production characteristics, resource endowments, have to be combined with household level data on soil quality and soil quality change. Beside the fact that this type of data is seldomly collected simultaneously, there is an additional problem with short-term data. Because the model deals with dynamics, for econometric analysis panel data are necessary. But only with very long time series is it possible to discern the mutual effects of soil quality and productivity on each other (Kruseman & van Keulen, this volume).There are so many random effects involved that the short-term changes are lost in the random variation.

In summary, three types of problems have been distinguished with respect to econometric estimation of household models. Functional form specification is difficult, and the signs of the coefficients are indeterminate, making it impossible to distinguish between true inference and spurious correlation. The data necessary to estimate the equations is difficult to gather, either because it covers different scientific realms, or because the values are elusive due to their subjective character. Finally, for capturing the effects of soil degradation very long time series are necessary.

8.4 Mathematical Programming Models: Advantages and Shortcomings

Although full econometric estimation of bio-economic farm household models is currently not feasible due to the lack of consistent data, there is, however, a large amount of data available in most developing countries related to separate components of the farm household model, that can be estimated econometrically. This section describes a procedure to link these components into a consistent framework. Since the household is assumed to maximise a utility function, application of mathematical programming techniques is most appropriate, as they allow inclusion of complex relationships and technology choice.

The basic structure of the farm household model was presented in Section 8.2. The production side is modelled using Leontieff production activities in terms of vectors of input-output combinations, that can be interpreted as points on an n-dimensional production function. The consumption is modelled with an econometrically specified utility function.

In contrast to other bio-economic models (Barbier, 1994; Dalton, 1996; Köbrich, 1997; Shiferaw, 1998), multiple objectives are considered in this model to account for consumer preferences (consumption utility) and sustainability criteria (macronutrient and organic matter balances). Specific weights are attached to each objective for different household types, based on the trade-offs under partial optimisation (Romero, 1993). The goal weights are calibrated through comparison of model results with empirical evidence (Bade *et al.*, 1997).

The combination of the two approaches is a farm household model in which an econometrically specified, non-linear, behavioural expenditure part is linked with a linear programming optimisation procedure of the production structure (Ruben *et al.*, 1997; Kuyvenhoven *et al.*, 1998). The use of a linear programming production framework and a direct expected utility function allows introduction of interactions between production and consumption, such as different buying and selling prices due to transaction costs (Omamo, 1995).

The structure of the bio-economic model is schematically presented in Fig 8.1. The model consists of six separate modules for (i) production activities, (ii) expenditures, (iii) prices, (iv) resource endowments based on a farm household stratification, (v) savings and investments, and (vi) objectives and goal weights.

The *production-activity module* describes the agro-ecological processes that determine production options for cropping, pasture, livestock and forestry activities. These options are 'generated' as specific combinations of inputs (seed, labour, nutrients) associated with certain output levels (Hengsdijk *et al.*, 1996) using a Technical Coefficient Generator (see chapter Kruseman & van Keulen, this volume). Nutrient and organic matter balances are derived simultaneously. Technical coefficients depend on soil and weather conditions, as well as farmers' strategies for the management of crop residues, soil conservation measures, and anti-erosion activities. Different technical coefficients are defined for currently applied farming practices (generally based on soil mining), as well as for alternative practices that guarantee more sustainable resource use (defined in terms of non-negative nutrient and organic matter balances).

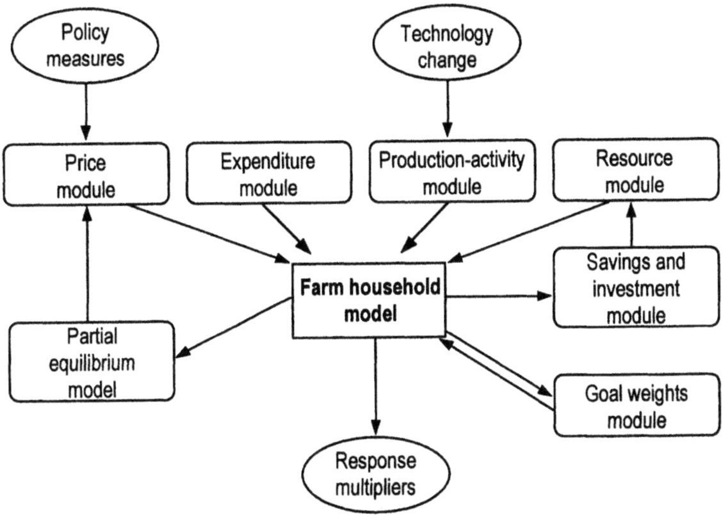

Figure 8.1: Structure of the modelling framework.

The interactions between various inputs that give rise to synergistic effects cannot be adequately captured with a continuous production function. Therefore, a series of discrete point-estimates are defined, where input efficiency is dependent on the availability of other production factors. Due to the complementarities between inputs, a non-homogenous n-dimensional production space results. In economic terms this can be interpreted as the existence of multiple production functions, whereby allocative efficiency is reached either through shifts along or between different production functions. Using discrete point estimates for technological choice enables reliance on linear programming techniques for optimisation.

The *expenditure module* is derived from a cross-sectional budget survey that enables the estimation of a direct utility function for different farm household income levels by assuming that the expenditures are the revealed preferences of households that choose amongst alternatives according to the marginal utility of several consumption categories. Through piecewise linearisation the derived concave utility function can be incorporated in the linear programming framework. This enables the direct evaluation of the impact of modifications in relative prices on farm household welfare.

The *price module* includes information on factor and product prices, based on historical price series. A price band that represents transaction costs accounts for the difference between market and farm gate prices. For partial analysis at farm household level, prices are considered exogenous, but at an aggregate, regional level interactions between farm households on factor and commodity markets are specified within the framework of a partial equilibrium analysis using metamodelling techniques (see Section 8.5).

Different farm household types are identified in the *resource module* according to their initial factor endowments, calculated from farm surveys and farming systems research. The *savings and investment module* defines an income-dependent fixed savings component and a household-specific transitory savings component. Fixed savings are a fraction of the expected income under normal weather conditions. Transitory savings are dependent on the difference between permanent and actual income. A small part of that difference is used for direct consumption, but a larger part is used for consumption smoothing over time. Transforming past savings into consumption can thus cover shortfalls in income. This implies that savings can be used either for investment in inputs, capital goods and land improving and sustainability enhancing measures (*e.g.* soil conservation measures), or for investment in near liquid assets to maintain income in adverse periods. The latter is based on the permanent income hypothesis and consumption smoothing behaviour (Rosenzweig & Wolpin, 1993), that is considered relevant for resource-poor households in fragile environments. Savings and investment behaviour in the past have resulted in current levels of capital accumulation, where near liquid assets (especially in cattle) are also used as a collateral for lending operations (Udry, 1990).

The procedure in the *goal-weighting module* merits special attention. The objective function developed in Section 8.2 can be rewritten in terms of discrete time steps and discrete weather conditions with specified probability of occurrence P_n with n referring to different states of nature:

$$u = \sum_t \sum_n (1+r)^{-t} P_n u(\mathbf{c}_{nt}, l_{nt}) \tag{8.10}$$

The dynamics of soil degradation and their effects on production are difficult to quantify, especially when weather risk is included. Hence the expectations of future production as a result of changing soil quality is subjective. Equation (8.10) can be split into two separate components, one relating to present consumption and the other to future consumption:

$$u = \sum_n P_n u(\mathbf{c}_{nt}, l_{nt}) + \sum_{t \neq 0} \sum_n (1+r)^{-t} P_n u(\mathbf{c}_{nt}, l_{nt}) \tag{8.11}$$

Since the second component depends on income generation and changes in soil quality and on subjective discount rates, it can be replaced by an indirect measure of potential future consumption. This gives:

$$u = \sum_n P_n u(\mathbf{c}_{nt}, l_{nt}) + j(r, \dot{s}, y) \tag{8.12}$$

The second component does not contain future income since it is notoriously difficult to estimate. Instead it contains present income, adjusted for the expected value of soil quality change over time. The simulation model can generate results for a completely myopic household that only considers present consumption, *i.e.* $j(r, \dot{s}, y) = 0$. The model can also generate results for a household that prefers future consumption potential to present consumption, *i.e.* $u = j(r, \dot{s}, y)$[2]. Both generated model results can be compared to empirical evidence and the relative weights of each component estimated. Estimating goal weights requires special techniques due to the fact that there is usually limited data available, and the empirical evidence used for verification is not always compatible. Maximum entropy econometrics offers tools in this case (Kruseman, 2000).

Using LP models to solve a household model does have some shortcomings next to its important advantages. One of the main problems of this approach is that it can only deal with weakly non-linear relationships[3].

Limiting the approach to weakly non-linear relationships has important implications for the production function. The use of Leontief production activities for specified input-output combinations, solves the problem of having to specify a full production function, but it also leads to very large sets of production activities, and it implies that marginal effects are more difficult to assess. Furthermore, scale effects cannot be dealt with adequately. In the model there is no difference whether an activity vector of inputs and outputs is applied to a square centimetre of land, or to the full arable crop area. The same holds true for livestock, where the herd is considered a continuous variable.

8.5 Metamodelling

Mathematical programming models, especially linear ones, are not always consistent in their outcomes. The problems of corner solutions and the lack of consistent partial derivatives leaves much to be desired in the analysis of model results. Fortunately there is a way out of these problems, through the use of metamodelling. A metamodel refers to a black-box representation of the relationship between exogenous variables and relevant endogenous variables of a specified model (Kleijnen, 1998). A metamodel of a mathematical programming model can be compared to the reduced form of a full household model. The bio-economic farm household simulation model results and the relevant exogenous variables are linked in a functional way. The resulting model does not explain the processes guiding decision making, but only gives the apparent relationships between inputs and outputs. The basic principle of metamodelling is fairly simple. Comparable to the case of econometric estimation of a farm household model using empirical evidence, the metamodel estimates a statistical relationship between exogenous and endogenous variables. The fundamental difference is the use of a simulation model to generate values for the endogenous variables.

---

[2] This does not mean that utility of current consumption equals zero, but only that consumption utility is not taken into consideration in the partial optimisation procedure.

[3] Well-behaved non-linear relationships satisfying strict convexity, without feedback mechanisms into other non-linear relationships.

Following Kleijnen & Sargent (1997) metamodels can be used for a number of purposes. They mainly serve to better understand the behaviour of the simulation model and by proxy the households it represents. The metamodel permits conducting sensitivity analysis, and 'what-if' studies. These 'what-if' studies are especially important in the assessment of technology change and policy instruments (Kruseman & Ruben, 1998; Kruseman, 1999).

Three application areas of metamodels are relevant in the analysis of technology choice and sustainable land use. The first is the use of metamodels to simplify technical relationships. The underlying goal is to obtain a better interface between the biophysical and economic components of the bio-economic modelling framework. The specification of the interlinked production functions for crops, pastures and livestock in terms of vectors of input-output combinations takes up a very large part of the bio-economic farm household model. If it is possible to reduce the size of the production matrix by estimating a functional relationship between the relevant inputs and outputs, this can be very useful[4].

The second area where metamodels are applied is in estimating response multipliers and elasticities. By estimating a statistical relationship between relevant exogenous parameters and endogenous variables a better understanding of the behaviour of farm households is gained. With linear programming models, results can be biased due to the frequent occurrence of corner solutions. The theoretical household model serves as a basis for defining the metamodel specification of household behaviour. As was discussed in Section 8.3, the functional form of the reduced-form equation cannot be derived due to the high complexity of the model. Therefore, a lot of attention is given to determining the most appropriate functional form. Testing for model specification in the metamodel, using a range of test statistics, is imperative.

The third area to which metamodelling is applied is the estimation of aggregate supply functions in partial equilibrium analysis. Traditionally two approaches have been used for the aggregation of household responses when this results in markets clearing prices. The first approach is myopic and just looks at the new equilibrium price and calculates new results with that price. The second approach is based on recursive, iterative procedures in which the previous step (myopic approach) is repeated until the price change of the equilibrium price drops below a tolerance level (Bade *et al.*, 1997; Roebeling *et al.*, 2000). For a single commodity or factor market this approach will work, although it can be time consuming, depending on the structure of the household model and whether the equilibrium is

[4] Preliminary tests with this type of model yield mixed results (Ruben & Ruijven, 1999). Although the metamodels decrease the simulation model size, performance in comparison to the input-output matrix is poor. The main reason lies in the difficulty to capture the synergistic effects of inputs especially in combination with discrete technology choice, in the context of multiple outputs (Kruseman & van Keulen, 2000). This type of metamodel is a predictive model and needs to be very robust to be used.

near a point of multiple near-optimal solutions. In the case of very different multiple near-optimal solutions, it might not be possible to calculate the equilibrium price in this way. Sensitivity analysis can point out whether this is the case. When more than one market is involved, the chances are small of reaching equilibrium through iterative procedures, because this implies strong non-linearities while the procedures only work for weak non-linearities. The usual difficulties with local and global optimal solutions can also occur.

Estimating response functions for each household (type) in the analysis with metamodels and linking those to the demand functions is a solution to this problem. The metamodel then captures the relevant responses to price incentives only. Unique solutions are now possible without resorting to tedious numerical methods.

8.6 Discussion and Conclusions

When dealing with non-separable farm household models applied to the analysis of soil degradation in developing countries, where data availability does not permit full econometric estimation of the model, simulation models based on mathematical programming techniques are very useful. An additional reason for using simulation models instead of full econometric analysis is that mathematical programming models are better suited to analyse technology change.

Data limitations in developing countries are very common. A complete household survey is seldomly available. When such a survey is available, the necessary data on soil quality have not been included. For policy assessment studies usually not enough time is available to collect all necessary information by conducting a multi-year survey. However, over the past decades vast amounts of data have been collected in developing countries for a variety of purposes. Taking advantage of that information is a prerequisite for applying a realistic modelling framework as an aid to policy makers. The simulation model introduced in this chapter is based on a theoretically sound set of equations. Each equation is matched with existing data and its parameters are estimated econometrically. Where econometric analysis is not possible, other quantitative methods are used to parametrise the equations.

A major advantage of mathematical programming models is the opportunity to combine economic and biophysical information in an integrated framework. For the analysis of technology change and sustainable land use, where both economic behaviour and biophysical processes play an important role, a methodology that accounts for both scientific realms is a necessary condition for a robust study. The use of Leontief input-output coefficients permits the combination of incompatible economic and biophysical production functions (Kruseman & van Keulen, this volume). When using continuous biophysical production functions (commonly called processes, Struif Bontkes, this volume), allocative efficiency is not guaranteed. When using economic production functions, synergy among inputs and interlinked relationships between outputs and environmental externalities are not adequately captured.

A disadvantage of mathematical programming models is the vast amount of results that can be generated with 'what-if' analyses. Especially when some of the parameters of the model are not known with great accuracy this can pose a problem. Arbitrarily choosing certain parameter values may result in biased answers, not only of parameter values, but also through the possible occurrence of multiple near-optimum solutions and the existence of corner solutions (especially with linear programming techniques).

One way of dealing with these shortcomings of mathematical programming is the application of metamodelling techniques in which the simulation model is reduced to a black box and the statistical correlation between input and output variables helps in the analysis of results. The structure of metamodels resembles the reduced-form equations of the theoretical household model, with the significant difference that the relationship can be estimated.

8.7 References

Askari, H. & J.T. Cummings (1976): Agricultural supply response: a survey of the econometric evidence. Praeger Publishers, New York

Bade, J., G. Kruseman, R. Ruben, P. Roebeling & H. Hengsdijk (1997): Farm household modelling in a regional setting: the case of Koutiala Cercle in Mali. DLV Report No. 6. AB-DLO/WAU, Wageningen

Barbier, B. (1994): Modélisation agronomique et économique de la durabilité d'un système agraire villageois. Ph.D. Thesis, CIRAD-SAR, Montpellier

Barnum, H.N. & L. Squire, (1979): A model of an agricultural household: theory and evidence. Johns Hopkins University Press, Baltimore

Becker, G.S. (1965): A theory of the allocation of time. The Economic Journal 75, 493-517

Benjamin, D. (1992): Household composition, labor markets and labor demand: testing for separation in agricultural household models. Econometrica 60, 287-322

Bulte, E.H. & D. van Soest (1999): A note on soil depth, failing markets and agricultural pricing. Journal of Development Economics 58, 245-254

Chayanov, A.V. (1923): The theory of peasant economy. Translated and commented by D. Horner, B. Kerblay & R.E.F. Smith (Eds.) (1966). Homewood, Ill.

Dalton, T.J. (1996): Soil degradation and technical change in Southern Mali. Ph.D. Thesis West Lafayette: Purdue University

De Janvry, A. & E. Sadoulet (1994): Structural adjustment under transaction costs. In: F. Heidhuess & B. Knerr (Eds.) Food and agricultural policies under structural adjustment. Peter Lang, Frankfurt

De Janvry, A., M. Fafchamps & E. Sadoulet (1991): Peasant household behaviour with missing markets: some paradoxes explained. The Economic Journal 101, 1400-1417

Goetz, S.J. (1992): A selectivity model of household food marketing behaviour in sub-saharan Africa. American Journal of Agricultural Economics 64, 444-452

Hengsdijk H., W. Quack, E.J. Bakker & J.J.M.H. Ketelaars (1996): A land use system and technology generator for the Koutiala region in South Mali. DLV Report no. 5., AB-DLO/PSS, Wageningen

Holden, S.T., J.E. Taylor & S. Hampton (1998): Structural adjustment and market imperfections: a stylized village-wide model with nonseparable farm households. Environment and Development Economics 4, 69-87

Kleijnen, J.P.C. (1998): Experimental design for sensitivity analysis, optimization, and validation of simulation models. In: J. Banks (Ed.) Handbook of simulation. John Wiley and Sons, New York

Kleijnen, J.P.C. & R.G. Sargent (1997): A methodology for the fitting and validation of metamodels in simulation. CentER Discussion Paper, no. 97116. Katholieke Universiteit Brabant, Tilburg

Köbrich, C.J. (1997): The construction and use of compromise programming models to measure the impact of development policies on the sustainability of peasant farming systems in central Chile. Ph.D. Thesis Department of Agriculture, Faculty of Agriculture and Food, University of Reading

Kruseman, G. (1999): Bio-economic modelling to assess the impact of policy incentives on sustainable land use: Do fertiliser prices and transaction costs matter? Paper presented at the NWO seminar Milieu, Economie en NMP4, Amsterdam, June 4, 1999

Kruseman, G. (2000): Bio-economic household modelling for agricultural intensification. PhD Thesis Wageningen University (in preparation)

Kruseman, G. & H. van Keulen (2000): Soil degradation and agricultural production: economic and biophysical approaches. This volume

Kruseman, G. & R. Ruben (1998): Marketing costs and incentives for sustainable land use: a farm household modelling approach for Southern Mali. Paper presented at the 57th seminar of the European Association of Agricultural Economists, "Agricultural markets beyond liberalisation", Wageningen, 23-26 September 1998

Kuyvenhoven, A., R. Ruben & G. Kruseman (1998): Technology, market policies and institutional reform for sustainable land use in southern Mali. Agricultural Economics 19, 53-62

Nerlove, M. (1979): The dynamics of supply: retrospect and prospect. American Journal of Agricultural Economics 61, 874-888

Omamo, S.W. (1995): Smallholder agriculture under market reform: the case of southern Siaya district, Kenya. Ph.D. Thesis, Stanford University

Penning de Vries, F.W.T., P.S. Teng & K. Metselaar (Eds.) (1995): Systems approaches for agricultural development. Proceedings of the international symposium on systems approaches for agricultural development, Bangkok, Thailand. Kluwer Academic Publishers, Dordrecht

Pinstrup-Andersen, P. (1994): World food trends and future food security. Food Policy report, IFPRI , Washington D.C.

Pol, F. van der (1993): Profits de l'Intensificacion. Royal Tropical Institute, Amsterdam

Roebeling, P.C., H.G.P. Jansen, A. van Tilburg & R.A. Schipper (2000): Spatial equilibrium modeling for inter-regional trade flows, land use and agricultural policy analysis. In: B.A.M. Bouman, H.G.P. Jansen, R.A. Schipper, H. Hengsdijk & A. Nieuwenhuyse (Eds.). Tools for land use analysis at different scales with case studies for Costa Rica. Kluwer Academic Publishers, Dordrecht

Romero, C. (1993): A research on Andalusian farmers' objectives: methodological aspects and policy implications. Paper presented to VII EAAE Congress, Stresa, Italy

Rosenzweig, M.R. & K.I. Wolpin (1993): Credit market constraints, consumption smoothing, and the accumulation of durable production assets in low-income countries: investments in bullocks in India. Journal of Political Economy 101, 223-244

Ruben R. & A. Ruijven. (1999): Technical coefficients for bioeconomic farm household models: a metamodelling approach with applications for southern Mali. Mimeograph, Wageningen University.

Ruben, R., G. Kruseman, H. Hengsdijk & A. Kuyvenhoven (1997): The impact of agrarian policies on sustainable land use. In: P.S. Teng, M.J. Kropff, H.F.M. ten Berge, J.B. Dent, F.P. Lansigan & H.H. van Laar (Eds.) Applications of Systems Approaches at the Farm and Regional Levels. Kluwer Academic Publishers, Dordrecht, 65-82

Sadoulet, E., A. De Janvry & C. Benjamin (1998): Household behavior with imperfect labor markets. Industrial-Relations 37, 85-108

Shiferaw, B. (1998) Peasant agriculture and sustainable land use in Ethiopia: economic analysis of constraints and incentives for soil conservation. Ph.D. Thesis, Department of Economics and Social Sciences. Agricultural University of Norway, As

Singh, I., L. Squire & J. Strauss (1986): Agricultural household models: extensions, applications, and policy. The Johns Hopkins University Press, Baltimore

Skoufias, E. (1994): Using shadow wages to estimate labor supply of agricultural households. American Journal of Agricultural Economics 76, 215-27

Struif Bontkes, T. (2000): Agricultural prices and land degradation in Koutiala, Mali: a regional simulation model based on farmers' decision rules. This volume

Udry, C. (1990): Credit markets in Northern Nigeria: credit as insurance in a rural economy. World Bank Economic Review 4, 251-270

Wolf, J., H. Breman & H. van Keulen (1991): Bio-economic capability of West-African drylands. Report 147, CABO-DLO, Wageningen

9 Agricultural Prices and Land Degradation in Koutiala, Mali: a Regional Simulation Model Based on Farmers' Decision Rules

Tjark Struif Bontkes

Department of Plant Sciences, Wageningen University and Research Centre
(Present address: IFDC-Africa, Lomé, Togo)

Abstract

To enhance sustainability of agricultural production and food supply, insight is required in the relevant agro-ecological and socio-economic processes at farm and regional level. In this chapter a modelling approach is presented that is suitable to provide a qualitative insight in the dynamics of agricultural development. The approach is used to develop a model of the Koutiala region in southern Mali that allows simulating the development of different farm types and their interactions. It is demonstrated how the model may help to identify discontinuities in agricultural development when conditions change, and how it may serve to explore effects of price policies on, for example, soil organic matter content.

9.1 Introduction

Land degradation constitutes a serious threat to food security in many countries of Sub-Saharan Africa (Stoorvogel & Smaling, 1990; van der Pol, 1992). Both ecological and economic processes play an important role in causing land degradation. These processes are interlinked: the environment sets limits to what the population can achieve, while humans, in turn, actively modify their environment to realise their objectives. Hence, the relationship between humans and their environment is regulated by feed-backs. This, however, is a slow process: it takes often many years before negative effects of human actions on the environment become apparent. Therefore, negative feed-backs that could induce corrective behaviour are scarce. Such systems can be considered as complex dynamic non-linear systems in which the human system and its environment co-evolve and do not reach a state of equilibrium (Ring, 1997).

It is obvious that blue-print solutions for such problem situations do not exist. This calls for adaptive management, in which policy makers test different policies and at the same time learn-by-doing (Holling *et al.*, 1998). As learning-by-doing requires time, and experiments are very costly or may even not be possible, the use of models may be helpful. Such models can be used to perform simulation experiments, increasing insight in the dynamics of the system.

Land degradation is the consequence of decisions taken at the farm level. These decisions can only partly be influenced by government policy, as other factors play a role as well, such as rainfall or world market prices. Moreover, processes at higher levels have also to be taken into account, as the role of common pastures, population growth and (food) prices cannot be analysed at farm level, requiring a multi-level analysis. This requires insight in the way humans react to variable and changing conditions. Boorsma (1990) distinguishes three approaches in modelling the behaviour of the farmer:

- econometric modelling, based on linear regression equations of a data set;
- mathematical programming;
- modelling decision processes based on a number of decision rules.

The econometric approach is based on statistical analysis of historical and / or cross-sectional data. Its advantage is that it provides a fairly accurate description of the behaviour of the system in the past. A disadvantage, however, is that it does not always provide insight in the processes that play a role and that it is not very suitable to deal with new phenomena (Wossink, 1993). Moreover, the approach requires the availability of sufficient data for estimation, preferably time series, which are rarely available in developing countries.

Mathematical programming is an approach, frequently applied in farm planning. This method allows determination of an optimal allocation of land, labour and capital, given a set of goals (*e.g.* maximisation of income and leisure and minimisation of risk) and constraints (*e.g.* labour and land) (Barnett *et al.*, 1982; Kruseman & Bade, 1998). Some authors have raised questions as to what extent such normative models are an appropriate way to represent decision making (Dent *et al.*, 1995). These authors argue that complexity, insufficient knowledge and inconsistencies of individual preferences and beliefs render normative methods unsuitable to describe individual decision making. Dent *et al.* (1995) propose therefore the use of decision rules in simulation models in which the behaviour of the farmer is considered as the outcome of the interplay between his 'disposition to act', his material resources (*e.g.* soil fertility) and the external context (*e.g.* prices). The 'disposition to act' of a farmer should not be viewed as unchangeable, but rather as cumulative due to his experiences (Shucksmith, 1993).

In this paper a modelling approach is presented that enables decision makers at the regional level to obtain insight in the processes related to the sustainability of the agricultural sector and the possible effects of their policies. Some important features of this approach are that it:

1. permits integration of ecological and economic processes at both the farm and regional level;
2. provides insight in the dynamics of the system;
3. is a suitable tool for 'learning by doing'.

To show how this approach can be applied, a regional model has been developed, representing the agricultural sector in Koutiala (Mali) (Struif Bontkes, 1999).

Koutiala has always been inhabited by people who grow crops such as sorghum and millet for their livelihood. Since World War II, cotton has been introduced in association with animal traction, and maize has become an important food and cash crop. This has commercialised farming, though food crop production for own consumption continues to play an important role. Income, derived from cotton production is invested in cattle, first for draught-animals, later also for the purpose of developing a savings account in which the animals are considered the capital and the calves the interest. As agriculture has become a profitable business, out-migration remained at a low level and the number of farms increased, expanding the area cultivated. The combination of an expanding cropping area and increasing herd size has led to overgrazing of the communal pasture land. Although farmers do use fertiliser on commercial crops, they do not apply it on their food crops. In addition, soil organic matter tends to decline, as decomposition of organic matter is not compensated by application of manure and crop residues.

The question is now which policy measures the government should take to prevent further soil degradation: what will happen without any intervention or exogenous disturbance? What will happen in case of external shocks? What can the government do to guide development in the desired direction? It is obvious that in the decision making process, the government should also take into account the human aspects, as human behaviour plays an important role in soil degradation and because the well-being of its subjects is a concern of the government.

9.2 Structure of the Model

Three soil types (sand, loam, and clay) and four farm types (A, B, C and D) are distinguished in the model. The latter division is based on household size, farm size, herd size, implements, crop choice, use of inputs and investment behaviour. The sizes decrease from A to D.

The model is a simulation model written in Vensim[1]. The model simulates the period 1980-2025, where the period 1980-1996 is used for calibration. Fig. 9.1 provides an overview of the relations at the farm level and Fig. 9.2 of those at the regional level. At the start of the simulation, the number of farms per farm type is known as well as their household sizes, the areas of sandy and loamy soil

[1] Ventana Systems, Inc., 149 Waverly Street, Belmont, MA 02178, USA

occupied, the number of animals per farm and the soil fertility characteristics per farm and soil type. Simulation proceeds by determining sequentially:

1. areas per crop and inputs (labour, biocides and organic and inorganic fertiliser) applied per farm type (*cf.* Fig. 9.1);
2. availability of nutrients for crop uptake (represented by soil fertility in Fig. 9.1);
3. crop production including production of the common pastures (*cf.* Fig. 9.1);
4. animal production based on the available quantity and quality of feed (*cf.* Figs. 9.1 and 9.2);
5. cereal prices (*cf.* Fig. 9.2);
6. sale or purchase of animals on the basis of farm incomes (*cf.* Figs. 9.1 and 9.2);
7. changes in the number of farms per farm type, their household sizes, the areas of sandy and loamy soil cultivated, the pasture area, the herd sizes per farm type and the soil fertility characteristics per farm and soil type (Fig. 9.2).

These then define the new situation at the start of the following year and the cycle starts from step 1 again.

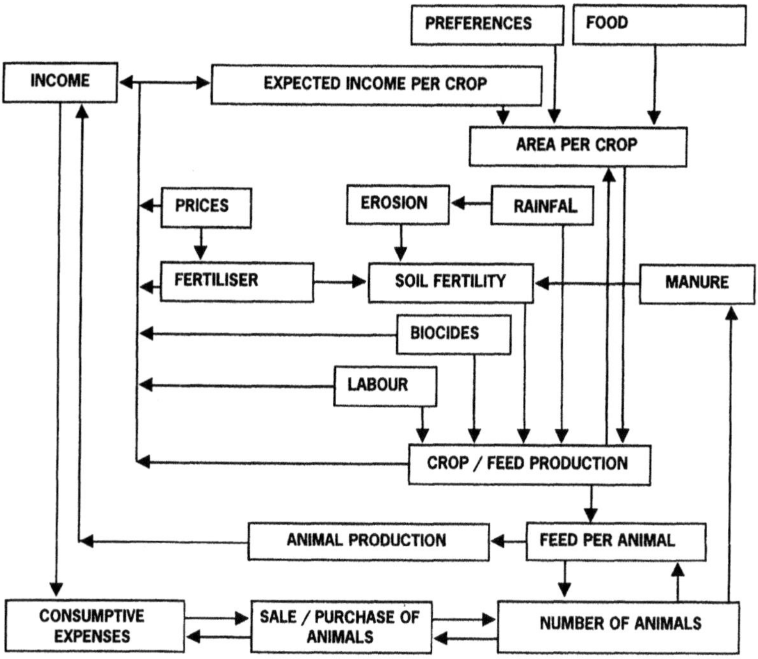

Figure 9.1: Overview of the relations at the farm level.

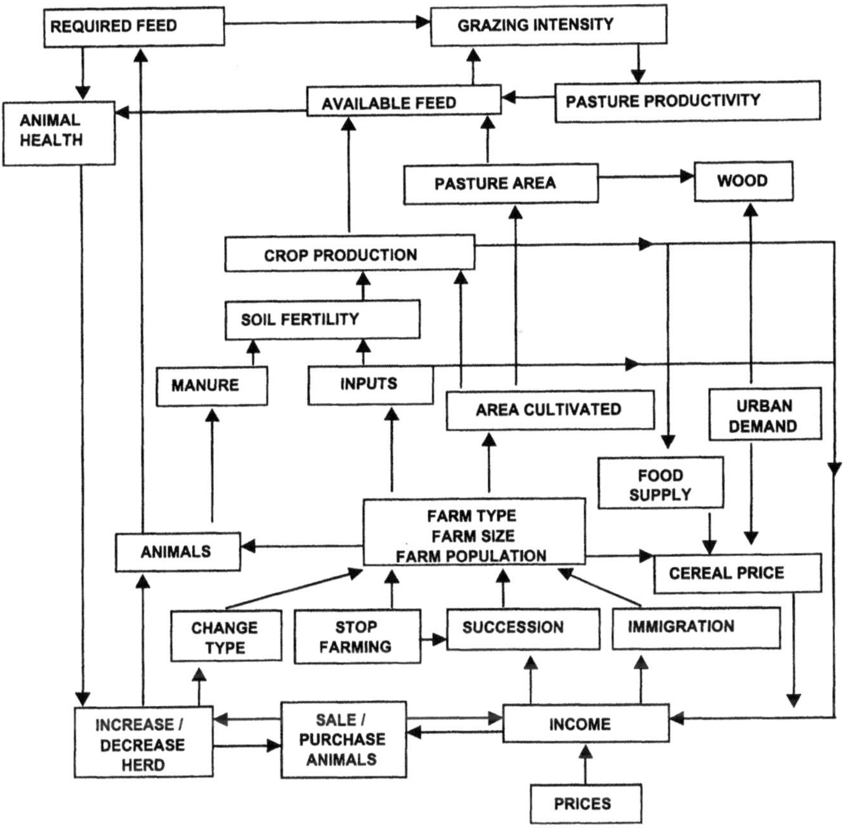

Figure 9.2: Overview of the relations at the regional level.

9.3 Detailed Description of Model Components

9.3.1 Choice of Crops and Inputs Applied per Farm Type

The decision on allocation of land to various crops differs per farm type and is represented by decision rules. Factors that play a role in this decision are food requirement, food preferences, expected yields, and economic surpluses per crop over the past three years. It is assumed that farmers first try to secure their food supply. In the model, millet and sorghum are combined and referred to as millet. As millet is the preferred cereal, farmers have a preference for millet as food crop and grow maize mainly as a cash crop. The planned share of millet (*cfmillet*) in the total diet is therefore described as a function of this preference (*cftaste*) and expected net incomes (per man-day) for millet and maize over the two preceding years:

$$cfmillet = \{net\ income\ millet\ /\ (net\ income\ millet + net\ income\ maize)\}^{cftaste} \quad (9.1)$$

cftaste changes over time to reflect a gradual shift in taste towards maize. On the basis of the shares of millet and maize in the household food supply, the required areas are determined by dividing the required amounts by the average yields over the past two years.

In spite of the wish of the farmers to produce sufficient food for household consumption, a highly profitable cotton crop could allow them to purchase cereals on the market. The extent to which the farmer is prepared to increase the area of cotton at the expense of the millet area is determined by the expected revenue of the cotton crop, the value of the millet production per ha at market prices, and a parameter representing the attitude of the farmers to risk. Large farmers are ready to take more risk than small farmers.

When the area for food crops has been determined, the areas that are used for growing cash crops (cotton, maize, groundnut and millet) are determined. The allocation per crop (*cfarea*) depends on average income per labour day over the preceding two years (*av income*) and on an adjustment factor expressing preferences (*cfpref*):

$$cfarea[crop] = \frac{av\ income[crop]^{cfpref[crop]}}{av\ income[cotton]^{cfpref[cotton]} + \ldots + av\ income[millet]^{cfpref[millet]}} \quad (9.2)$$

Cotton is a preferred crop, as credit for inputs is readily available, delivery of inputs is guaranteed and crop prices are announced prior to the time of sowing. *cfpref* also differs per farm type: small farmers are more risk averse, resulting in lower preferences for cotton and maize.

Application of animal manure depends on its availability and is crop-specific. It is assumed that cotton and maize receive animal manure up to an amount equivalent to 30 kg N per ha; the remainder is applied to other crops. Farmers apply chemical fertiliser on cotton and maize. The quantity of fertiliser applied on cotton is related to the crop-fertiliser price ratio. To estimate the effect of that ratio on fertiliser application, the response of cotton and maize to different levels of compound fertiliser has been simulated for the period 1980 to 1994, using actual rainfall data. The relationship between crop-fertiliser price ratio and the fertiliser application rate that produced the highest net return per ha was established, resulting for cotton in:

$$fertiliser\ on\ cotton = (272 * price\ cotton\ /\ price\ compound) - 100 \quad (9.3)$$

For maize no relationship was found between the crop-fertiliser price ratio and the optimal level of compound fertiliser. This was caused by the sensitivity of maize to variations in rainfall: in years of high rainfall optimum application of compound fertiliser was very high, while in years of low rainfall it was very low. The application of compound fertiliser on maize is therefore assumed to be constant: 50 kg per ha. Biocides are only applied to cotton, at a rate that is slightly below the recommended rate.

9.3.2 Availability of N and P for Crop Uptake

The amount of N available depends to a large extent on soil organic matter content. Soil organic matter decreases due to decomposition and erosion losses, and increases due to crop residues and application of manure. The nitrogen, mineralised during decomposition of organic matter, becomes available for uptake or for re-incorporation into organic matter or may be lost through volatilisation or leaching. Similarly, mineralised phosphorus is available for uptake or for re-incorporation into organic matter or into the inorganic phosphorus pool (Jones *et al.*, 1984).

Phosphorus is also released from inorganic phosphorus pools that further change through erosion losses and incorporation of phosphorus from other sources. Erosion is determined according to Roose (1977), and depends on rainfall, topography, crop, texture and soil conservation measures. As soil acidity affects the availability of N and P, soil pH has also been included in the model. Changes in soil pH are calculated by taking into account the production and consumption of protons as a result of the various soil processes, and the buffering capacity of the soil (Helyar *et al.*, 1990).

9.3.3 Crop Production, Including Production of the Common Pastures

Crop yield is determined as the minimum of the maximum yield of the crop and that dictated by water supply and the availability of nitrogen and phosphorus, corrected for labour (*cflabour*) and the effect of pests and diseases (*cfpests*):

$$yield = cflabour * cfpests * MIN(maximum\ yield,\ water\text{-}limited\ yield,\ N\text{-}limited\ yield,\ P\text{-}limited\ yield) \quad (9.4)$$

Labour requirements are calculated per month and depend on area per crop, farm activities during that month and availability of tools, such as plough and sowing equipment. As labour availability is critical for timely land preparation, sowing and weeding, the effect of labour shortage is based on the availability and requirements during the months of May, June and July. If available household labour falls short of the requirement, the larger farm types are assumed to hire external labour up to 25% of the available household labour. The effect of the availability of labour is crop-specific to express its sensitivity to late sowing or competition of weeds. The effect of pests and diseases is also crop-specific.

Maximum yields have been estimated on the basis of the highest reported yields in the area (van Duivenbooden, 1992). Water-limited yields are determined by the availability of water per growth stage (Doorenbos & Kassam, 1979).

9.3.4 Animal Production Based on Available Quantity and Quality of Feed

Quantity and quality of feed intake determine animal production. Feed intake depends on the amount of the different feed types available and their quality, represented by their N content: the higher the availability and the higher the N-content, the higher feed intake. The model calculates growth of heifers on a monthly basis, according to the method as described in Breman & De Ridder (1991). Death rates, calving rates and milk production are derived from the annual growth rate of heifers.

9.3.5 Cereal Prices

While prices of cotton are exogenously determined, cereal prices are assumed to be affected by local production and demand. When the quantity of cereals that is produced in excess of the requirement of the farm households (surplus production) exceeds the food requirement of the non-farm and the food deficit farm households (food requirement), only a quantity that is equivalent to the quantity that can be purchased by these households (demand) is taken to the market. In that case the minimum price plus costs of transport determine the price. With increasing food requirement, a larger area is required for production, and costs of transportation increase. Improving market access may reduce transportation costs.

If food requirement exceeds surplus production, cereal prices will increase, until demand equals surplus production. Following Colman & Young (1989), the market demand function for a product (*e.g.* cereals) can be specified as:

$$demand_c = f\,(price_c, price_i, M, POP, ID) \qquad (9.5)$$

where,

$demand_c$: demand for cereals;
$price_c$: price of cereals;
$price_i$: prices of alternative products;
M: per caput income;
POP: market population;
ID: index of income distribution.

As demand equals surplus production, and population size determines food requirement, the following equation can be derived:

$$price_c = f\,(surplus\ production\ /\ food\ requirement,\ income\ distribution) \qquad (9.6)$$

Assuming that in case of food shortage people obtain either the full requirement or nothing, the fraction of the population able to purchase cereals (*cfpopulation*) is equivalent to *surplus production / food requirement*. Hence, the fraction of the population that is able to purchase cereals is determined by price and income distribution and is represented by:

$$cfpopulation = 1 / (1+e^{X * (price - Y)}) \tag{9.7}$$

where X is a measure of income inequality and Y the price at which half of the population is able to satisfy its food requirement by purchasing locally produced cereals. The price then follows from:

$$price = Y+(\ln ((1\text{-}cfpopulation)/ cfpopulation))/X \tag{9.8}$$

This implies that, if the fraction of the population that can be fed with the surplus production (*cfpopulation*), income inequality (X) and the price at which half of the population is able to satisfy its food requirement (Y) are known, the cereal price can be determined. The parameters in this equation have been determined by fitting the curve of maize price development to empirical data. The millet price is set to 1.2 times the price of maize.

9.3.6 Sale or Purchase of Cattle on the Basis of Farm Incomes

It is assumed that the primary objective of farmers is to satisfy the food requirements of their household members and to provide them with at least a minimum amount of cash for purchasing salt, sugar, soap, kerosene and clothes. These requirements per household member increase with increasing farm size, going from the small to the large farm type. If food production is insufficient, millet is purchased at the market at a higher price. If available cash falls short of the requirement, the farmer sells one or more of his animals, and if that is not sufficient, he tries to secure a loan. After satisfying the food and cash requirements, surplus income is used to purchase more luxury items and additional cattle. The investment behaviour of the farmers varies among farm types.

9.3.7 Changes in the State Variables

So far, processes have mainly been described at farm level, except for the price formation of cereals. The next step is to aggregate the results of these processes to the regional level. This cannot be done by simply adding the number of all farms per farm type, as the number of farms per farm type may change due to population dynamics, migration, changes in farm size and transition of farms from one type to another. Changes in area cultivated, herd size and available pasture may affect farm development. If a farmer is able to expand his herd size, he may move to a larger farm type. On the other hand, if he is losing animals, he may move to a smaller farm type. Whether a farmer, retiring after 30 years of farming, is succeeded by his son depends on the relative income, *i.e.* the ratio of expected farm income and the income that can be earned outside agriculture. When relative income is high, the number of sons willing to start farming may exceed the number of farmers who retire. Farms may then be split in two smaller farms, or one son may succeed his father while the other starts a new farm. If relative income is high, immigrants are attracted.

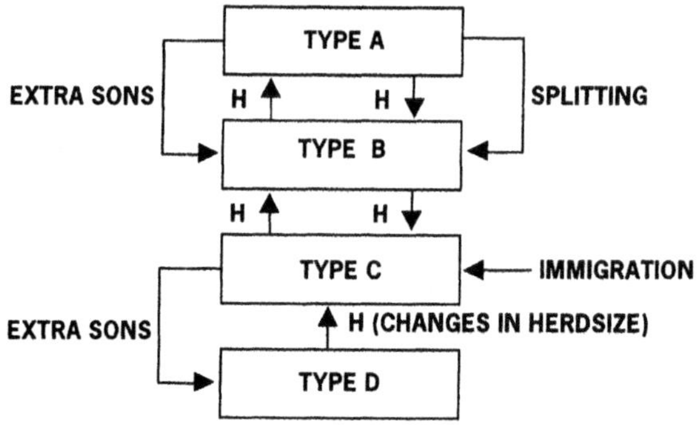

Figure 9.3: Overview of changes in the number of farms per farm type.

To determine the new household size per farm type, the total population for all farms per farm type is determined. Subsequently the change in total population per farm type is determined:

$$dtpop = tpop\ increase + tpop\ change\ in - tpop\ change\ out - tpop\ retire \qquad (9.9)$$

where:

dtpop:	change in total population per farm type;
tpop increase:	net increase due to births and deaths;
tpop change in:	population coming in with the farms making the transition to the farm type;
tpop change out:	population leaving the farm type by transition to other farm types;
tpop retire:	population leaving the area due to retirement.

By dividing the new total population per farm type by the new number of farms per farm type, the new household size per farm type is determined.

When a farmer moves to another farm type, he takes his land with him. The size of the area he takes with him is determined by the household size and the standard area per person. If that area is larger than the average farm size of the farm type that he leaves, he will expand his farm by withdrawing land from the pasture area. If he moves to a smaller farm type, the surplus land is turned into pasture land. The new average farm sizes per farm type are now determined by adding or subtracting the areas of the farmers that enter or leave the farm type and dividing the total area by the number of farms per type.

Soil fertility characteristics change from year to year depending on soil type and farm management. As no individual plots are distinguished, average soil fertility characteristics are determined per farm and soil type. First the changes in soil fertility characteristics are determined per farm type, crop and soil type, which are averaged per farm and soil type to yield values for the farm level.

When a farmer changes to another farm type, he brings (part of) his land with him (plus possibly some pasture land). The soil characteristics of the land he takes with him may differ from those of the farm type he enters. The new average soil fertility characteristic (*e.g.* labile N) of a particular soil type (*e.g.* loamy soil) of a particular farm type (*e.g.* B) for the whole region is determined by summing the labile N content of all loamy soils of the B-type farms and divide this by the total area of these soils. The total labile-N content in the loamy soils of the B-type farms at a particular time (t) depends on the value of N at the preceding time (t-1), and the rate of change (*tot dNlabile*):

$$total\ N\ labile[B,loamy]_t = total\ N\ labile[B,loamy]_{t-1} + tot\ dNlabile[B,loamy]_{t-1} \qquad (9.10)$$

The change in total labile-N content in the loamy soils of all B-type farms is determined by:

- the change in labile N per ha loamy soil, the area of loamy soil per farm and the number of B-type farms;
- the area of loamy soil and its labile-N content, coming in with farms from the A and the C type;
- the area of loamy soil and its labile-N content, leaving with farms quitting the B type;
- the area of loamy soil and its labile-N content withdrawn from the pasture land.

The changes in average herd size per farm type are determined similarly to the changes in household size per farm type, taking into account that farms that move out of a class transfer a different number of animals than the average number of the type they leave.

9.4 Some Illustrations of Simulation Results

9.4.1 Base Run

The model simulates the dynamics of the system over the period 1980 – 2025, using actual annual rainfall data up to 1996 and average rainfall data over the remaining period. Fig. 9.4 shows a decreasing soil organic matter content when no additional measures are taken, leading to decreasing millet yields, while maize yields increase as a result of application of fertiliser and animal manure. Only in

1984 maize yields are low because of the drought in that year, to which maize is very sensitive. The increase in maize yields from 1995 onwards is caused by the increasing availability of animal manure due to the increased herd size of the A farms. This is a result of the devaluation in 1994, leading to higher cotton prices, more income and hence increased investment in cattle.

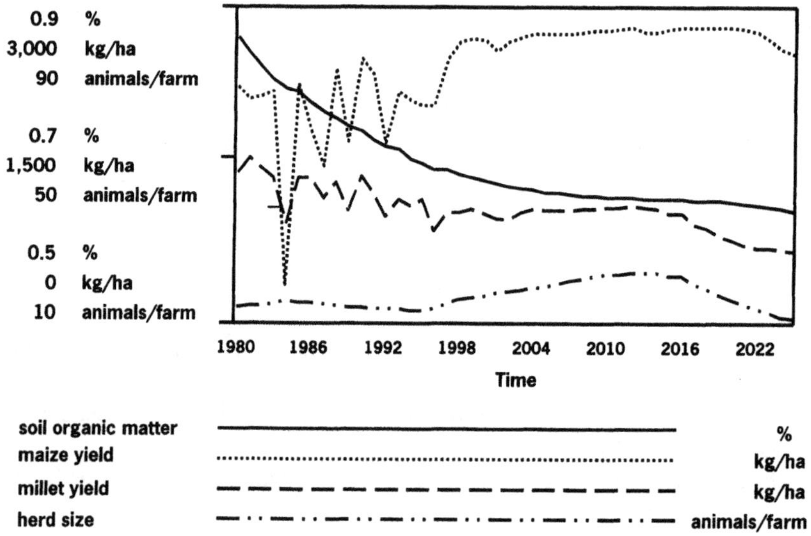

Figure 9.4: Simulated development of soil organic matter content, yields of maize and millet and herd size for A farms.

The increased availability of animal manure allows the farmers to apply part of it to millet fields, stabilising millet yields. From 2013 onwards, millet yields decrease again, due to a reduction in manure availability, caused by decreasing herd size of the A farms. This reduction also affects maize yields after some time. The reduction in herd size is caused by increasing animal death rates and decreasing calving rates, as farmers continue to invest their surplus income in cattle, resulting in a strong increase in herd size, while at the same time the number of farms increases, reducing the pasture area and hence available feed. Due to the decreasing herd sizes, A farms become B farms and B farms become C farms.

9.4.2 Policy Experiment

Though prices of cotton and fertiliser are determined at the national level, room for manoeuvre is limited, as these prices are strongly influenced by the world market. The ratio between the prices of both products is an important determinant for the total area of cotton: in the eighties when fertiliser prices increased while cotton prices remained stable, the share of cotton in the total cultivated area decreased. In addition farmers reduced the amount of fertiliser applied. In this experiment the effects of two price changes from 1997 onward for cotton and compound fertiliser (for cotton as well as for maize) have been explored:

- a 20% price increase of cotton from FCFA 155 to FCFA 185 per kg;
- a 20% price reduction of compound fertiliser from FCFA 201 to FCFA 161 per kg.

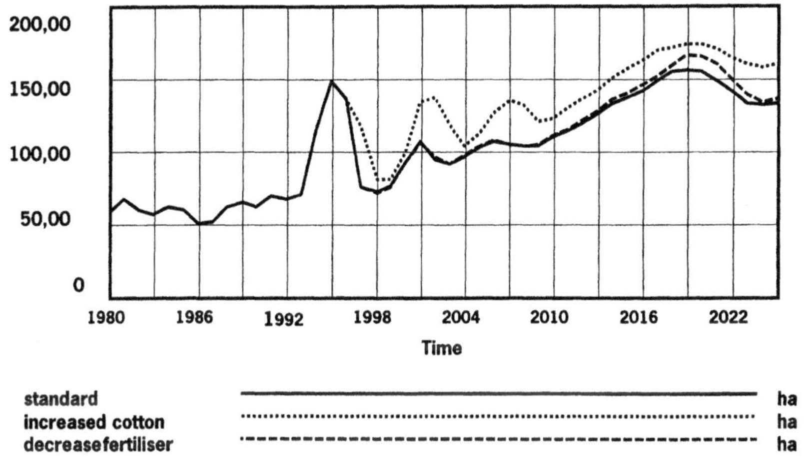

Fig. 9.5: Effect of changes in prices of cotton and fertiliser on total cotton area.

Fig. 9.5 suggests, as may be expected, that an increased price of cotton has a positive influence on the area grown with cotton. The increase in cotton area, immediately after the price increase of cotton, causes a reduction in the area of millet and maize, increasing cereal prices. This increase renders cereal production more attractive, causing an increase in the area cultivated with cereals, resulting in an increased cereal production and hence in a lower price. At the end of the simulation period the cotton area decreases. This is caused by decreasing cereal yields, due to soil organic matter depletion, forcing farmers to increase their area of food crops to assure an adequate food supply to their households. A change in the price of fertiliser has a smaller effect on the cotton area, as this also renders maize more attractive.

Table 9.1 shows the effect of an increase in cotton price on net income per capita and on the C balance per ha sandy soil for the different farm types over the medium and long term. Although the soil organic matter content continues to decline, the results show that increasing the cotton price has a positive influence on this development, especially for large farms.

Table 9.1: Response multipliers (average percentage change per year) for net income per capita and C balance for the four farm types at a 20 % increase of the cotton price in 1998.

	net income per capita		C balance	
farm type	1998-2002	1998-2025	1998-2002	1998-2025
A	12.9	10.4	4.9	11.1
B	12.7	8.9	2.7	5.1
C	11.2	7.7	-0.3	1.0
D	6.9	2.5	0.4	0.8

Table 9.2 shows the effect of a reduction in the price of fertiliser on net income per capita and on the C balance for the different farm types over the medium and long term. Again, the effect is a reduction in the decline of soil organic matter on the larger farms.

Table 9.2: Response multipliers (average percentage change per year) for net income per capita and C balance for the four farm types at a 20 % reduction of the fertiliser price in 1998.

	net income per capita		C balance	
farm type	1998-2002	1998-2025	1998-2002	1998-2025
A	0.2	0.8	1.4	6.1
B	0	1.0	0.7	5.4
C	-0.3	-0.2	0.1	-0.1
D	-0.4	-0.6	0	0.1

9.5 Discussion

The purpose of this study is to develop an approach to regional land use modelling that satisfies a number of requirements that are not, or only partly, met by other approaches. These requirements pertain to the integration of ecological and socio-economic aspects, the possibility to include decision making at farm level and represent it in a descriptive rather than a normative way, and to obtain insight in the dynamics of the system rather than formulating a static picture.

The model presented here integrates ecological and socio-economic aspects: farm management decisions affect the quality of the resource base of the farmer, and that quality in turn affects farm management. In addition, the model integrates the farm and the regional level: the behaviour of the farmer influences cereal prices, land availability and quality, which influence the behaviour of the farmer in the following period.

An important difference between the present approach and approaches based on mathematical programming exists with respect to the integration of ecological and socio-economic aspects. Mathematical programming approaches use input-output relationships (technical coefficients) that are derived from separate biophysical simulation models. This results in a model that can only deal with discrete input-output relationships: a particular combination of inputs always results in a particular combination of outputs. In the present approach, dynamics of ecological processes are fully integrated in the model and take into account changes in soil fertility characteristics, resulting in continuous input-output relationships.

In the present study, decision making by farmers is simulated by using decision rules. These rules are mainly based on subjective interpretations of interviews and literature, and to a lesser extent on socio-economic research and they are therefore rather arbitrary. This can be considered as a shortcoming of the present study, but not of the approach, as it is quite well possible to develop decision rules based on solid research (Gladwin, 1989). An advantage of the use of decision rules is that they allow more flexibility in the modelling of decision processes by taking into account a large variety of considerations. Decision rules are also relatively easy to validate with stakeholders. A limitation of the use of decision rules is that this approach is not suitable to select the 'best' alternative production system out of a multitude of alternative production systems. Another difficulty is that decision rules are based on the existing situation, whereas in reality decision rules may change under changing circumstances. This can be anticipated to some extent by running the model and identify where present decision rules are unlikely to be continued and replace them with decision rules that are more likely to be used under the new circumstances (Struif Bontkes, 1999). Mathematical programmers are faced with similar problems, as they usually introduce a number of restrictions in their models to prevent selection of corner solutions and to tune the model to the present situation.

Questions may be raised with respect to the predictive capacity of such models. Is it possible to make reliable predictions on agricultural development of a large region, comprising many farms of different types? Finding an answer to this question is further complicated by lack of tested theories, relevant to the situation, and by paucity of reliable longitudinal data that are required to construct and validate a model. Moreover, unpredictable events, such as drought, devaluation, changes in world market prices of cotton and/or political changes constitute inherent sources of uncertainty. The model should therefore be considered as a tool for analysis providing relevant insights, rather than an instrument that enables the decision maker to predict the future with some certainty.

As, by definition, a model is unable to capture the real world and as the real world continuously changes, the model should be repeatedly subjected to a process of testing. In this process, model predictions are compared with real world data, followed, if necessary, by adjustment of the model. This implies that a model should be considered as an evolving representation of the real world. The validity of such models will be further enhanced if developed in continuous interaction with stakeholders, including farmers, researchers and policy makers.

By involving various stakeholders in the process of model building, it also has a consensus-building function (Costanza, 1996; Vennix, 1996). In this way it can help to build mutual understanding and maintain a meaningful dialogue between the groups. Modern simulation software packages, such as STELLA and Vensim allow interactive model building by first developing a graphical structure of the model which can then be translated into the final model.

Such models may be helpful to improve understanding of the dynamics of the system, allowing decision makers to improve the quality of their decisions. The model may help to discover discontinuities in behaviour when conditions change, as shown above. Similarly the model may be useful in discovering undesirable trends and permit the exploration of effects of various policies or the identification of topics for agricultural research that may contribute to avoid or remedy problems in the future.

9.6 References

Barnett, D., B. Blake & B.A. McCarl (1982): Goal programming via multidimensional scaling applied to Senegalese subsistence farms. American Journal of Agricultural Economics 64, 720-727

Boorsma, A. (1990): Mogelijkheden voor akkerbouw in de Veenkoloniën. PhD thesis, Groningen State University

Breman, H. & N. de Ridder (Eds.) (1991): Manuel sur les pâturages des pays sahéliens. Karthala, Paris

Colman, D. & T. Young (1989): Principles of agricultural economics. Markets and prices in less developed countries. Wye Studies in Agricultural and Rural Development. Cambridge University Press, Cambridge

Costanza, R. (1996): Ecological economics: reintegrating the study of humans and nature Ecological Applications 6, 978-990

Costanza, R., L. Wainger & N. Bockstael (1997): Integrated ecological economic systems modelling: theoretical issues and practical applications. In: R.Costanza (Ed.) Frontiers in ecological economics: transdisciplinary essays. Elgar, Cheltenham

Dent, J.B., G. Edward-Jones & M.J. McGregor (1995): Simulation of Ecological, Social and Economic Factors in Agricultural Systems. Agricultural Systems 49, 337-352

Doorenbos, J. & A.H. Kassam (1979): Yield response to water. FAO, Rome

Duivenbooden, N. van (1992): Sustainability in terms of nutrient elements with special reference to West-Africa. CABO-DLO, Wageningen

Gladwin, C.H. (1989): Modelling farmers' decisions to change: using cognitive science in the design of agricultural technology. In: D. Groenfeldt & J.L. Moock (Eds.) Social science perspectives on managing agricultural technology. IIMI

Helyar, K.R., P.D. Cregan & D.L. Godyn (1990): Soil acidity in new South Wales - Current pH values and estimates of acidification rates. Australian Journal of Soil Research 28, 523-537

Holling, C.S., F. Berkes & C. Folke (1998): Science, sustainability and resource management. In: F. Berkes, C. Folke & J. Colding (Eds.) Linking social and ecological systems: management practices and social mechanisms for building resilience. Cambridge University Press; Cambridge

Jones, C.A., C.V. Cole, A.N. Sharpley & J.R. Williams (1984): A simplified soil and plant phosphorus model: I. Documentation. Soil Science Society of America Journal 48, 800-805.

Kruseman, G. & J. Bade (1998): Agrarian policies for sustainable land use: bio-economic modelling to assess the effectiveness of policy instruments. Agricultural Systems 58, 465-481

Pol, F. van der, (1992): Soil Mining, an unseen contributor to farm income in Southern Mali. Royal Tropical Institute, Amsterdam

Ring, I. (1997): Evolutionary strategies in environmental policy. Ecological Economics 23, 237-249

Roose, E. (1977): Erosion et ruissellement en Afrique de l'Ouest: vingt années de mesures en petits parcelles expérimentales. Travaux et documents de l'ORSTOM 78

Shucksmith, M. (1993): Farm household behaviour and the transition to post-productivism. Journal of Agricultural Economics 44, 466-478

Stoorvogel, J.J. & E.M.A. Smaling (1990): Assessment of soil nutrient depletion in sub-Saharan Africa, 1983-2000. Winand Staring Centre for Integrated Land, Soil and Water Research (SC-DLO), Wageningen

Struif Bontkes, T.E. (1999): Modelling dynamics of agricultural development: a process approach – the case of Koutiala (Mali). Wageningen Agricultural University, Wageningen

Vennix, J.A.M. (1996): Group model-building: facilitating team learning using system dynamics. John Wiley & Sons, Chichester

Wossink, A. (1993): Analysis of future agricultural change: a farm economics approach applied to Dutch arable farming. PhD thesis, Wageningen Agricultural University, Wageningen

10 Integrating Site-Specific Biophysical and Economic Models to Assess Trade-offs in Sustainable Land Use and Soil Quality

John M. Antle[1] **& Jetse Stoorvogel**[2]

[1] Department of Agricultural Economics and Economics, Montana State University, Bozeman, U.S.A

[2] Laboratory of Soil Science and Geology, Wageningen University and Research centre, Wageningen, the Netherlands

Abstract

In this paper we describe the sustainability of a production system in terms of trade-offs between present and future outcomes such as crop productivity. Based on the observed site-specificity of the relationship between soil characteristics and productivity, we hypothesise that to accurately assess the sustainability of an agricultural production system we must measure the interactions between farmers' management decisions and soil quality on a site-specific basis and then aggregate the resulting relationships. The analysis of linked, site-specific economic and bio-physical models shows that soil quality at a point in space and time depends on both fixed site characteristics and on the history of farmers' management decisions. Aggregation then shows that both on-farm productivity and potential productivity are functions of parameters that define the distributions of physical and economic variables in a region. Using this result, we show that measurements of sustainability based on 'representative' data such as mean soil characteristics will generally be biased and thus may lead to inaccurate assessment of a system's productivity and sustainability.

10.1 Introduction

This paper develops a conceptual framework and analytical models that can be used to address the impacts of economic policy and technology choice on the sustainability of agricultural land use as indicated by changes in indicators of soil quality. Impacts of changes in policy or technology can be described as trade-offs between current and future outcomes (*e.g.* output, income, health, environmental quality, *etc.*), as illustrated in Fig. 10.1[1]. The degree of sustainability of the system is defined here as the inverse of the absolute elasticity of the trade-off curve between present and future outcomes. Thus, as the trade-off curve's slope approaches zero, so that there is no impact of present activities on future outcomes, the system is fully sustainable; as the trade-off curve's slope approaches minus infinity the system becomes absolutely unsustainable in the sense that any attempt to increase present activities forces future outcomes to zero. Fig. 10.1 also illustrates that both sustainable and unsustainable systems may, in principle, achieve either a low or high level of productivity. Clearly, systems that exhibit both a high level of productivity and a high degree of sustainability are preferred over systems with low productivity and low sustainability. In many cases, however, society may have to choose between systems that present various trade-offs between present and future outcomes[2]. The goal of integrated assessment of agricultural production systems is to quantify these trade-offs so that both private and public decision makers may make informed choices between present and future outcomes.

We hypothesise that to quantify soil quality accurately, and to understand the interactions between soil quality and farmer behaviour, analysis must be conducted on a site-specific basis. Thus, the modelling approach developed in this paper links bio-physical attributes of land and farmer decision-making at a spatial scale where sufficiently homogeneous land units can be defined for the relevant soil quality attributes. This approach shows that farmers' management decisions are determined by the joint distribution of bio-physical and economic factors in the producer population. This joint distribution evolves over time in response to the various factors that influence farmers' decision making, including climate, prices, technology, and policies. The outcomes of these site-specific interactions are statistically aggregated to the scale relevant to policy analysis.

Aggregate economic analysis often uses the construct of the 'representative agent'. Analogously, aggregate analysis of the impacts of policy on land use and soil quality utilise data for 'representative soils'. An implication of the analysis in this paper is that use of representative soil quality indicators may result in aggregation bias. Such bias may lead to inaccurate assessments of the productivity and sustainability of agricultural production systems.

---

[1] Here the tradeoff curves are illustrated as straight lines. More generally, these relationships are represented as distributions, and tradeoff curves can be defined *e.g.* in terms of the means and other moments of these distributions as discussed in Antle *et al.* (1998).

[2] The consideration of tradeoffs between present and future outcomes raises the question of discounting future outcomes. Such discounting as deemed appropriate by the analyst is assumed to be incorporated into future values.

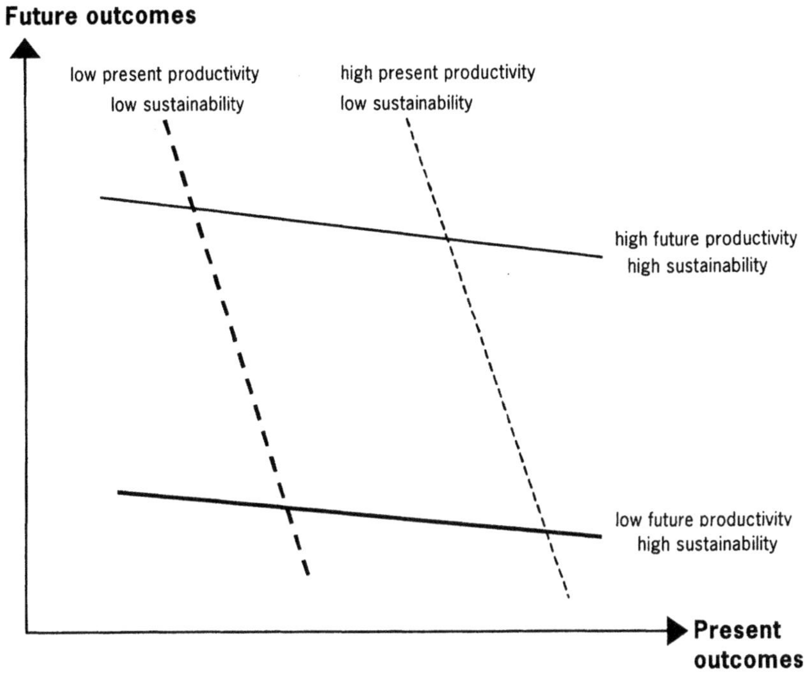

Figure 10.1: Trade-offs between present and future outcomes with low and high levels of productivity and sustainability.

We begin with a discussion of various concepts of soil quality in the soil science literature, and how these concepts can be quantified. Section 10.2 presents an economic model of site-specific land use and management, derives the implications of this model for the properties of functions describing the behaviour of soil quality over space and time, and then considers the implications for aggregation and analysis of trade-offs between current and future productivity. We conclude with a discussion of how the site-specific approach can be implemented, using a system of integrated bio-physical and economic models.

10.2 Defining and Measuring Soil Quality

With a large part of the soils of the world being degraded (Oldeman *et al.*, 1991), soil quality is receiving increasing attention. Soil degradation is often linked to its adverse impact on agricultural productivity and sustainability (Lal *et al.*, 1998). Soil quality is extremely difficult to define in general terms since its definition typically depends on soil use. Soil quality may have implications for environmental quality, for example, through its effects on the risk of pesticide leaching (Bouma, 1997). One can also express soil quality in relation to urban environments (Huinink, 1998). The major productive use of soils is for agriculture and it is logical, therefore, to focus on the effects of soil quality on agricultural productivity (Bouma *et al.*, 1998). Soils are difficult to evaluate without considering other environmental conditions (*e.g.* topography, climate, *etc.*) and therefore many researchers prefer the concept of land quality rather than soil quality. Pieri *et al.* (1995) define land quality as '*the condition of land relative to the requirements of land use, including agricultural production, forestry, conservation and environmental management*'. This definition also includes the conditions of land relative to land use.

Soil characteristics are interrelated and as such one can not evaluate one single soil characteristic. Land qualities, *i.e.* complex attributes of land that affect its suitability for uses in a distinct way, on the other hand, provide better insight in soil behaviour (FAO, 1976). However, in many cases land qualities can only be estimated by modelling the interaction of a number of land characteristics, which makes them more difficult to quantify. Increasingly, simulation models that can quantify these interactions are being used to assess the effects of changes in soil quality on agricultural productivity. Results of crop models can be interpreted as an estimate of potential yield or the inherent productivity of a site, and thus can be viewed as a summary statistic representing the effects of soil quality on productivity. Most crop simulation models, however, can only deal with a limited number of factors and processes. The interactions between pesticide use, crop performance and the soil biological parameters are still an unknown field that needs to be investigated. Most simulation models were developed to simulate the effects of soil water and soil nutrients on crop growth under specific conditions. As a consequence, extensive calibration and validation exercises are necessary if these models are to be applied under other conditions. Mutsaers and Wang (1999) critically discuss the usefulness of crop growth simulation models, arguing that they are not yet capable to estimate crop productivity with sufficient robustness to varying conditions. Consequently, whether or not crop simulation models can be used to generate indicators of soil quality remains an open question.

Sustainable land use is defined as land use where soil quality stabilises at an acceptable level. A certain level of degradation – *i.e.* reduction of soil organic matter and top soil depth – is tolerable and in many cases inevitable. We must consider various components of production systems, including substitution (trade-offs) among various dimensions of soils that affect productivity, as well as trade-offs between soil quality and other properties of a system. A loss of soil quality can often be compensated by external inputs.

Changes in soils quality can be highly site-specific, with changes taking place at the field or sub-field scale in response to land use and management practices. To illustrate, we consider the following example from the production system of northern Ecuador. In this system, the movement of topsoil takes place as a result of the use of tractors ploughing on steep slopes. As Fig. 10.2 shows, the process of cultivating potatoes and other crops on steeply sloped hillsides mechanically transports soil down-slope on each field, so that the topsoil is eventually removed from the up-slope part of the field. Research shows that each year in an average field 1.2 % of the topsoil is completely removed (Veen, 1999). As topsoil depth decreases, productivity declines slowly at first, but beyond a critical depth productivity declines rapidly and approaches zero (Fig. 10.3). The issue that we want to address in this paper is how to quantify this type of site-specific change in soil conditions on productivity. Clearly, if average soil depth per field or in the region were used to estimate productivity, one would conclude that there would be little impact on total productivity. For example, using the SUBSTOR potato simulation model contained in the DSSAT system (Tsuji *et al.*, 1994), the loss of production associated with average soil depth was only 0.8% per year, whereas the loss measured on a sub-field scale in a typical field ranged from 97% in the upper part where most topsoil was lost, to a gain of 7% in the lower parts of the field where sediments accumulated.

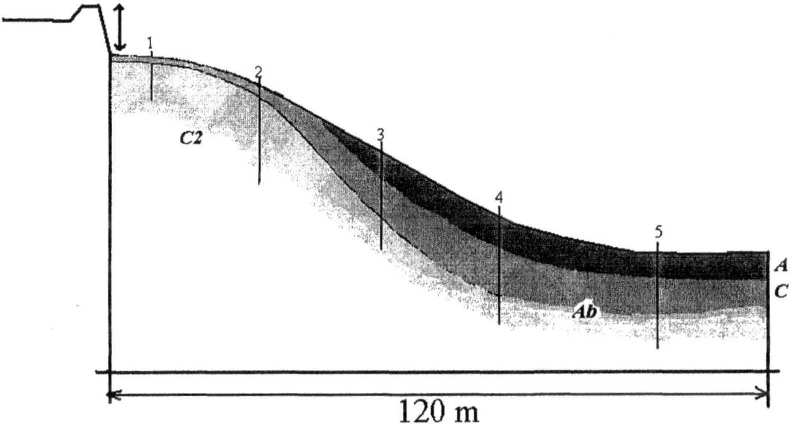

Figure 10.2: The effect of mechanical erosion on a farmer's field (source: Veen (1999)).

These data suggest that it is essential to consider the spatial variability in soils within fields to detect this loss in productivity. Representative soils for fields or soil mapping units will not reveal the loss in productivity since the average profile does not necessarily change as a result of mechanical erosion. Our observation of these types of site-specific processes provide the basis for our hypothesis that accurate estimation of the productivity and sustainability of production systems generally needs to account for spatial variability in soil quality that occurs at the field or even sub-field scale. Therefore, it follows that in analysis of the inter-relationships between policy and sustainability, we need to carefully consider the scale of analysis. At a regional scale, model-based analysis may lead to the erroneous conclusion that mechanical erosion and other site-specific processes have little impact on the sustainability of a production system, whereas analysis conducted at the field or sub-field scale and then aggregated to the regional level, may result in a very different estimate of a system's sustainability.

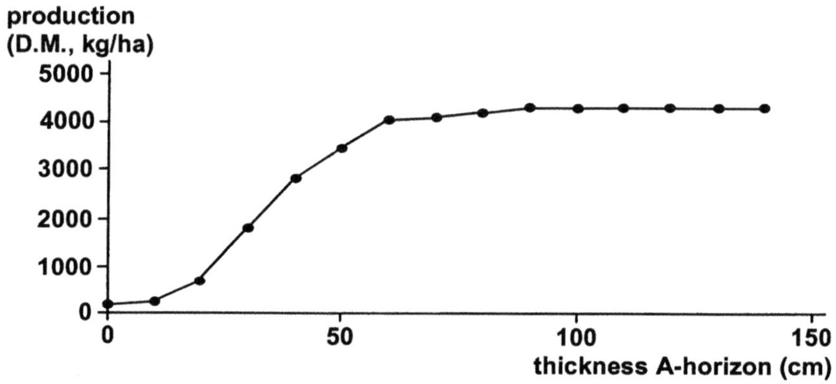

Figure 10.3: The effect of differences in the thickness of the fertile A-horizon on the dry matter production of potatoes as simulated with DSSAT.

10.3 An Economic Model of Site-Specific Land Use and Management

This section develops a static theoretical economic model of site-specific farm management that can be linked to site-specific soil quality data discussed in the preceding section. This model is a generalisation of ones discussed in the literature on agriculture–environment interactions (*e.g.* Hochman & Zilberman (1978); Just & Antle (1990); Antle & Just (1991); Antle & Capalbo (1991); Opaluch & Segerson (1991); Wu & Segerson (1995); Antle (1996); Hardie & Parks (1997); Antle *et al.* (1998)).

A production process at site i for crop j in period t is described by a technology set $\boldsymbol{v}_{ijt} \in \boldsymbol{T}(\boldsymbol{z}_{ijt}, \boldsymbol{e}_{it})$ where $\boldsymbol{v}$ is a vector of variable inputs, $\boldsymbol{z}$ is a vector of allocatable quasi-fixed factors of production and other fixed effects, and $\boldsymbol{e}$ is a vector of environmental characteristics of the site. The technology set $\boldsymbol{T}$ is specific to a particular site through the $\boldsymbol{z}$ and $\boldsymbol{e}$. For convenience we also assume that the form of the technology is independent of time, although the model can be generalised to account for technological change. A stochastic generalisation could be developed in which output is represented as a probability distribution conditional on $\boldsymbol{z}$ and $\boldsymbol{e}$.

We assume that the production function for each production activity corresponding to $\boldsymbol{T}(\boldsymbol{z}_{ijt}, \boldsymbol{e}_{it})$ is non-joint in inputs, following the discussion of nonjointness of agricultural production technology in Just *et al.* (1983). The individual production function can be written as $q_{ijt} = f(\mathbf{v}_{ijt}, \mathbf{z}_{ijt}, \mathbf{e}_{it})$ with corresponding expected profit function $\pi_{ijt} = \pi_j(p_{ijt}, \mathbf{w}_{ijt}, \mathbf{z}_{ijt}, \mathbf{e}_{it})$ where p_{ijt} is expected output price and $\mathbf{w}_{ij}$ is a vector of input prices. Define $\delta_{ijt} = 1$ if the jth crop is grown at location i at time t and zero otherwise. If the land is not in crop production then it is in a conserving or other productive use that earns a return π_{cit}. Letting $\delta_{cit} = 1 - \sum_j \delta_{ijt}$, the land use decision is defined as solving:

$$\max_{(\delta_{i1t}, \cdots, \delta_{int})} \sum_{j=1}^{n} \delta_{ijt} \pi_j \left(p_{ijt}, \mathbf{w}_{ijt}, \mathbf{z}_{ijt}, \mathbf{e}_{it}\right) + \delta_{cit} \pi_{ct} \tag{10.1}$$

The solution takes the form of a discrete step function

$$\delta^*_{ijt} = \delta(\mathbf{p}_{it}\ \mathbf{w}_{it}, \mathbf{z}_{it}, \mathbf{e}_{it}, \pi_{cit}), \tag{10.2}$$

where $\mathbf{p}_{it}$ is a vector of the p_{ijt} and likewise for the other vectors. We shall use the notation $\boldsymbol{\delta}^*_{it}$ to denote the vector of the δ^*_{ijt} for $j = 1,...,n$. Using Hotelling's lemma, the quantity of planned production on the ith land unit is given by:

$$q_{ijt} = \delta^*_{ijt}\, \partial\pi_j(p_{ijt}, \mathbf{w}_{ijt}, \mathbf{z}_{ijt}, \mathbf{e}_{it}) / \partial p_{ijt} = q_{ijt}(\mathbf{p}_{it}, \mathbf{w}_{it}, \mathbf{z}_{it}, \mathbf{e}_{it}, \pi_{cit}) \tag{10.3}$$

Variable input demands are likewise given by:

$$\mathbf{v}^*_{ijt} = -\, \delta^*_{ijt}\, \partial\pi_j(p_{ijt}, \mathbf{w}_{ijt}, \mathbf{z}_{ijt}, \mathbf{e}_{it}) / \partial \mathbf{w}_{ijt} = \mathbf{v}_{ijt}(\mathbf{p}_{it}, \mathbf{w}_{it}, \mathbf{z}_{it}, \mathbf{e}_{it}, \pi_{cit}) \tag{10.4}$$

The solution to (10.1) applies to a given site (a farmer's field), and is based on the assumption that each field can be managed independently. Various conditions may cause the management of distinct fields to be interrelated. One consideration discussed by Just, Zilberman and Hochman (*op. cit.*) is constraints on allocatable fixed inputs such as land or capital. In this case, problem (10.1) would be embedded in a choice problem for all land units subject to an adding-up constraint on the allocatable fixed inputs (elements of z). Whether or not an adding-up restriction is appropriate depends on how farmers utilise machinery and other quasi-fixed inputs. In many cases rental markets are available for machinery and for custom operations such as pesticide and fertiliser applications. In this case, services of capital (*e.g.* tractor hours) may be interpreted as variable inputs rather than fixed. The same can be said for the total area of land managed by farmers. Given land rental markets, the total number of acres managed in a growing season may be a choice variable rather than a constraint.

Another factor leading to interdependence of management decisions across land units is risk. We can add a stochastic term to each production function and assume that these stochastic elements are jointly distributed across land units. Then under the assumption that farmers are risk averse and choose production activities to maximise expected utility or some other risk-based criterion, it can be shown that production decisions are inter-related across land units. *E.g.* if farmers maximise a linear mean-variance criterion, land use decisions will be inter-related if the choice of production activities affects the covariance matrix of returns across land units.

The dynamics of crop rotations can be readily incorporated into the above model. For example, a crop/fallow rotation is incorporated by specifying the profit function to include the appropriate costs and returns (Antle & Capalbo, 2000). For simplicity of exposition, we proceed with the analysis using the simple static model (1) in which decisions are independent across sites, while recognising that resource constraints and risk could lead to spatial interdependence, and that crop rotations could result in temporal interdependence.

10.4 Site-Specific Management and Soil Quality

Analysis of the inter-relationships between land use and soil quality necessarily involves introduction of temporal variation brought about by the dynamics of the production system, *i.e.*, the time-dependent relationships between land use and soil quality. The vector $\boldsymbol{e}_{it}$ can be disaggregated into soil quality $\boldsymbol{s}_{it}$ which is site-specific and may vary with time, and a vector $\boldsymbol{c}_i$ of site characteristics (*e.g.* climate, soil type, slope) that do not vary with time. Climate is defined in terms of the parameters of distributions of weather events; these parameters do not vary with time (assuming no climate change) whereas weather events (temperature and precipitation) are realisations of climate that do vary with time.

The discussion above shows that soil quality at a point in space and time is a function of: site characteristics that are fixed in time (*e.g.* texture, slope); the soil quality realised in the previous time period; land use and management in the previous time period; and random realisations of weather and other stochastic events ε_{it}. Thus,

$$\boldsymbol{s}_{it} = \boldsymbol{s}(\boldsymbol{c}_i, \boldsymbol{s}_{it-1}, \delta_{it-1}, \boldsymbol{v}_{it-1}, z_{it-1}, \varepsilon_{it-1}) \tag{10.5}$$

Recursive substitution of this relationship shows that:

$$\boldsymbol{s}_{it} = \boldsymbol{s}(\boldsymbol{c}_i, \boldsymbol{s}_{i0}, \delta_i^{t-1}, \boldsymbol{v}_i^{t-1}, z_i^{t-1}, \varepsilon_i^{t-1}) \tag{10.6}$$

where $\delta_i^{t-1} = (\delta_{it-1}, \delta_{it-2}, \ldots, \delta_{i1})$ and other variables are defined similarly. Thus, soil quality at a point in space and time is a function of the following factors: fixed physical site characteristics; initial soil quality; histories of land use and crop management, production technology, weather, and other stochastic events. Further substitution of Equations (10.2) and (10.4) into (10.6) gives

$$\boldsymbol{s}_{it} = \boldsymbol{s}(\boldsymbol{c}_i, \boldsymbol{s}_{i0}, \boldsymbol{p}_i^{t-1}, \boldsymbol{w}_i^{t-1}, z_i^{t-1}, \pi_{ci}^{t-1}, \varepsilon_i^{t-1}) \tag{10.7}$$

demonstrating that soil quality at a point in space and time can also be described as a function of the histories of prices and any related policy variables that affect economic returns.

To investigate the implications of this relationship for farmer behaviour, we consider the simple land-use decision problem in Equation (10.1) solved subject to Equation (10.7). Observe that because soil quality evolves over time as a function in part of farmer's land use and management decisions, a fully rational decision maker would consider the impacts of decisions in period t on economic returns in future periods t+1, t+2, *etc*. For simplicity of exposition, however, we initially assume farmers are myopic and do not take into account the impacts of their current decisions on their future well-being. This assumption also approximates the situation typical of many developing countries where farmers face high rates of interest, so that the present value of expected returns more than a few periods into the future is approximately zero.

Under the assumption that land-use decisions are based on Equation (10.1), we obtain the land use decision rules defined in Equation (10.2). Replacing $\boldsymbol{e}$ with $\boldsymbol{s}$ and $\boldsymbol{c}$ in (10.2), and combining (10.2) with (10.7) gives:

$$\delta^*_{ijt} = \delta\,(\boldsymbol{p}_{it},\, \boldsymbol{w}_{it},\, \boldsymbol{z}_{it},\, \boldsymbol{s}(\boldsymbol{c}_i\,,\, \boldsymbol{s}_{i0},\, \boldsymbol{p}_i^{t-1},\, \boldsymbol{w}_i^{t-1},\, \boldsymbol{z}_i^{t-1},\, \pi_{ci}^{t-1},\, \boldsymbol{\varepsilon}_i^{t-1}),\, \boldsymbol{c}_I,\, \pi_{cit}). \tag{10.8}$$

Similarly, management decisions in Equation (10.4) can be expressed as:

$$\boldsymbol{v}^*_{ijt} = \boldsymbol{v}_{ijt}(\boldsymbol{p}_{it},\, \boldsymbol{w}_{it},\, \boldsymbol{z}_{it},\, \boldsymbol{s}(\boldsymbol{c}_i\,,\, \boldsymbol{s}_{i0},\, \boldsymbol{p}_i^{t-1},\, \boldsymbol{w}_i^{t-1},\, \boldsymbol{z}_i^{t-1},\, \pi_{ci}^{t-1},\, \boldsymbol{\varepsilon}_i^{t-1}),\, \boldsymbol{c},\, \pi_{cit}). \tag{10.9}$$

Thus, the dynamics of soil quality combined with the land-use decision-making of farmers gives the result that both land use decisions and the state of soil quality at a point in time and space are functions of: current prices and related policy factors that affect returns to economic activity in the present period; site-specific physical conditions; the history of prices, policy parameters, weather, and other events that affected land use and soil quality in the past.

A model with foresight can be formulated as follows. From Equation (10.1) let:

$$\pi_{it} = \textstyle\sum_j \delta_{ijt}\, \pi_j(\boldsymbol{p}_{ijt},\, \boldsymbol{w}_{ijt},\, \boldsymbol{z}_{ijt},\, \boldsymbol{e}_{it}) + \delta_{cit}\, \pi_{cit}\,. \tag{10.10}$$

Letting the interest rate be r_t , the land-use problem with foresight over period (t,T) is then defined as:

$$\max_{\{\delta_{it}\}} \sum_{k=t}^{T} \left(1/1 + r_k\right)^k \pi_{ik} \qquad \text{s.t. } (10.5) \tag{10.11}$$

The solution to this problem (*e.g.* derived by recursive solution) can be shown to be a function of current and future expected prices and other predetermined variables. If price expectations are modelled as functions of past prices, then it follows that the reduced form solution to (10.11) is a function of the same variables contained in Equation (10.8). However, the structure of the solution and the implied time paths for the solution variables are in general different. The solution with foresight yields land use and management decisions that maintain a level of soil quality over time that yields a higher present value of expected returns than the myopic solution. It should be emphasised, however, that the solution with foresight still can yield a solution in which soil quality can be declining over time (McConnell, 1983).

10.5 Aggregation and Policy Analysis

We assume that policy makers deal with trade-offs between present and future outcomes, such as agricultural production, farm and consumer incomes, environmental quality, *etc.* Here, for simplicity, we focus on the trade-off between present and future agricultural productivity as suggested by Lynam & Herdt (1992), and to simplify matters we use an individual crop's yield as a measure of productivity rather than a more complete multi-output, multi-factor measure of total factor productivity.

Let $\boldsymbol{e}_{i0} = (\boldsymbol{c}_i, \boldsymbol{s}_{i0})$ be distributed in the population of land units with cumulative distribution function $G(\boldsymbol{e}_{i0}|\theta)$, where time zero is when cultivation of the land began. The analysis of Section 10.4 shows that land use, management, and soil quality are functions of weather events and the initial condition of the soil, and are therefore jointly distributed random variables. The joint distribution of these random variables evolves over time according to the laws of motion derived above (*i.e.* Equations (10.7), (10.8) and (10.9)). In addition to climate and soil variation, farmers' behaviour is a function of spatially and temporally varying output and input price $\boldsymbol{p}$ and $\boldsymbol{w}$, and fixed factors of production z. Spatial and temporal variation in all of these factors induce variation in farmer behaviour. This variation is represented by the joint distribution $F(\boldsymbol{p}, \boldsymbol{w}, z|\phi_t, \psi_t)$, where ϕ_t is a time-dependent parameter representing factors causing autonomous or exogenous shifts in distributions of prices and fixed factors over time (*e.g.* technological or demographic change in output markets), and ψ_t is a policy parameter that shifts price distributions.

For purposes of deriving the distribution of the aggregate outcomes, it is sufficient to observe that an aggregate quantity, such as total production, is the sum of random variables over all land units in the region and therefore is itself a random variable. Because land use, management, soil quality, and agricultural output at each site are each generally a non-linear function of the underlying random variables, and because these random variables may be both spatially and temporally related (*e.g.* weather events and prices may be correlated over space and time), the form of this joint distribution generally is unknown, even if the distributions of the underlying random variables are known. Nevertheless, it is possible to derive the moments of the aggregate distributions. For the purposes of this discussion, we focus on the first moments of these distributions, with obvious generalisations to higher moments.

Because we are defining productivity in terms of output per unit of land, let the size of a land unit be $a_{i\,t}$ hectares, and let the cumulative distribution of land units in the region be $A(a_{it})$. Using Equations (10.2) and (10.3), expected yield realised by farmers in the region is then :

$$\mu_{jt}(\theta, \phi_t, \psi_t) = \iiint \{q^*_{ijt}(p_{it}, w_{it}, z_{it}, e_{it}, \pi_{cit})/a_{it}\} \, dG(e_{i0}|\theta) \, dF(\mathbf{p}, \mathbf{w}, \mathbf{z} \mid \phi_t, \psi_t) \, dA \,. \tag{10.12}$$

Letting the potential yield at a site be $y^p(e_{it})$, the expected potential yield in the region is:

$$\mu^p_{jt}(\theta, \phi_t, \psi_t) = \iint y^p(e_{it}) \, dG(e_{i0}|\theta) \, dF(\mathbf{p}, \mathbf{w}, \mathbf{z} \mid \phi_t, \psi_t) \,. \tag{10.13}$$

A key result from Equations (10.12) and (10.13) is that *both* expected yield and expected potential yield in the region are functions of the same sets of parameters that define the distributions of physical and economic variables in the region.

Trade-offs between current and future expected productivity can now be defined using (10.12) and (10.13). A trade-off curve is defined as the set of outcomes generated by varying one parameter and holding other parameters constant. Here we generate trade-offs by letting price parameters vary within a specified range Φ, under the assumption that expected productivity is monotonic in these price parameters. For example, a trade-off curve for expected on-farm productivity between periods t and $t+n$ can be defined as:

$$M(\theta, \Phi, \psi_t, n) \equiv \{(\mu_{jt}, \mu_{jt+n}) \mid \phi_t \in \Phi\} \,. \tag{10.14}$$

We can now utilise this definition of a trade-off curve to quantify the concept of sustainability and to assess the effects of policy on trade-offs between current and future productivity, *i.e.* the effects of policy on the sustainability of the system. Using (10.14), define the elasticity of the trade-off between present and future productivity as $E(\theta, \Phi, \psi_t, n) \equiv dln\ \mu_{jt+n} / dln\ \mu_{jt} \leq 0$. The system's sustainability is then defined as:

$$S(\theta, \phi_t, \psi_t, n) \equiv 1/|E(\theta, \Phi, \psi_t, n)| \,. \tag{10.15}$$

The impacts of policy on the sustainability of the system is now calculated simply as $\partial S(\theta, \phi_t, \psi_t, n)/\partial \psi_t$.

To illustrate the difference between the analysis of productivity trade-offs derived from site-specific models and models based on 'representative' soils, we observe that both on-farm and potential crop productivity exhibit a highly non-linear response to factors such as soil depth (as discussed above in relation to Fig. 10.3). Beginning with adequate soil depth for a crop's genetic potential to be realised, one finds that productivity initially declines at a low rate with a decrease in soil depth, but as soil depth falls below a critical level productivity declines rapidly, as in Fig. 10.4. Therefore, estimates of on-farm productivity or potential productivity based on a 'representative' set of environmental conditions defined as:

$$\mu^e_{jt}(\theta) \equiv \int e \, dG(\theta) \,. \tag{10.16}$$

will generally overstate expected productivity in the population based on site-specific information. That is, if output and yield are concave functions of e then by Jensen's inequality,

$$\mu^P_{jt}(\theta, \phi_t, \psi_t) \leq y^P(\mu^e_{jt}(\theta)) \,. \tag{10.17}$$

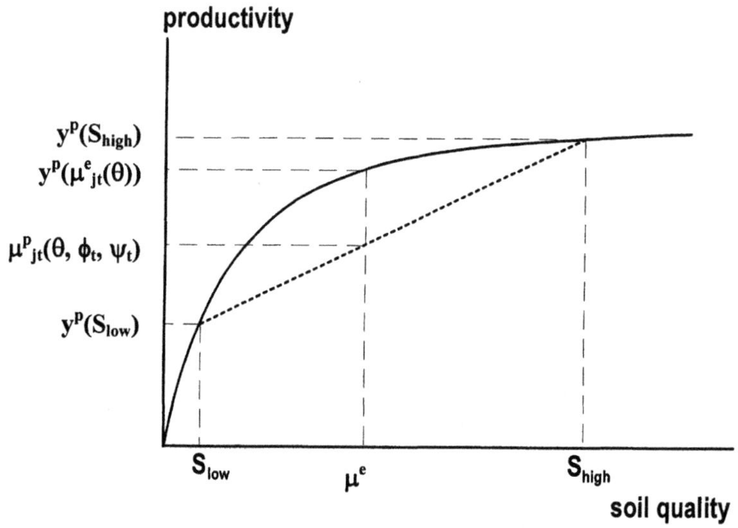

Figure 10.4: Site-specific soil quality, mean soil quality, and productivity measurement.

Equation (10.17) is illustrated in Fig. 10.4 for two soil qualities, low and high. As the figure shows, the degree to which productivity based on mean soil characteristics overestimates expected productivity depends on the non-linearity of the production function and on the degree of heterogeneity of soil characteristics in the population of fields in the region[3]. Therefore, the position as well as the slope of the trade-off curve defined in (10.14) will be different depending on whether $\mu^P_{jt}(\theta, \phi_t, \psi_t)$ or $y^P(\mu^e_{jt}(\theta))$ is used. Thus, the measured sustainability of the system, as given in Equation (10.15), will depend on whether it is based on a representative soil or derived from an aggregation of site-specific data.

[3] See Heerink (1994, Ch. 2) for a related analysis of aggregation bias in the context of measurement of income inequality.

10.6 An Example: Mechanical Erosion

We now consider the example of mechanical erosion, *i.e.* the process of downhill soil displacement that occurs as steeply sloped fields are cultivated. Mechanical erosion, rather than hydric or water-caused erosion, is a principal cause of soil and productivity loss in some regions of the world, such as the Equatorial Andes. The crops such as potatoes are cultivated on steeply-sloped hillsides of deep volcanic soils with high infiltration rates. These soils, with mostly low-intensity precipitation, are hardly susceptible to water erosion. However, continuous potato cultivation gradually displaces soil down-slope. This displacement is accelerated when tractors with mouldboard ploughs are used. These tractors are driven directly downhill because of the steep slopes. To represent this process, let soil depth be determined by the following process:

$$s_{ijt} = \lambda_{ijt}\, exp(\varepsilon_{it})\, s_{ijt-1},\ \lambda_{it} = 1 - (\alpha + \beta v_{ijt} + \Sigma_j\, \gamma_{ijt}\, \delta_{ijt}) \tag{10.18}$$

where parameter α represents the rate of soil loss independent of land use and management, βv_{ijt} is the component of soil loss associated with management practices (*e.g.* use of tractors), and the terms $\gamma_{ijt}\, \delta_{ijt}$ capture the effect of alternative land uses (*e.g.* crops, pasture). A richer specification could incorporate management and crop interactions. From (10.18) it follows that:

$$s_{ijt} = \prod_{k=1}^{t} \lambda_{ijk} \exp\left(\sum_{m=1}^{t} \varepsilon_{im}\right) s_{i0} \tag{10.19}$$

To further simplify the analysis, replace λ_{ijk} with the average annual rate of mechanical erosion λ_{ij} and let $u_{it} = exp(\Sigma_m \varepsilon_{im})$ so that Equation (10.19) becomes:

$$s_{ijt} = (\lambda_{ij})^t\, u_{it}\, s_{i0}\,. \tag{10.20}$$

The impact of a discrete change in policy parameter ψ_t on soil depth can now be expressed as:

$$\Delta s_{ijt} / \Delta \psi_t = t\, (\Delta \lambda_{ij} / \Delta \psi_t)(\lambda_{ij})^{t-1}\, u_{it}\, s_{i0} \tag{10.21}$$

where,

$$\Delta \lambda_{ij} / \Delta \psi_t = (\Delta \lambda_{ij} / \Delta\, \delta_{ij})(\Delta \delta_{ij} / \Delta \psi_t)|_v + (\Delta \lambda_{ij} / \Delta v_{ij})(\Delta v_{ij} / \Delta \psi_t)|_\delta\,. \tag{10.22}$$

In this last expression, δ_{ij} and v_{ij} are used to denote the land use pattern and management associated with the time horizon over which the average erosion rate λ_{ij} was calculated. Equation (10.22) demonstrates that the impact of policy on soil depth operates through the impacts of policy on land use and management, and through the impacts of land use and management on mechanical erosion. Aggregation analysis conducted along the lines in Section 10.5 would show that these impacts of policy computed on a site-specific basis would not be equivalent to impacts conducted with representative data.

10.7 Conclusions

In this paper we describe the sustainability of a production system in terms of trade-offs between present and future outcomes such as crop productivity. Based on the observed site-specificity of the relationship between soil characteristics and productivity, we hypothesise that to accurately assess the sustainability of an agricultural production system we must measure the interactions between farmers' management decisions and soil quality on a site-specific basis and then aggregate the resulting relationships.

The analysis of linked, site-specific economic and bio-physical models show that soil quality at a point in space and time depends on both fixed site characteristics and on the history of farmers' management decisions. Aggregation then shows that both on-farm productivity and potential productivity are functions of parameters that define the distributions of physical and economic variables in a region. Using this result, we show that measurements of sustainability based on 'representative' data such as mean soil characteristics will generally be biased and thus may lead to inaccurate assessment of a system's productivity and sustainability.

Site-specific modelling for analysis of agricultural sustainability can be carried out using GIS-based models and data that are integrated at the requisite spatial scale for accurate assessment of production-environment interactions. By simulating these interactions over a statistically representative sample of the land units in a region, for various price, technology, and policy scenarios, the results can then be aggregated and used to construct trade-off curves for policy analysis such as those illustrated in Fig. 10.1. Stoorvogel *et al.* (2000) present a general framework and software for this type of integrated biophysical and economic modelling.

10.8 References

Antle, J.M. (1996): Methodological issues in assessing the potential impacts of climate change on agriculture. Agricultural and Forest Meteorology 80, 67-85

Antle, J.M. & S.M. Capalbo (1991): Physical and economic model integration for measurement of the environmental impacts of agricultural chemical use. Northeastern Journal of Agricultural and Resource Economics 20, 68-82

Antle, J.M. & S.M. Capalbo (2000): Econometric-process models for integrated assessment of agricultural production systems. Research Discussion Paper No. 40, Trade Research Center, Montana State University

Antle, J.M. & R.E. Just (1991): Effects of commodity programs on resource use and the environment. in commodity and resource policies. In: R.E. Just & N. Bockstael (Eds.) Agricultural systems. Springer-Verlag Publishing Co., New York, 97-128

Antle, J.M., C.C. Crissman & S.M. Capalbo (1998): Tradeoffs in policy analysis: conceptual foundations and disciplinary integration. In: C.C. Crissman, J.M. Antle & S.M. Capalbo (Eds.) Economic, environmental, and health trade-offs in agriculture: pesticides and the sustainability of Andean potato production. Kluwer Academic Publishers, Boston

Bouma, J. (1997): Soil environmental quality: a European perspective. Journal of Environmental Quality 26, 26-31

Bouma, J., N.H. Batjes & J.J.R. Groot (1998): Exploring land quality effects on world food supply. Geoderma 86, 43-59

Food and Agriculture Organization (FAO) (1976): A framework for land evaluation. FAO Soils Bulletin 32, FAO, Rome

Hardie, I.W. & P.J. Parks (1997): Land use with heterogeneous land quality: an application of an area base model. American Journal of Agricultural Economics 79, 299-310

Heerink, N. (1994): Population growth, income distribution, and economic development: theory, methodology, and empirical results. Springer-Verlag, Berlin

Hochman, E. & D. Zilberman (1978): Examination of environmental policies using production and pollution microparameter distributions. Econometrica 46, 739-760

Huinink, J.T.M. (1998): Soil quality requirements for use in urban environments. Soil & Tillage Research 47, 157-162

Just R.E. & J.M. Antle (1990): Interactions between agricultural and environmental policies: a conceptual framework. American Economic Review 80, 197-202

Just, R.E., D. Zilberman & E. Hochman (1983): Estimation of multicrop production functions. American Journal of Agricultural Economics 65, 770-780

Lal, R., L.M. Kimble, R.F. Follett & C.V. Cole (1998): The potential of US cropland to sequester C and mitigate the greenhouse effect. Ann Arbor Press, Chelsea, MI

Lynam, J.K. & R.W. Herdt (1992): Sense and sustainability: sustainability as an objective in international agricultural research. In: J.L. Moock & R.E. Rhoades (Eds.) Diversity, farmer knowledge, and sustainability. Food Systems and Agrarian Change Series. Cornell University Press, Ithaca and London

McConnell, K.E. (1983): An economic model of soil conservation. American Journal of Agricultural Economics 65, 83-89

Mutsaers H.J.W. & Z. Wang (1999): Are simulation models ready for agricultural research in developing countries? Agronomy Journal 91, 1-4

Oldeman, L.R., R.T.A. Hakkeling & W.G. Sombroek (1991): World map of the status of human-induced soil degradation: an explanatory note. International Soil Reference and Information Centre, Wageningen

Opaluch, J.J. & K. Segerson (1991): Aggregate analysis of site-specific pollution problems: the case of groundwater contamination from agricultural pesticides. Northeast Journal of Agricultural and Resource Economics 20, 83-97

Pieri, C., J. Dumanski, A. Hamblin & A. Young (1995): Land quality indicators. World Bank Discussion Papers #315, The World Bank, Washington, D.C.

Stoorvogel, J., J.M. Antle, C.C. Crissman, & W. Bowen (2000): The trade-off model: a policy decision support system for agriculture. Quantitative Approaches in Systems Analysis, Wageningen

Tsuji, G.Y., G. Uehara & S. Balas (1994): DSSAT v3. University of Hawaii, Honolulu, HI

Veen, M. (1999): The development of land use and land management and their effect upon soils in processes of mechanical erosion and compaction. M.Sc thesis Wageningen Agricultural University, Wageningen, The Netherlands & CIP, Quito, Ecuador

Wu, J. & K. Segerson (1995): The impact of policies and land characteristics on potential groundwater pollution in Wisconsin. American Journal of Agricultural Economics 77, 1033-1047

11 Effects of Land Degradation in a Diversified Economy with Local Staple and Labour Markets: A Village-Town CGE Analysis from Mexico

George Dyer[1], **Antonio Yunez-Naude**[2] **& J. Edward Taylor**[1]

[1] Department of Agricultural and Resource Economics, University of California, Davis, U.S.A

[2] Centro de Estudios Economicos, El Colegio de Mexico, Mexico

Abstract

Land degradation has potentially complex effects on diversified economies typical of rural areas in less-developed countries. Market linkages transmit the impacts from the directly affected households to others in the economy. When high transactions costs inhibit trade in food, labour, capital, or other goods, shifts in consumer or input demands influence prices and unleash general-equilibrium effects in local economies. These effects are missed by the extensive research on first-round effects of agricultural policies on household-farms, including those in imperfect market environments. Economy-wide models, including computable general equilibrium (CGE) models, are designed to capture the second and higher-round feedback of exogenous shocks to the macro economy. Our research uses an adaptation of CGE techniques presented in Taylor & Adelman (1996) to explore the impacts of land degradation on village economies, and on migration for an area in rural Mexico characterised by local markets for staples and labour. Our village-town CGE blends microeconomic household-farm models with economy-wide modelling. Results suggest that second round effects, which are not captured by household models, can be important.

11.1 Introduction

The causes of land degradation lie at the farm level as much as outside the farm. The consequences of land degradation extend beyond the farm, affecting other economic actors in complex ways ignored by most economic models. Other chapters in this book analyse links between economic and biophysical processes at different levels: from the household to the national level, and beyond. Some of these processes occur in regional economies, including villages and towns, where most of the world's population lives. Local general-equilibrium feedback, particularly in regional economies facing high transportation and transaction costs, may reshape the impacts of environmental changes on rural households. It may even transmit an important part of these impacts to households not directly affected by the environmental change. In spite of their potentially critical influence on how policy, market, and environmental changes affect rural households, local economic relationships are poorly understood and usually ignored by researchers and policy makers.

This paper applies micro economy-wide modelling techniques (Taylor & Adelman, 1996) to illustrate the potential impact of land degradation on household incomes in a regional, village-town economy in Mexico. Our model incorporates a series of agricultural household models into a local general-equilibrium framework for (a) a rural *municipio* or county seat — the town —, and (b) the surrounding villages that, together with the town, make up the *municipio*. The town and village models are then linked spatially through trade. This makes it possible to capture both the direct and indirect impacts in the villages and town, where transaction costs may differ, as well as to trace the transmission of effects between the two. We estimated the model using original survey data, gathered in conjunction with El Colegio de Mexico's PRECESAM (Program for the Study of Economic Change and Sustainability in Mexican Agriculture). We use the village-town model to illustrate the effects of land degradation on the local economy, particularly effects extending across local economic activities — agricultural and non-agricultural — and thus only evident at the local level. We describe how these effects vary across different market scenarios, defined by the assumption of either fixed or endogenous wages and staple prices. Despite the fact that land degradation is evident in this area of Mexico, we lack data on its nature and causes. While assuming that land degradation is given, *i.e.* exogenous to our model, we explore the model's sensitivity to the nature and causes of land degradation by using two different specifications for it.

11.2 The Model and Data

Land degradation has potentially complex effects on diversified economies typical of rural areas in developing countries, where agricultural households are simultaneously engaged in a 'portfolio' of farm and non-farm activities. Market linkages transmit the impacts from the directly affected households to others in the rural economy (suppliers of inputs and consumer and investment goods to directly affected households). When high transactions costs hamper trade with the outside in food, labour, capital, or other goods, shifts in consumer or input demand influence prices and unleash general-equilibrium effects in local economies. These effects are missed by the extensive research focusing on first-round effects of agricultural policies on household-farms (*e.g.* Singh *et al.*, 1986), including those in imperfect market environments (de Janvry *et al.*, 1991). Economy-wide models, including computable general equilibrium (CGE) models, are designed to capture the second and higher-round feedback of exogenous shocks to the macro economy. However, national CGE models abstract from local economies, and they do not provide the detail needed to reliably uncover the full impact of changes on diversified rural economies. Our village-town CGE blends microeconomic household-farm models with economy-wide modelling. It offers an approach complementary to micro household-farm models and aggregate CGE models for policy analysis, providing insights that cannot be obtained by the other two approaches.

Consider the effect of a change in an exogenous variable Z (*e.g.* a government income transfer or price of a village tradable) on an endogenous variable Y (*e.g.* production, income of a household group, or migration). Let P denote a vector of local input and output prices. The full impact of the change in Z on Y is given by:

$$dY/dZ = \partial Y/\partial Z + \partial Y/\partial P\, dP/dZ \tag{11.1}$$

The first term represents direct income effects, an economy-wide analogue to the partial effects in a neo-classical household-farm model in which all prices are held constant. The second term represents the indirect, general-equilibrium effects of the shock through endogenous local prices. If all goods and factors are tradable at both the village and household levels (that is, all prices are given to the local economy), the second term drops out. In this case, a micro agricultural household model may be sufficient to estimate local production, marketed-surplus, and income effects of land degradation. However, if some goods (*e.g.* labour, output) are village non-tradables, the second term in Equation (11.1) is non-zero. Market linkages resulting from endogenous prices may modify the effects of land degradation on rural economies. In rural Mexico, if alternative local labour-intensive activities absorb labour released from staple production, the first-round impacts on employment and migration may be dampened or, as we shall see later, even reversed.

11.2.1 The Village-Town Model

The building block of our village-town CGE model is a series of models of agricultural households engaged in crop production, but also potentially in a portfolio of other economic activities, including migration. The model in GAMS is available on the web (//www.agecon.ucdavis.edu/Faculty/Ed.T/Taylor.html). It consists of four equation blocks: (1) a household-farm production block, (2) a household-farm income block, (3) an expenditure block, and (4) a set of general equilibrium closure equations.

The household-farm economy includes four production sectors and a migration activity. Production in each of the sectors is carried out with four factors: labour, animal capital, physical capital and land. Household-farms are assumed to maximise utility, defined on consumption goods and savings. On the production side, this implies maximising net farm income from each activity, given market prices for inputs and outputs, as well as land and capital constraints. Capital and land are fixed in total supply, but land may be mobile across sectors. Their prices (rental rates) are endogenous, equal to their marginal value products in local production. Since capital is immobile, its rental rates differ across activities. Wages may be determined exogenously outside the village-town economy or endogenously, through the interplay of supply and demand.

The household-farm sector in our model consists of two groups: subsistence households and commercial households. Household-farm income is the sum of wage income, net income from household-farm production activities, and migrant remittances, which are a function of labour allocated to migration. The expenditure block includes the consumption demand for products and manufactured goods produced in the villages, town, or elsewhere in Mexico; savings, including investments in physical and human capital (schooling); taxes, including a 5% levy on maize production; and some household-to-household transfers.

The general-equilibrium closure equations include local market-clearing conditions for factors and goods, a village-town savings-investment balance, and a trade balance equation. For goods and factors for which the village-town is a price-taker (*i.e.* tradables), the market-clearing conditions determine net village-town marketed surplus. For non-tradables, they determine local prices. The savings-investment balance constrains investments in physical and human capital to be self-financed, *i.e.* out of local savings. The trade equation constrains the value of village-town exports of goods and factors to equal the value of village-town imports, minus net borrowing from outside (analogous to foreign savings in national CGE models). It represents the redundant equation in our village-town CGE system. Production technologies are specified as Cobb-Douglas, and consumption demands are modelled using a linear expenditure system (LES) approach. This functional form imposes restrictions on the model, but it also facilitates the illustration of local economic processes.

The village-town CGE model solves for local equilibrium prices of non-tradables (rental rates on fixed land and capital factors and, in two of our model-closure scenarios, local wages and village staple prices), and net marketed surplus for tradables. In this analysis, tradables refer to goods for which the village-town economy engages in trade and is a price taker in national markets. Non-tradables include goods for which local trade is possible but for which prices are endogenously determined through the interaction between local supply and demand. Local trade in goods and factors can occur within the villages, within the town, and between villages and town. This latter form of trade defines village-town interactions; all goods and factors can be traded freely between villages and town, except as stated in the different scenarios.

11.2.2 Data

The village-town model was estimated using data from a 1993 survey of 196 households in Michoacan, Mexico, including 53 in a town, or county seat, and 143 in the surrounding villages that, together with the town, make up the *municipio* (analogous to a U.S. county). Most land in the survey area is *ejido* (reform-sector) land. The range of cultivated landholdings in our sample is from 0 to 19 ha, and the average is 2.8 (sd=3.9; 0.4 ha for subsistence households (n=115, sd=0.67) and 6.3 ha for commercial households (n=81, sd=3.9)). Factor inputs and the prices of physical capital and hired labour were observed directly. Implicit land rents were estimated econometrically from regressions of income from land-based activities (crops and livestock) on household-farm asset holdings. Family labour value-added was calculated as the value of production minus the costs of intermediate inputs, hired labour, physical capital, and implicit land rents. Household-farm savings and expenditure shares were estimated from expenditure survey data. Remittance elasticities were estimated from a regression of the log of households' migrant remittance receipts (net of amortised migration costs) for households that participated in migration on the log of their labour allocated to migration, controlling for family migration networks and correcting for potential sample selectivity bias.

Incomes in the surveyed households are highly diversified. 64% of households in the sample received income from staple production, including imputed income from subsistence production. Most of these staple-producing households also received income from several other sources, and staples accounted for only 6% of their total income. 46% of households received income from Mexico-to-U.S. migration, but the majority of these also received income from staple, cash-crop, and non-agricultural production.

11.3 Simulations

We performed three sets of experiments on alternative model-closure scenarios. Experiments 1 and 2 explore the adjustments undertaken by farmers to cope with two different forms of land degradation. In experiment 1, we assume that all land degrades, but only staples are sensitive to the degradation: staple productivity decreases exogenously by 10%. In experiment 2, we assume that production of staples degrades some of the land, henceforth preventing its use in any activity: 10% of land in staple production is exogenously retired from all uses. Although both exogenous changes would result in a 10% reduction in staple production in the absence of farmer response, the rationale for, and the implications of, the two experiments are different. Experiment 1 applies where loss of the original forest cover results in soil erosion under any agricultural use, but only fast growing annual crops — in this case the staples sector — show the effect of land degradation. Experiment 2 applies where the cultivation of perennial crops and grasses in the cash-crop and livestock sectors precludes soil erosion, but growth of annual crops degrades land, preventing its further use in any agricultural sector. These two specifications are intended as examples of the many forms of land degradation that can actually take place.

Four model scenarios were considered for these two experiments: (*a*) a perfect-markets neo-classical specification, in which the village-town micro-economy is a price taker in all markets with the exception of capital and land (which are in fixed supply); (*b*) an endogenous wage scenario, in which wages are determined by the interplay of village-town labour supply, on one hand, and local labour demand and migration, on the other. Scenarios (*c*) and (*d*) are identical to (*a*) and (*b*), respectively, except that the staple market in the villages is closed (*i.e.* constrained to be local), resulting in an endogenous village staple price. This corresponds to a world in which transaction costs for staples are low for the town (*i.e.* the town is a price taker in outside markets) but high for the surrounding villages (*i.e.* village households face high costs of transacting with the town, and beyond). Rural Mexico is characterised by a patchwork of staple surplus-producing areas and others producing little or no surplus for national markets. In the latter areas, the price of staples is likely to be endogenous. Our market closure specifications, which assume either perfectly closed or perfectly open markets, are somewhat restrictive. A more flexible specification with endogenous prices would assume open markets, but less than perfect substitution between local and foreign goods[1].

In experiments 1 and 2, farmers can respond to land degradation by changing the allocation of land and labour among agricultural activities[2]. When certain agricultural activities are more sensitive to land degradation than others, degradation itself may impose restrictions on the mobility of land across activities. This is common in rural Mexico, where farmers commonly grow pastures on land that is too degraded to grow staples. Experiment 3 explores the ramifications of environmental restrictions on land reallocation across activities, by simulating an exogenous 10% increase in the price of staples when (a) land is mobile, and (b) land mobility across sectors is constrained by degradation[3]. Scenarios (c) and (d) are identical to (a) and (b), respectively, except for the assumption of an endogenous local wage. In all scenarios village staple markets are closed, to illustrate how policies may indirectly affect village households. Tables 11.1 through 11.3 report the results of our experiments.

---

[1] See *e.g.*, de Melo & Robinson (1989).

[2] Land is mobile across crop and livestock sectors; labour is mobile across all sectors; capital is fixed.

[3] Land used for livestock or other crops cannot be converted to staple production.

Table 11.1: Percentage effects of a 10% decrease in staple productivity under alternative market closure rules[1].

Labour scenario:	Fixed wage[2]		Endogenous wage	
Staple scenario:	Fixed price	Endogenous price in village	Fixed price	Endogenous price in village
Village				
Output				
Staple	-32.8	-9.1	-32.7	-9.0
Cash crop	4.5	1.2	4.7	1.3
Livestock	4.1	-0.5	4.1	-0.5
Non agricultural	0.0	0.0	1.2	0.3
Staple Price	n/a	11.3	n/a	11.3
Labour Demand	-2.0	0.1	-1.3	0.2
Wage	n/a	n/a	-0.1	0.0
Migration	0.0	0.0	0.4	0.1
GDP	-3.3	-0.1	-2.7	0.0
Household real income				
Commercial	-2.0	-0.8	-1.6	-0.7
Subsistence	-1.7	-2.0	-1.4	-1.9
Capital investment	-1.8	-0.1	-1.4	0.0
Town				
Output				
Staple	-43.8	-43.8	-43.6	-43.8
Cash crop	3.5	3.5	4.5	3.7
Livestock	2.4	2.4	2.3	2.3
Non agricultural	0.0	0.0	8.8	2.2
Labour Demand	-1.6	-1.6	2.8	-0.5
Wage	n/a	n/a	-0.1	0.0
Migration	0.0	0.0	0.4	0.1
GDP	-2.6	-2.6	2.6	-1.3
Household real income				
Commercial	-2.0	-2.0	2.3	-0.9
Subsistence	-1.1	-1.1	1.1	-0.6
Capital investment	-1.2	-1.2	1.3	-0.6

[1] Land can be destined to any agricultural activity either in the town or in the villages, but its supply is fixed in each location.

[2] Village-town is a price taker in labour market and/or Lewis surplus-labour assumption.

Table 11.2: Percentage effects of a 10% retirement of land under alternative market closure rules[1,2].

Labour scenario:	Fixed wage[3]		Endogenous wage	
Staple scenario:	Fixed price	Endogenous price in village	Fixed price	Endogenous price in village
Village				
Output				
Staple	-0.9	-0.7	-0.9	-0.6
Cash crop	-1.8	-1.9	-1.8	-1.9
Livestock	-1.7	-1.8	-1.7	-1.8
Non agricultural	0.0	0.0	0.1	0.1
Staple Price	n/a	0.1	n/a	0.1
Labour Demand	-0.2	-0.2	-0.1	-0.1
Wage	n/a	n/a	0.0	0.0
Migration	0.0	0.0	0.0	0.0
GDP	-0.9	-0.8	-0.8	-0.8
Household real income				
Commercial	-0.6	-0.6	-0.5	-0.5
Subsistence	-0.4	-0.4	-0.4	-0.4
Capital investment	-0.5	-0.5	-0.4	-0.4
Town				
Output				
Staple	-0.9	-0.9	-0.8	-0.8
Cash crop	-2.8	-2.8	-2.7	-2.7
Livestock	-1.9	-1.9	-1.9	-1.9
Non agricultural	0.0	0.0	1.0	0.9
Labour Demand	-0.4	-0.4	-0.1	-0.1
Wage	n/a	n/a	0.0	0.0
Migration	0.0	0.0	0.0	0.0
GDP	-0.9	-0.8	-0.8	-0.8
Household real income				
Commercial	-0.6	-0.6	-0.2	-0.2
Subsistence	-0.3	-0.3	-0.1	-0.1
Capital investment	-0.4	-0.4	-0.1	-0.1

[1] Land can be destined to any agricultural activity either in the town or in the villages, but its supply is fixed in each location.

[2] Equivalent to 10% of the land originally on staple production is retired.

[3] Village-town is a price taker in labour market and/or Lewis surplus-labour assumption.

Table 11.3: Percentage effects of a 10% increase in staple price under alternative market closure rules[1,2].

Labour scenario:	Fixed wage[3]		Endogenous wage	
Staple scenario:	Fixed price	Endogenous price in village	Fixed price	Endogenous price in village
Village				
Output				
Staple	0.0	0.0	-0.1	-0.1
Cash crop	0.0	0.0	-0.1	0.0
Livestock	0.0	0.0	0.0	0.0
Non agricultural	0.0	0.0	-0.5	-0.3
Staple Price	0.0	0.0	0.0	0.0
Labour Demand	0.0	0.0	-0.3	-0.2
Wage	n/a	n/a	0.1	0.0
Migration	0.0	0.0	-0.2	-0.1
GDP	0.0	0.0	-0.3	-0.1
Household real income				
Commercial	0.0	0.0	-0.2	-0.1
Subsistence	0.0	0.0	-0.2	-0.1
Capital investment	0.0	0.0	-0.2	-0.1
Town				
Output				
Staple	53.1	18.8	52.8	18.8
Cash crop	-3.3	0.0	-3.7	-0.1
Livestock	-4.1	-0.1	-4.0	-0.1
Non agricultural	0.0	0.0	-3.8	-2.0
Labour Demand	2.6	1.3	0.7	0.4
Wage	n/a	n/a	0.1	0.0
Migration	0.0	0.0	-0.2	-0.1
GDP	4.0	2.2	1.8	1.2
Household real income				
Commercial	1.7	0.4	-0.1	-0.5
Subsistence	0.4	-0.3	-0.5	-0.7
Capital investment	1.9	1.1	0.8	0.5

[1] Mobile land can be destined to any agricultural activity either in the town or in the villages, but its supply is fixed in each location; immobile land is fixed in each activity.
[2] Staple prices in the village are endogenous.
[3] Village-town is a price taker in labour market and/or Lewis surplus-labour assumption.

11.4 Results

11.4.1 Experiment 1: Decrease in Staple Productivity

Staple output contracts with the 10% decline in staple productivity in all market-closure scenarios. When the villages have an open staple market, production of staples drops 33% in the villages and 44% in the town, as land and labour are reallocated to more profitable agricultural sectors. Other crop output and livestock sales increase between 2 and 5%. If the wage is fixed, employment drops as the decrease in labour demand in the staples sector exceeds increases elsewhere. Real income of all households decreases. If the labour market is strictly local, the endogenous wage decreases with the drop in staple productivity. This stimulates production in other sectors, especially in the town's labour-intensive non-agricultural activities, where households earn most of their income. The real income of town households increases, but that of village households still contracts.

When the villages have a local staple market, town activity responds to the decline in productivity exactly the same way as in the previous case, but the villages experience a local scarcity of staples that pushes up the staple price by 11%. Village staple production falls by 9%, less than in the exogenous staple-price scenario. In response to the high staple price, more land and labour are allocated to village staple production, despite its reduced productivity. This restricts the expansionary effect on other village sectors; cash crops grow by only 1%, while livestock output decreases by 0.5%. Moreover, the wage does not decline even when it is endogenous, restricting the response of the non-agricultural sector. Real income decreases for all households in the region. The intensification of staple production in response to land degradation potentially creates a vicious circle, resulting in greater productivity loss in the future.

11.4.2 Experiment 2: Decrease in Land Availability

When land is taken out of production, all agricultural activities contract. Since land is mobile across local agricultural activities, the outcome is not sensitive to the type of land retired; remaining land is always reallocated in the same way. Staple output decreases by around 1%, while output of other crops and livestock decreases between 2 and 3%. If the village staple market is closed, local scarcity of staples drives up the staple price; village staple output decreases relatively less, and other village agricultural activities relatively more, than when the market is open. When the labour market is strictly local, a modest wage decline drives non-agricultural output slightly upwards. In all scenarios, changes in household income are negative, but due to the possibility of reallocating land into the most profitable activities, the changes are always smaller, in absolute-value terms, than in the corresponding scenarios in experiment 1.

11.4.3 Experiment 3: Increase in World Market Price for Staples

In all scenarios, the high price for staples stimulates staple production in the town but not in the villages, where a separate and closed staple market exists. All other sectors contract — including village sectors when the local wage is endogenous — as factor marginal value products rise in the town's staple sector. The town's gross domestic product (GDP) increases in all scenarios, but real effects on households differ in each case. When land is mobile across sectors, local land and labour are drawn into staple production in the town, which increases by more than 50%, while other agricultural sectors in the town contract by 3 - 4%. Real income for town households increases with a higher staple price when the wage is fixed, but village households are not affected, as village economies are isolated from the price policy. When the labour market is strictly local, the endogenous wage rises, extending the impact of the policy to the town's labour-intensive non-agricultural sector — which contracts by 4% — and to the village economies; even village staple output contracts. In this case, real income decreases for both town and village households, despite the rise in GDP. This is due to staple-price inflation and a slight decrease in migration and therefore, in migrant remittances.

When land cannot be moved among agricultural activities, the staple-price policy has a smaller effect on the region's output, but the final effect on households depends again on the nature of the labour market. In general, the restriction on land mobility limits the growth of town GDP to 1 - 2%, compared with 2 - 4% when land is mobile. The town's staple output rises by a relatively modest 19% — all attributable to greater labour use — which now does not impact other agricultural sectors significantly. GDP grows most when the wage is fixed, but only commercial households experience a rise in real income, as subsistence households — net buyers of staples — lose in real terms. When the wage is endogenous, the town's non-agricultural activity decreases due to a higher wage, which draws labour from the village to the town. All households experience a decrease in real income, as even the 1% increase in town GDP is not enough to outweigh the decrease in migrant remittances.

11.5. Conclusions

The high degree of diversification in Mexico's small-farm economy buffers households from the negative impacts of land degradation. An agricultural household model can partially capture the role of diversification. Under the perfect markets specification, the percentage results from our village-town CGE model are the same as those from a model of the average household; namely, households respond to land degradation by shifting factors among different agricultural activities. The loss of staple productivity entails the largest redistribution of factors, away from staple production and into other agricultural activities undertaken by the household. Retirement of land from all production induces smaller changes in factor reallocation, as land and labour are pulled out of all agricultural activities. Although retirement of land has a more limited effect on village-town GDP than does a decrease in staple productivity, in the perfect markets scenario, all households lose income when land is retired. Commercial households, which produce an agricultural surplus, are most affected under both types of land degradation.

The agricultural household model fails to capture indirect, albeit potentially important effects of land degradation on local and regional economies. The village-town economy may react to the land degradation through changes in endogenous prices for staples and/or wages, which induce second round adjustments in households. Use of our village-town model to capture these effects suggests that none of the results from the household model alone can be generalised to these situations. When the staple market in the villages is closed, the effect of land degradation is shaped by endogenous price changes, which may pull factors into, rather than out of, staple production, despite declining staple productivity. This potentially creates a vicious circle of more intensive staple cultivation and increasing land degradation. The price increase for staples also shifts negative impacts of land degradation from commercial to subsistence households, which must buy staples to cover their production deficit. When the wage is endogenous, the impact of land degradation extends to the non-agricultural sector, raising the incomes of at least some town households. Thus, when second round effects are taken into account, it is not possible to conclude that land degradation is always detrimental to all households.

Second round effects are also relevant in policy making. Our last experiment can be used to illustrate the danger of not considering these effects. For example, suppose that future land degradation is checked, but the effects of past degradation endure: reduced production of staples has created a food deficit in the country, and farmers can no longer grow staples in degraded areas. The government attempts to implement a policy to assist farmers while reducing the deficit in staples, through a 10% increase in the staple price. The price policy will not reach remote villages with closed staples markets; only town households are targeted, since they have low transportation costs and thus can increase the supply of staples to the national market. If we based our predictions of the policy's impact on a household model alone, we would conclude that staple output would increase by 19%. The policy might create a distortion in rural economies by changing relative prices, but its effect on other agricultural sectors would be small due to land degradation, which would preclude land from going back into staple production. Therefore, we would predict that the policy would induce intensification of staple production rather than displace other agricultural activities. Only the town's livestock sector would decline slightly, because of higher feed prices, and village economies would not be affected at all. In the towns, subsistence households would pay a higher price for staples, but this would certainly benefit surplus producers.

However, our village-town model's findings tell a different story. By failing to account for the existence of regionally restricted labour markets in our stylised rural economy, we would overlook how intensive staple production in towns pulls labour away from agriculture in the villages, from local non-agricultural activities, and from migration, to which rural households increasingly turn to supplement their incomes. The policy would in fact create a small income loss for most households, aggravating income losses from land degradation. The greater the price linkages between households benefitting directly from the policy and others in the rural economy, the more complex the potential ramifications of the policy.

We have tried to illustrate the importance of local economic forces in the study of farm level processes and the possibilities of regional CGE models to study these forces. Regional CGE models are not presented here as a substitute for household models or models at other levels, but rather as a complement. Finally, we have treated land degradation as exogenous throughout the analysis. This reflects a lack of information about land degradation processes in the region and not a restriction of regional CGE models. It underscores the imperative for future interdisciplinary research on the implications of environmental changes for rural economies.

11.6 References

De Janvry, A., M. Fafchamps & E. Sadoulet (1991): Peasant household behavior with missing markets: some paradoxes explained. The Economic Journal 101, 1400-1417

De Melo, J. & S. Robinson (1989): Product differentiation and the treatment of foreign trade in computable general equilibrium models of small economies. Journal of International Economics 27, 47-67

Singh, I., L. Squire & J. Strauss (Eds.) (1986): Agricultural household models, extensions, applications and policy. The World Bank and The Johns Hopkins University Press, Washington, D.C.

Taylor, J.E. & I. Adelman (1996): Village economies: the design, estimation, and use of villagewide economic models. Cambridge University Press, Cambridge

12 Agricultural Production and Erosion in a Small Watershed in Honduras: a Non-Linear Programming Approach

Bruno Barbier

CIAT International Center of Tropical Agriculture

Abstract

This paper describes a multi-period non-linear programming model of agricultural production that includes soil erosion and its effect on crop production within a small watershed. The purpose of the model is to predict the likely effect of different decisions taken by the local community of the watershed on land uses, soils and incomes. The model has been applied to a small watershed in the hillsides of Central Honduras. Results show how erosion will affect local production in the long term. It also shows how farmers are likely to cope with the problem. In the model, the effect of erosion is calculated as a function of soil loss through decreasing soil depth. To better describe the spatial dimension of the problem, the watershed is divided in homogenous land units of different sizes. These land units are specified according to altitude, slope and farm type. The land units are inter-related through transport time, access to road, and water transfer. The main actual agricultural activities are available to the model, including land conservation techniques. Each land unit starts with a different stock of soil volume, land use area, crop productivity, and human population. These variables are modified over time by both natural and human activities.

12.1 Introduction

Classic books on farm modelling usually do not include erosion (Hazell & Norton, 1986; Boussard & Daudin, 1988; Mc Carl & Spreen, 1997). One reason is that most of these books focus on temperate regions, where erosion is less of a priority. Models representing tropical farming systems, however, need to include erosion, since its effect on yields explains their performance to a large extent. Not only tropical conditions induce more severe erosion, but socio-economic conditions of most tropical regions hamper farmers' capacities to control it.

The effect of erosion on yield is not easy to measure. The first reason is that the biophysical models describing erosion are not very accurate. Methods derived from the Universal Soil Loss Equation (USLE) (Wischmeier & Smith, 1978) and the WEPP equation (Flanagan & Nearing, 1995) have been widely used, but both methods remain controversial. They produce rough estimates, rather than exact predictions.

It is even more difficult to assess the effect of erosion on yields. Plant growth models such as EPIC, APEX, SWAT or CROPSYST include the effect of erosion on yields, but these models also are based on the controversial USLE.

When data on erosion produced under different land uses are available, through measurement or biophysical models, it is possible to introduce erosion in mathematical programming models in various ways. A first method consists of comparing the total erosion produced under different scenarios (Kruseman *et al.*, 1995). Another method consists of restricting total erosion in the model and examining the change in other variables (Barbier, 1988; White, 1988; Estrada *et al.*, 1999). These models compare the off-site effects of erosion and not its on-site effect. There are a limited number of mathematical programming models that include the on-site effect of erosion. When they do, the erosion effect is described as a direct function of eroded soil (Parikh, 1991). For instance, one ton of eroded soil means x tons of yield loss. However, this effect of erosion is not always considered the most damaging to agricultural production. The most damaging effect of erosion might be that the volume of arable soil progressively decreases, increasing the difficulties for plant roots to develop through harder and poorer soil layers. This process is usually slow, but many consider it quasi irreversible (Morgan, 1995). Modelling of this effect of erosion on yields requires modelling the dynamics of the volume of soils and determination of the progressive effect of insufficient soil volume on yields. Soil loss and associated yield reduction is the type of erosion we modelled and that we present in this paper.

The model presented in this paper was applied at the watershed level, but only to take into account the water problem. The erosion equations that are presented in this paper can also be applied in farm level models. First, we present the model structure and then we briefly present its application to a small watershed in Honduras.

12.2 Model Structure

The objective function of the model maximises aggregate utility of the whole watershed over a multi-year planning horizon. Utility is defined as the discounted value of future monetary net income plus the value of livestock, plus the value of the stock of wood in the forest, plus the opportunity costs of leisure. The discount rate is set to the real interest rate in the area, but can be changed in a sensitivity analysis. Costs include the cash expenses for farm production, capital depreciation and interests paid on loans. The traditional constraints included in mathematical programming models are land, labour and capital. We added water, output market and erosion constraints.

12.2.1 Land Limitation

The first land limitation is the area of the watershed. 'A watershed is the total area above some point on a stream or river that drains to that point' (Quinn *et al.*, 1995). The model presented here should be applied to small watersheds (a few thousand hectares) well connected to the market. The reason is that the model represents a 'price taker', meaning that output prices are exogenous. If the watershed is large or if it is not well connected to the market, the production of the watershed is likely to have an impact on prices, making alternative scenarios less realistic (Taylor & Adelman, 1996).

The second land limitation in the model is the partitioning of the watershed in different land units, identified on the basis of different relevant criteria. In the Honduran model, altitude, slope and farm size were selected as criteria, because they explain a large part of the difference in land uses, productivity and erosion.

Two types of land units were distinguished. First the *altitude/farm* land unit that we call Land Management Unit (LMU). For instance, the Honduran model distinguishes three altitude classes and two groups of farms. Altitude classes are the area above 1100 meters, that below 800 meters and the intermediate area. The two groups of farms are characterised by their size: those with less than 10 ha and those with more. Three altitude classes and two groups of farms produce 6 LMUs. Each LMU can be considered one large farm, and is characterised by an initial stock of socio-economic attributes, such as number of persons, cows and mules, machinery and capital. The attributes of each LMU change endogenously over time. Each LMU is characterised by a specific distance to the main road, leading to different transportation times between LMUs.

The second level of land units adds a slope criterion to the altitude and farm criteria LMUs. We call the resulting units Land Production Units (LPU), since the crop production functions are specified at this level. Classes are slopes of less than 10%, slopes of over 25% and intermediate slopes. Hence, a division in three altitude classes, two farmer groups and three slope classes results in 18 LPUs. Each LPU can have different combinations of land use such as forest, pasture, fallow, crops, and land conservation structures.

12.2.2 Water Limitation

Water is necessary for domestic use, irrigation and livestock. The needs are specified in the model for each group of farms at each altitude. Available spring water is specified for each altitude and access time to streams for each farmer group. In addition, unused water from one sector runs out to the next sector located at a lower altitude. This allows analysis of the effect of increasing water use by one sector on the sector located downstream.

12.2.3 Labour Limitation

A fraction of the watershed population is available as labour force. Birth and mortality rates are treated as exogenous to the model. However, population density remains endogenous to the model, because emigration flows can affect the size of the population. Permanent or temporary migration is limited to a fraction of the population on an annual basis.

The watershed population is distributed per LMU, meaning that a sub-population exists for each *altitude/farm* type unit. If necessary, these sub-populations exchange labour within the watershed, which is paid per day at the official wage. However, commuting between sectors consumes additional time.

12.2.4 Cash Limitation

The model chooses the optimal way to manage cash flow during each period in order to maximise monetary net income over the whole planning period. In the cash balance, any cash not reinvested or consumed at the end of one season is saved for the next season. Monetary net income at the end of the year also determines family monetary consumption in the next year. The model has the option of borrowing money from the bank or intermediaries to buy inputs. Loans with interests have to be repaid before the end of the year.

12.2.5 Market Limitation

The model represents a price taker, meaning that output prices are exogenous. For some crops the model also represents a 'quantity taker', which means that some production levels are limited by an upper bound. Farmers that produce vegetables have special ties with a few traders. If this restriction is not included, the model is likely to produce very special crops for which there is no market.

12.2.6 Crop Production Modelling

The crop production function calculates the average response to different production factors for each LPU. To simplify the presentation, the equations below will not specify the indices for a LPU. The production functions apply to each type of crop and each year of the planning horizon. Total production ($TPROD_{c,t}$) of crop c in period t equals the potential yield ($yield_c$) multiplied by the cropped area ($CROP_{c,t}$), plus the effect of fertilisers ($FERT_{c,d,t}$), minus the effect of erosion ($EROSION_{c,e,t}$), minus the effect of insufficient soil depth ($SOILDEF_{c,f,t}$). The effects of erosion and fertilisation are non-linear and are approximated by linear segments. The indices *d, e* and *f* indicate the limits of the segments on the x-axis. All endogenous variables are given in capitals.

$$TPROD_{c,t} = yield_c \cdot CROP_{c,t} + \sum_{d=1}^{D} effnpk_{d,c} \cdot FERT_{c,d,t} - \sum_{e=1}^{E} efferos_{e,c} \cdot$$

$$EROSION_{c,e,t} - \sum_{f=1}^{F} (effdef_{c,f} \cdot SOILDEF_{c,f,t}) \quad (12.1)$$

Land use change includes expansion ($NEWCROP_{c,t}$) or contraction ($CUTCROP_{c,t}$) of the initial cropped areas ($CROP_{c,t-1}$).

$$CROP_{c,t} = CROP_{c,t-1} + NEWCROP_{c,t} - CUTCROP_{c,t} \quad (12.2)$$

The same type of equation applies to the non-cropped area ($LAND_t$):

$$LAND_t = LAND_{t-1} + NEWLAND_t - CUTLAND_t . \quad (12.3)$$

Because there are only two land uses, the *NEWLAND* variable exactly equals *CUTCROP* and *CUTLAND* equals *NEWCROP*. We kept two land uses in this explanation to keep it more understandable. However, one can add a third or a fourth land use within each LPU. For instance, the model would be more realistic if the variable *LAND*, representing the non-cropped area, would be replaced by two variables, such as forest and pastures.

12.2.7 Erosion

In each LPU, erosion is described as a basic erosion level, minus the effect of conservation structures and fertilisation. The potential erosion level equals the area under a certain type of crop ($CROP_{c,t}$), multiplied by an average volume of eroded soil per hectare ($eros_c$) without conservation structures or fertilisers. The total 'erosion mitigating' effect of conservation structures equals the area of conservation structures ($CONSER_t$) multiplied by the effect of one unit area of conservation structure on the volume of eroded soil ($effcons_c$). The total 'erosion mitigating' effect of fertilisation equals the fertiliser quantity ($FERT_{c,d,t}$) in tons multiplied by the volume of soil not eroded as the result of application of one ton of fertiliser ($effert_{d,c}$).

$$\sum_{e=1}^{E} EROSION_{c,e,t} = eros_c \cdot CROP_{c,t} - effcons_c \cdot CONSER_{c,t} -$$

$$\sum_{d=1}^{D} effert_{d,c} \cdot FERT_{c,d,t} \quad (12.4)$$

In each LPU, two initial volumes of topsoil are distinguished, one below crops ($VOLCROP_{c,t}$) and one below the remaining land (forest, grassland and soil conservation structures, $VOLAND_{c,t}$). To simplify the explanations we do not consider erosion in the non-cropped area. Soil volume under the crop decreases by erosion ($\Sigma_c EROSION_{c,t}$). However, land use areas change between years.

If in year t the crop area returns to non-cropped area ($CUTCROP_{c,t}$) a fraction of the soil volume below crops ($VOLCROP_t$) will be transferred to the volume of soil below the non-cropped area ($VOLAND_{c,t}$). The transferred volume equals the per hectare volume under crops of the previous year ($VOLCROP_{t-1}/\Sigma_c\ CROP_{c,t-1}$) multiplied by the transferred area ($CUTCROP_{c,t}$). Similarly, if new land is brought under crops ($NEWCROP_{c,t}$) that will bring new soil volume to the existing volume of soil under crops ($VOLCROP_t$). This transferred volume equals the per hectare volume of non-cropped land of the previous year ($VOLAND_{t-1}/LAND_{t-1}$) multiplied by the area brought under crops ($NEWCROP_{c,t}$).

$$VOLCROP_t = VOLCROP_{t-1} - \sum_{c=1}^{C}\sum_{e=1}^{E} EROSION_{c,e,t} - (VOLCROP_{t-1} / \sum_{c=1}^{C} CROP_{c,t-1}) \cdot \sum_{c=1}^{C} CUTCROP_{c,t} + (VOLAND_{t-1} / LAND_{t-1}) \cdot \sum_{c=1}^{C} NEWCROP_{c,t} \tag{12.5}$$

We follow the same procedure for 'other uses'. If farmers plant crops on other land ($CUTLAND_t$) the new area carries the volume of soil to the 'other land use' ($VOLAND_{t-1}/LAND_{t-1}$).

$$VOLAND_t = VOLAND_{t-1} - (VOLAND_{t-1} / LAND_{t-1}) \cdot CUTLAND_t + (VOLCROP_{t-1} / CROP_{t-1}) \cdot NEWLAND_t \tag{12.6}$$

Within each LPU, the two volumes of soil, the one below crops and the one below 'other uses' evolve over time, depending on of the management chosen by the model.

If the soil volume under crops ($VOLCROP_t$) decreases below a threshold value per hectare ($soilvol_f$), a deficit will appear ($SOILDEF_{c,f,t}$). An index d allows partitioning of the variable $SOILDEF$ in different partial volumes having a different impact on yield. For instance, when the deficit is small the effect on yield will be small.

$$VOLCROP_t \geq soilvol_f \cdot \sum_{c=1}^{C} CROP_{c,t} + \sum_{c=1}^{C} SOILDEF_{c,d,t} \tag{12.7}$$

Why is the deficit ($SOILDEF_{c,f,t}$) crop-specific? The reason is that the production function is crop-specific and that the deficit ($SOILDEF_{c,f,t}$) affects each crop differently. However, in the preceding equation, $SOILDEF$ is summed over c, because the soil deficit for the LPU is not crop specific. Without further specification, the model will 'allocate' the deficit to the least productive crops. For instance, a less productive crop, such as sorghum, will be cultivated only to take on the deficit $SOILDEF_{'sorghum',d,t}$, while a more productive crop, such as maize, would not take on any deficit ($SOILDEF_{'maize',d,t}$) and thus maize production would not be affected even if the soil volume ($VOLCROP_t$) in this land unit is small. To avoid this allocation, we added the following equation in which *soilfix* represents a maximum deficit per hectare for any crop:

$$SOILDEF_{c,d,t} \leq soilfix \cdot CROP_{c,d,t} \,. \tag{12.8}$$

As explained in the production function, the model can build conservation structures ($CONSER_{c,t}$) to control erosion. Existing soil conservation structures deteriorate (*depline*) due to climatic factors, can be removed by farmers ($CUTCONS_{c,t}$) and may be maintained or extended by farmers ($NEWCONS_{c,t}$):

$$CONSER_{c,t} = depline \cdot CONSER_{c,t-1} - CUTCONS_{c,t} + NEWCONS_{c,t} \,. \tag{12.9}$$

The conservation structure conserves an area not exceeding the cropped area:

$$CONSER_{c,t} \leq CROP_{c,t} \,. \tag{12.10}$$

12.3 Application to a Small Watershed in Honduras

The model has been applied to a small watershed called La Lima in the hillsides of Central Honduras. More details about this study can be found in Barbier & Bergeron (1999). The objective was to compare different development scenarios. First, the model was validated on the basis of the real land use from 1975 to 1995. Simulated and observed land uses were similar. Secondly, we ran the model recursively until 2075 to examine the effect of increasing population pressure and erosion on the watershed in the long term. Erosion values were derived from the biophysical model EPIC (Williams, 1994). In this paper we focus on the results related to erosion problems.

Figure 12.1 shows the changes in soil volume over time in one of the 18 land units of the watershed. The soil volume under crops decreases rapidly because of erosion. Soil volume in the non-cropped area decreases slowly. The reason for the reduction in soil volume in the non-cropped areas is not erosion, as that was assumed absent. The soil loss is due to the transfer of land from the cropped area to the non-cropped area, as the cropped area gradually decreases. In other words, each hectare withdrawn from cropping transfers a volume of soil to the non-cropped area. Since this volume is smaller than the average volume per hectare of land in the non-cropped area, the average volume in the non-cropped area decreases.

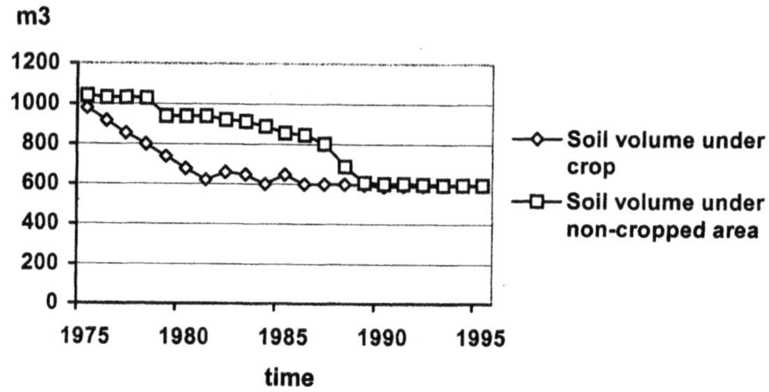

Figure 12.1: Evolution of the simulated arable soil volume in one land unit.

After 1989, the soil volume remains constant under both land uses, because in that year the model built a conservation structure in the cropped area, resulting in reduced soil loss. The model reproduces the reality that farmers return eroded plots to fallow when yield loss becomes excessive. When the whole area has reached a minimum soil volume, farmers start to construct conservation structures.

The results illustrated in Fig. 12.1 refer to the LPU with the steeper slope; the results for the whole watershed were more favourable. The results confirmed that the current cost of erosion in the watershed is low, because most soils are deep and application of fertilisers reduces erosion and compensates for the loss of nutrients. This low cost explains the absence of land conservation structures in the watershed.

However, the model predicts that erosion might become a problem in the future. As population increases, farmers might expand cropping on the slopes, and erosion is likely to intensify. When soils will become shallow, yields will be reduced and farmers will have to return some land units to pasture and later they will have to invest in land conservation structures.

Contrary to conventional wisdom, the intensification process in the watershed does not reduce erosion. Conventional wisdom says that expansion of labour-intensive crops, such as vegetables, usually reduces the area of other less labour-intensive crops, such as maize. In La Lima, however, the maize area did not decrease because it has become a valuable cash crop. The continued cultivation of maize and the expansion of vegetables during the rainy season in fact have increased total erosion.

Results also suggest that the benefits of land conservation structures will remain small for a few decades, because the opportunity cost of labour will remain high. The model predicts that some investments will occur during low activity periods and in areas where the opportunity cost of labour is low. Irrigation possibilities that result in high labour costs throughout the year, are a strong constraint to the construction of land conservation structures.

The results of the model confirm that the relation between population pressure and soil quality is U-shaped. Areas with low or high population densities are likely to have less erosion, areas with medium population density more. In low population density areas, farmers tend to crop less erodible soils. When population increases, farmers cultivate steeper land, that is more erodible. When the soil becomes too shallow, farmers tend to reclaim new land instead of investing in land conservation structures. Only when reclaiming new land becomes more costly, farmers start to adopt conservation practices.

12.4 Discussion and Conclusion

The model presented in this paper is an original method to incorporate erosion in mathematical programming. The model describes the decreasing volume of soil within one land unit until it reaches some threshold below which yields decline more rapidly. The model more realistically simulates the problem of sustainability than the usual methods that do not include the relation between yields and remaining soil volume. The principle of the model can be used for other natural resources. The method has been applied to soil organic matter (Barbier, 1998) and soil nutrients (Barbier & Hazell, 1999).

There are some limitations to the modelling approach though. First, the LPUs are a rough aggregation of dozens or hundreds of plots of various sizes and slopes. The resulting bias can be reduced by increasing both the number of LPUs and the criteria to select the LPUs. For instance, the criterion soil type was missing from our analysis. The main limitation for increasing the number of LPUs is computer capacity. Hence, in due course it will become easier to add more LPUs.

Another limitation of the approach is the maximisation of the utility of the whole watershed. Furthermore, comparing the watershed to one large farm and considering that the factor market between farms is functioning properly might be considered far-fetching assumptions, especially in Latin America, where relations between large land owners and small farmers are often conflictive. However, we observed that the results of the model were close to reality. Even if the labour, land and capital markets were not perfect, they were functioning properly.

12.5 References

Barbier, E. (1988): The economy of farm level adoption of soil conservation measures in upland Java. World Bank. Environment Division Working Paper 11

Barbier, B. (1998): Induced innovation and land degradation: results from a bioeconomic model of a village in West Africa. Agricultural Economics 19, 15-25

Barbier, B. & G. Bergeron (1999): Impact of policy interventions on land management in Honduras: results of a bioeconomic model. Agricultural Systems no 60, 1-6

Barbier, B. & P. Hazell (1999): Implications of Population Growth and Declining Access to Transhumant Grazing Areas for the Sustainability of Agropastoral Systems in the Semi-Arid Areas of Niger. IFPRI/ILRI proceedings of the international Symposium on 'property rights, risk and livestock development' held in Feldafing, Germany in September 1998

Boussard, J.M. & J.-J. Daudin (1988): La programmation linéaire dans les modèles de production. Masson, Paris

Estrada, R.D., O. Chaparro & B. Rivera (1999): Utilización de los modelos de simulación para evaluación ex-ante. Instrumentos metológicos para la toma de decisiones en el manejo de los recursos naturales. Centro Internacional de Agricultura Tropical, Cali

Flanagan, D.C. & M.A. Nearing (1995): USDA - Water Erosion Prediction Project: Hillslope profile and watershed model documentation. NSERL Rep. 10, USDA-ARS-NSERL, West Lafayette

Hazell, P. & R. Norton (1986): Mathematical programming for economic analysis in agriculture. Mc Millan, New York

Kruseman, G., R. Ruben, H. Hengsdijk & M.K. van Ittersum (1995): Farm household modelling for estimating the effectiveness of price instruments in land use policy. Netherlands Journal of Agricultural Science 43, 111-123

McCarl, B.A. & T.H. Spreen (1997): Applied mathematical programming using algebraic systems. Texas A&M. ttp://agrinet.tamu.edu/mccarl/agec642/newb.pdf

Morgan, R.P.C. (1995): Soil erosion and conservation. Second edition. Cranfield University.

Parikh, K.S. (1991): An operational definition of sustainable development. Indira Gandhi Institute of Development Research, Bombay

Quinn, R.M, K.N. Brooks, P.F. Ffolliott, H.M. Gregersen & A.L. Lundgren (1995): Reducing resource degradation: Designing policy for effective management. Working Paper no. 22. University of Minnesota, Department of Forest Resources.

Taylor, E. & I. Adelman (1996): Village economies. Cambridge University Press, New York

White, D.C. (1988): The economics of sediment management using an aggregate model of a large agricultural watershed. PhD dissertation. University of Illinois at Urbana-Champaign

Williams, J.R. (1994): The EPIC model. Temple, Texas, U.S.A.: U.S. Department of Agriculture, Agricultural Research Service, Economics Research Service, and Soil Conservation Service

Wischmeier, W.H. & D.D. Smith (1978): Predicting rainfall erosion: A guide to conservation planning. Agriculture Handbook No. 537. U.S. Department of Agriculture, Washington D.C.

13 Alternative Approaches to the Economics of Soil Nutrient Depletion in Costa Rica: Exploratory, Predictive and Normative Bio-Economic Models

Hans G.P. Jansen[1], Robert A. Schipper[2], Peter Roebeling[2], Erwin H. Bulte[3], Huib Hengsdijk[4], Bas A.M. Bouman[5] & André Nieuwenhuyse[6]

[1] Agricultural Economics Research Institute (LEI), Wageningen University and Research centre, The Hague, the Netherlands

[2] Development Economics Group, Department of Economics and Management, Wageningen University and Research centre, Wageningen, the Netherlands

[3] Department of Economics, Tilburg University, Tilburg, the Netherlands

[4] Laboratory of Theoretical Production Ecology, Department of Crop Science, Wageningen University and Research centre, Wageningen, the Netherlands

[5] International Rice Research Institute (IRRI), Makati City, Philippines

[6] ZONISIG project, La Paz, Bolivia

Abstract

This chapter discusses three methodologies to analyse land use with respect to income generation and sustainability in the Atlantic Zone of Costa Rica: the explorative SOLUS methodology, the predictive UNA-DLV methodology, and a normative optimal control approach. Each methodology may be classified as bio-economic; the first two are single period optimisation models at the regional and farm level, respectively, while the third is a dynamic model at the plot level. The SOLUS and UNA-DLV methodologies incorporate a multi-market structure for commodities produced by a wide spectrum of land use systems with specified technologies. For each of these systems, technical coefficients are generated related to discounted benefits and costs, and to sustainability indicators, including soil nutrient depletion. The optimal control model is confined to a pasture plot with cattle, property of a single farmer, and analyses the inter-temporal trade-off between soil nitrogen depletion and net benefits. Apart from assessing the 'economic benefit - sustainability' trade-off, all methodologies are capable of analysing the effects of technological change. After providing some example results, the chapter concludes with a comparison of the relative merits of each methodology.

13.1 Introduction

Degradation of soil resources is an important issue in many developing countries and associated irreversible productivity losses represent an increasing threat to agricultural production (Blaikie & Brookfield, 1987; Thampapillai & Anderson, 1994). In this chapter three methods are discussed to address this issue: the SOLUS methodology, the UNA-DLV methodology, and an optimal control approach[1]. The nature of these methods can be classified as *explorative, predictive* and *normative*, respectively (Table 13.1). In the next section each method is elaborated. Next, the region of application, the Atlantic Zone (AZ) of Costa Rica will be introduced, followed by the presentation of a few example results. The chapter ends with some conclusions.

13.2 Methods

13.2.1 The SOLUS Methodology

The *explorative* SOLUS (Sustainable Options for Land USe) methodology (Fig. 13.1) makes use of a bio-economic agricultural sector model which allows exploration of the aggregate effects of certain policy measures (*e.g.* the introduction of new technologies) on both efficiency and non-economic and (environmental) sustainability objectives at the regional level, both with a relatively long time-horizon (20-30 years; Bouman *et al.*, 1998b; 1999a).

The heart of the SOLUS methodology is REALM (Regional Economic and Agricultural Land use Model; Fig. 13.1), an LP agricultural sector model. REALM selects the optimal combination of production systems by maximising the economic surplus generated by the agricultural sector in the AZ.

---

[1] These methodologies were developed and applied within the Research Program on Sustainability in Agriculture (REPOSA) in Costa Rica, in collaboration with the UNA-DLV program (1996-1998) of the *Universidad Nacional* (National University of Costa Rica) and DLV (*Duurzaam Landgebruik en Voedselvoorziening*, Sustainable Land Use and Food Security) research program of Wageningen Agricultural University and the Research Department of the Ministry of Agriculture, Nature Management and Fisheries (DLO-LNV), Wageningen, the Netherlands. The REPOSA program (1986-1999) was an interdisciplinary research and education program focusing on land use analysis at different scales. The main collaborating partners included Wageningen Agricultural University, the Tropical Agricultural Research and Higher Education Center (CATIE; *Centro Agronómico Tropical de Investigación y Enseñanza*) and the Costa Rican Ministry of Agriculture and Livestock (MAG; *Ministerio de Agricultura y Ganadería*).

Table 13.1: Methodologies for land use analysis at different scales.

Nature of methodology (name)	Scale	Principal objective	Characterisation and methods	Key outputs
Explorative (SOLUS)	Region	Exploration of technological and policy options for land use under combined biophysical and socio-economic constraints	Bio-economic (expert systems, LP, econometrics, GIS)	Technological options & trade-off analysis & aggregate policy effects
Predictive (UNA-DLV)	Farm & region	Short-term prediction of the effects of agricultural policies on farmers' land use decisions	Bio-economic (expert systems, LP, econometrics, GIS)	Technological options & farmers' reactions & (partial and aggregate) policy effectiveness
Normative	Plot	Computation of optimal soil nutrient stocks in the long run	Bio-economic (optimal control theory)	Privately and socially optimal soil nutrient management strategies

Information on different production systems (technologies) is generated by two so-called Technical Coefficient Generators (TCGs)[2]. These TCGs are essentially expert systems that quantify input-output coefficients of a large number of land use systems (or agricultural production technologies) at the plot level. In order to capture the economic surplus, REALM is formulated with a multi-market structure (Sadoulet & De Janvry, 1995). Depending on the type of market, the economic surplus consists either of the sum of the consumer and producer surpluses (domestic markets) or of the producer surplus only (export markets). By maximising the sum of these economic surpluses, demand and supply in all markets are equilibrated (Takayama & Judge, 1971).

Below, the major features of REALM as part of the SOLUS methodology are explained. The REALM model itself, including the complete formulation in GAMS (Brooke *et al.*, 1992) on a CD-ROM, can be found in Schipper *et al.* (2000).

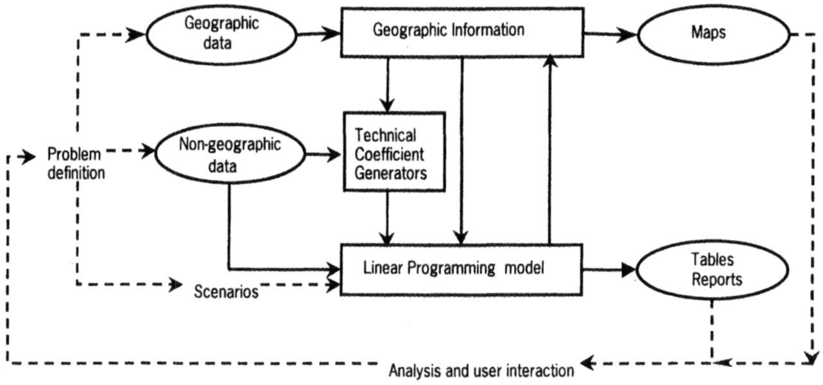

Figure 13.1: Structure of the SOLUS methodology. Rectangles represent models/tools; ovals are data; names without boxes are activities; solid lines are flows of data; dotted lines are flows of information.

Supply from the AZ competes with supply from other regions within Costa Rica (domestic markets) and with supply from other countries (export markets). For a number of products (*i.e.* banana, palm heart and plantain), the production of the AZ constitutes a significant part of the national and/or world supply. Therefore, prices of these products are likely to be influenced by the supply originating from the AZ, *i.e.* to be endogenous. Hazell and Norton (1986) show a method to

[2] PASTOR (Pasture and Animal System Technical coefficient generatOR; Bouman *et al.*, 1998a) and LUCTOR (Land Use Crop Technical coefficient generatOR; Hengsdijk *et al.*, 1998); for a general discussion of TCGs, see Kruseman & Van Keulen, this volume.

incorporate variable prices in LP models by linearising demand functions around an observed quantity and price. In a regional model, demand functions faced by producers rarely coincide with national demand functions. Product demand may be met by products purchased in other regions, while demand for local products may also arise in other regions. Kutcher (1972; 1983) therefore, argues that the demand function faced by the local producer is determined by the national demand function as well as by supply functions from other regions of the country, and depends on the relative importance of local production in total national production. The regional demand elasticity η_r can be expressed as follows (Hazell & Norton, 1986):

$$\eta_r = \eta \frac{1}{K} - \sigma_{nr} \frac{1-K}{K} \qquad (13.1)$$

where η represents the national demand elasticity, K is the share of region r in national production, and σ_{nr} the supply elasticity from other regions. Equation (13.1) shows that the demand elasticity faced by the local producer approaches the national demand elasticity when local production approximates national production, while the demand elasticity faced by the local producer approaches infinity when local production forms a relatively insignificant part of total national production. Geurts *et al.* (1997) and van der Valk (1999) estimated national own-price elasticities of demand. Cross price elasticities of demand were assumed to be zero, while elasticities of supply from other regions were set at 0.7, in line with Mamingi (1997) and Sadoulet & De Janvry (1995).

In REALM, product prices depend on geographical location within the AZ due to variation in the distance to markets and the quality of the roads. This variation in product prices was addressed by dividing the AZ into sub-regions, each with its own specific transport costs calculated on the basis of a regression model that relates unit transportation costs to geographical distance and road quality (Jansen & Stoorvogel, 1998), to the most relevant market (depending on the type of product and final destination). The sub-regions are the result of a GIS-overlay of three zonification maps based on equal transport costs (Bouman *et al.*, 1998b). The first map depicts the transport costs of agricultural products to the road in the Southwest of the AZ, where the products leave the AZ en route to the domestic market. The second map refers to livestock products shipped to the same destination. The third map refers to the transport of export products to Limón harbour in the Southeast of the AZ. To keep the size of the model within limits, while still distinguishing meaningful transport zones, 12 'iso-transport cost' sub-regions were delineated, as shown in Fig. 13.2(b). In the LP model, farm-gate prices were calculated per sub-region by subtracting the transport costs from the product prices in the respective market outlets of the AZ. No significant spatial variation was found of prices for agricultural inputs (*e.g.* seed, fertiliser and biocides).

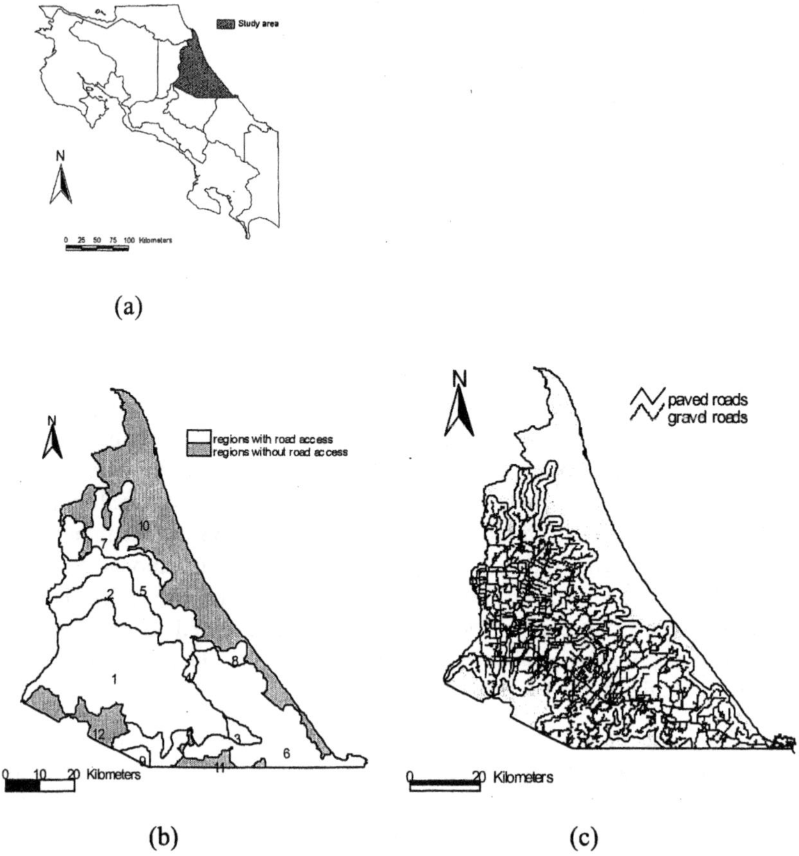

Figure 13.2: Maps of location of the northern part of the Atlantic Zone in Costa Rica (a), 12 sub-zones (b) and road infrastructure (c).

All production originating from the AZ competes for limited land and labour resources. Land availability in each of the 12 sub-regions of the AZ (Fig. 13.2) is specified for three soil groups (see Section 13.3), using map overlaying in GIS. Soil data were available from Stoorvogel & Eppink (1995) and Nieuwenhuyse (1996).

To introduce competition for labour between the agricultural sector in the AZ and other sectors and regions, potential labour availability in each sub-region of the AZ is divided into three components: the existing agricultural labour force in that sub-region, the existing agricultural labour force in other sub-regions of the AZ, and labour from outside the agricultural sector within the AZ or from outside the AZ. In each sub-region, the first labour component is the cheapest, working at a fixed wage without transaction costs. Additional labour can be attracted from other sub-regions, albeit at certain additional (transaction) costs, which depend on their relative location. Finally, labour in excess of the total regional labour pool can be attracted from elsewhere, but at higher wages according to an upward sloping labour supply function. Similarly to agricultural products, the regional labour supply elasticity, ε_r, for each sector/region r is given by:

$$\varepsilon_r = \varepsilon \frac{1}{M} - \theta_{nr} \frac{1 - M}{M} \tag{13.2}$$

where ε is the national labour supply elasticity, θ_{nr} the labour demand elasticity in the remainder of the economy, and M the share of agricultural labour in region r in the national labour market. A labour supply elasticity of 0.2 has been assumed at the national level, in line with other studies (Bosworth *et al.*, 1996). With regard to the labour demand elasticity in the remainder of the economy, -0.5 would be a plausible approximation (Bosworth *et al.*, 1996). Using Equation (13.2) and a labour share of 0.05, an ε_r of 13.5 for the labour supply elasticity in the AZ agricultural sector would thus be a reasonable approximation. Such a high elasticity implies a very gently upward sloping labour supply function, if more labour would be required than presently available (*i.e.* after the horizontal part of the labour supply function).

REALM is a one-year model, but production systems in the TCGs are defined for different periods of time. Therefore, for production systems with a life span exceeding one year, annuity costs and benefits were calculated using the capital recovery factor and a discount rate[3]. Assuming constant prices over time,[4] discounting occurred in quantitative rather than monetary terms, *i.e.* was applied, for example, to production (in kg or other physical units). After calculating such 'present quantities', the latter were multiplied by their respective prices to obtain present values.

Land use activities include actual and alternative land use systems, which were generated using two TCGs (see above) for 13 land utilisation types: eight crops (banana, black beans, cassava, maize, palm heart, pineapple, plantain and natural forest for sustainable timber extraction) and five pasture types (three fertilised improved grasslands, a grass-legume mixture, and a mixture of natural(ised) grasses). These land utilisation types were combined with three major land units identified in the northern AZ (see Section 13.3), each subdivided into mechanisable

[3] On the basis of Zuñiga (1996), the discount rate was set to 7% which is considered to be a reasonable approximation of the opportunity cost of capital under the conditions that exist in Costa Rica in the late 1990s.

[4] Constant prices normally are a basic assumption of the linear programming models used in land use analysis. However, in models where one cannot assume an infinite demand elasticity for a product (as in REALM), one implicitly has to assume constant prices over time, even though these prices are determined endogenously by the model.

and non-mechanisable sub-units. Actual land use systems are quantified based on data derived from descriptions of the current best farm practices in the AZ, while alternative systems were generated using a target-oriented approach (see Kruseman & van Keulen, this volume), under the condition that soil nutrient balances must be kept in equilibrium (no changes in soil nutrient stocks for N, P and K). For alternative crop systems, different technology levels were generated by combining levels of mechanisation, fertiliser use, and crop protection. Weeding, fertilisation levels and stocking rate determine pasture technology, apart from pasture type. A total of 1352 crop land use systems and 1756 pasture land use systems were generated. Activities were distinguished for cattle breeding and cattle fattening, each at four animal growth rates and associated feed requirements. Finally, five feed supplements were defined.

Technical coefficients were generated on a 'per ha' basis and include, next to sustainability indicators (see below), yields, labour requirements and costs of inputs. Technical coefficients are either averages per year (labour use) or annuities of the present value over the life span of the land use systems (yield and input costs). All systems are characterised by specific quantitative combinations of physical inputs and outputs, thus representing fixed input-output technologies.

For analysing sustainability and environmental scenarios, the model allows for specific restrictions regarding eight sustainability and environmental parameters. Based on the relevant sustainability issues in the AZ, the following technical coefficients of biophysical sustainability were implemented: the depletion of soil nutrient stock (*i.e.* a negative change in stock) for nitrogen (N), phosphorus (P) and potassium (K); N losses to the environment; volatilisation and (de)nitrification of N as proxy for greenhouse gas emissions; and the use of biocides, expressed as the total amount of active ingredients (BIOA) and by an ordinal biocide index (BIOI; Jansen *et al.*, 1995) that takes into account not only the active ingredients used but also their degree of toxicity and their persistence in the environment. For each sustainability parameter, restrictions can be formulated on its total value. Summation over the relevant land use systems can be implemented at the land unit level, the sub-regional level or the level of the AZ. The REALM model can then analyse the effects of such restrictions on the economic surplus.

13.2.2 The UNA-DLV Methodology

The UNA-DLV *predictive* methodology incorporates farm(er) behaviour and can be used to analyse ('predict') the likely short-term (say, less than 5 years) effects of policy measures (including technical change) on farmers' land use decisions. The methodology is based on individual farm models for four representative farm types in the AZ of Costa Rica, though it can also be used for regional analysis of farmers' policy reactions, since the results of the individual farm type models are linked with product market equilibrium conditions (Roebeling *et al.*, 2000; Fig. 13.3). Objective functions and resource endowments differ by farm type, while technical coefficients of different crop and livestock production systems and technologies are generated by the two expert systems also used in SOLUS (Section 13.2.1). These coefficients are incorporated in separate farm models of a bio-economic type.

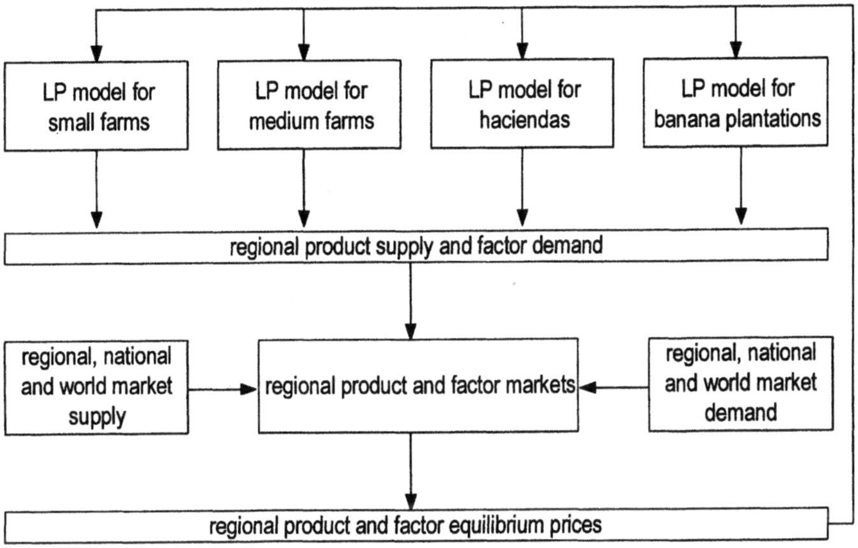

Figure 13.3: Multiple farm structure of the UNA-DLV projective land use analysis methodology.

In the context of the UNA-DLV methodology, models were constructed for the following representative types of producers in the AZ of Costa Rica: households with small farms (<20 ha), households with medium farms (20-50 ha), extensive beef cattle farms or *haciendas* (>50 ha), and banana plantations (>200 ha) (Table 13.2). These representative farm types were identified according to dominant land use and perceived objectives, based on the latest agricultural census (DGEC, 1987).

Models for small and medium farms are based on the farm household modelling approach described by Singh *et al.* (1986). They link a LP optimisation procedure of the production structure to an econometrically specified expenditure module (Roebeling *et al.*, 2000). First, long- and medium-term farm investment decisions are made on the basis of expected prices to maximise *net farm income* (see below), subject to credit constraints, production technologies, consumption requirements, and initial farm characteristics and resource availability.

Table 13.2: Average farm resource availability.

Farm type	Farm area (ha)	Labour (days)	Savings (US$)	Cattle (AU*)	Number of farms
Small	8.9	491.9	281.8	4.1	6480
Medium	39.2	412.7	773.5	48.3	1690
Hacienda	170.4	570.0	5524.9	188.1	803
Banana	226.3	-	-	-	83

-: not restricted;
*: and Animal Unit is a hypothetical animal of 400 kg liveweight, fed at maintenance;
Source: Roebeling *et al.*, 2000.

Subsequently, short-term production and consumption decisions are taken – in a non-separable model – on the basis of actual prices, maximising *utility* (see below), subject to net farm income and farm resource availability. The UNA-DLV methodology acknowledges the potentially important role of credit for the adoption of soil conservation practices, by assuming that formal credit (12 % interest rate) is limited to a fraction of the farm household's land value (25%) and value of the cattle stock (20%), while short-term informal credit (47% interest rate) is limited to a fraction of its crop production value (10%). Farm households allocate their available labour between off-farm (wage labour) and on-farm activities, with the latter determined by crop and livestock production activities selected by the models.

Optimisation takes place for income and utility objectives. Utility *UTIL*, is obtained through the consumption of on-farm production as well as leisure (Q_j^{cons}), and purchased products (Q_j^{buy}). Utility functions[5] are derived from the National Household Income and Expenditure Survey (DGEC, 1990), using a negative exponential utility function for basic food crops and an exponential utility function for other food products, non-food products and leisure. The objective utility function, Z_1, is given by:

$$Z_1 = UTIL = \sum_j U_j^{max}\left(1 - e^{-\rho_j\left(Q_j^{cons} + Q_j^{buy}\right)}\right) + \sum_j \rho_j \left(Q_j^{cons} + Q_j^{buy}\right)^{\sigma_j} \tag{13.3}$$

where U_j^{max} denotes the maximum attainable utility with commodity j, ρ_j is the conversion factor of consumption to utility, and σ_j is the exponent of consumption commodity j. The net farm income objective, *NFI*, is defined as the returns from marketed production (Q_j^{sold}) and off-farm employment (O^{off}), net of the costs

[5] Direct utility functions are used, applying the methodology proposed by Kruseman *et al.* (1997), with estimation of utility functions on the basis of Engel curves.

related to the use of fixed and variable inputs (I_i), capital resources (C_b) as well as consumption (Q_j^{cons}), corrected for the expected monetary value of nutrient losses (B_n):

$$Z_2 = NFI = \sum_j p_j Q_j^{sold} + w^{off} O^{off} - \sum_i p_i I_i - \sum_b p_b C_b - \sum_j p_j (Q_j^{cons} + Q_j^{buy}) - \sum_n p_n B_n \tag{13.4}$$

where w^{off} is the off-farm wage rate, and p refers to the prices related to commodities j, inputs i, capital sources b, and the reservation price related to nutrients n. The production and consumption components of the model are calculated separately, and final model results are obtained by their weighted sum, with weights calibrated to reflect actual land use. Transaction costs are taken into account to determine farm-gate prices (Roebeling *et al.*, 2000).

The dynamic LP model for *haciendas* evaluates technologies for pasture-based beef cattle production, according to an income objective (see below), and subject to resource and cash flow constraints. Dynamic properties include multi-period livestock activities, and a savings and investment module within the 10-year planning horizon (Roebeling *et al.*, 1998). Restrictions are defined by the *hacienda*'s initial resource availability (Table 13.2), while the availability of resources can change over time due to investment in land and/or cattle. Formal credit (10% interest rate) is limited to a proportion (25%) of the land and cattle value of the previous year.

Economic decision-making on extensive *haciendas* is guided by a long-term income objective, consisting of the net returns from beef production and the terminal value of the land (Roebeling *et al.*, 1998). The latter refers to the returns derived from the ownership of assets in the long run, obtained over and above the returns obtained from the productive use of these assets. According to van Hijfte (1989), land is the major asset and considered the most important long-term investment (I_s^{land}) for *hacienda* owners. Net returns refer to the present value of beef production (Q_j), net of expenditures on variable inputs (Q_i), investments in cattle (I_j^{cattle}) and land (I_s^{land}), capital costs (C_b), and tax payments (τ_r) over the *hacienda*'s resource value (R_r). The present value of the long-term income objective (LY) over the 10-year (y) planning period is given by:

$$LY = \sum_y e^{-iy} \left\{ \begin{array}{l} \sum_s p_{s,y=10} I_{sy}^{land} + \sum_j p_j Q_{jy} - \sum_i p_i Q_{iy} - \sum_b p_b C_{by} - \sum_j p_j I_{jy}^{cattle} - \\ \sum_s p_{sy} I_{sy}^{land} - \sum_r \tau_r R_{ry} \end{array} \right\} \tag{13.5}$$

where i is the time discount rate, and p represents prices related to animal classes j, inputs i, capital sources b, and land units s. Tax levies (τ_r) are differentiated for the resources (r) land and cattle. The *hacienda* model makes use of expected market prices, taking transaction costs into account.

The banana plantation model is a linear programming model that analyses the production side to determine technology choice. Banana production technologies are evaluated according to a profit objective, subject to availability of certain resources. The availability of land is restricted, while capital availability is assumed to be unlimited (see Table 13.2). Labour and inputs can be obtained at fixed market prices. Banana plantations are guided by a profit objective (Roebeling *et al.*, 2000), with profit defined as the returns from banana export, $Q_{j=ban}$, sold at world market prices, $p_{j=ban}$, net of the costs related to the use of fixed and variable inputs, I_i, and invested capital, $C_{b=own}$. Mathematically, the profit objective (*PROFIT*) can be expressed as:

$$PROFIT = \sum_{j=ban} p_j Q_j - \{\sum_i p_i I_i + \sum_b p_b C_b\} \tag{13.6}$$

where p_i is the price of fixed and variable inputs i, and p_b the price related to the use of operating capital sources b. Model optimisation takes place using actual prices, while accounting for transaction costs in determining prices at the farm gate level.

Even though single farm decisions do not affect product and factor prices, total product supply and factor demand of all farms in the region may affect equilibrium prices within the region. This implies that prices, while exogenous at the individual farm level, may become endogenous at the regional level. In the UNA-DLV methodology, partial model results for the different farm types are used to determine total regional product supply and factor demand through weighted aggregation, based on the number of farms per farm type (see Table 13.2). In the application for the AZ of Costa Rica, such aggregation takes place on the output side, to determine market-clearing prices. Regional product supply is confronted with the product demand function faced by producers in the region (see Section 13.2.1) to determine regional equilibrium prices for each product. These newly determined equilibrium prices form the input of subsequent partial model runs for each farm type. This procedure is repeated until product prices deviate less than 1% from corresponding prices determined in the previous iteration.

13.2.3 An Optimal Control Model

While the SOLUS and the UNA-DLV methodologies allow quantification of trade-offs between socio-economic and biophysical objectives at the regional level, they both apply static, single-period LP models. These LP models are supported by land use options as calculated by TCGs. Such land use options should ideally represent equilibrium conditions and indeed the alternative land use options with 'zero soil nutrient balances' represent such conditions. On the other hand, the actual land use options that are also part of the TCGs are basically non-stable systems whose input-output combinations are subject to change over time. In general, soil nutrient balances in actual land use systems are negative, implying that the soil is being 'mined'. This suggests a need for dynamic models keeping track of stocks over time to fully understand the developments in the study region.

The basic LP model can be extended somewhat to introduce a time component in the conceptual analysis (see *e.g.* Bunce, 1942; Walker, 1982; McConnell, 1983), although it should be noted that relatively few examples exist of such dynamic analyses for developing countries. For the AZ of Costa Rica, a time component involving declining yields of natural pastures has been simulated in a recursive sequence of running a static LP model with decreasing pasture yields in a step-wise manner (Bouman & Nieuwenhuyse, 1999). Their results suggest absence of a short-term incentive for farmers in the AZ to replace soil N mining on natural pastures by more sustainable systems. Recursive programming methods have also been used by Barbier (1995) in a study of long-term land degradation in West Africa which paints a gloomy picture of irreversible land degradation and decreasing incomes. Kuiper (1997), in a study for one settlement in the AZ of Costa Rica, found that multi-period programming models in which the magnitude of soil nutrient depletion is used to reduce productivity levels of land use systems in the next period, have a tendency to use sustainable and unsustainable (*i.e.* soil mining) technologies in earlier and later periods, respectively. Generally, these results suggest that profit-maximising farmers have no incentive to replace current soil mining practices by more sustainable systems. However, true dynamic modelling of soil nutrient depletion requires a dynamic (rather than static) framework, as offered by optimal control theory.

In line with the arguments above, suppose that *current* profits may be increased if farmers decide to 'mine the soil', *i.e.* extract more nutrients from the soil than are added. This implies that nutrient stocks decline and that future productivity is jeopardised. The farmer thus faces a trade-off between current production (profits) and future production (profits). His optimal choice will depend on a number of considerations, including expected future prices of output, fertiliser prices and the discount rate. McConnell (1983) has demonstrated that, under certain conditions, farmers find it in their interest to mine and deplete the soil (and indeed, that from an economist's perspective, soil mining may be socially optimal as it may maximise the present value of all future net benefits). The relevant trade-off that needs to be studied is therefore between the marginal benefit of current nutrient extraction and the marginal cost of extraction as measured by foregone future production possibilities. This is exactly what a normative optimal control model does[6].

---

[6] See Leonard & Van Long (1992) for an extensive discussion of the virtues, shortcomings and techniques of optimal control.

The optimal control approach explicitly focuses on balancing current and future productivity. Rather than observing that (one–period) profit maximising behaviour may conflict with the desire to keep N stocks constant, we now model optimal choices of farmers taking into account that current soil mining negatively impacts on future productivity. A central issue of dynamic models is the concept of user cost, which is defined as the loss in future benefit resulting from the extraction of a natural resource in the current period. However, focusing on the inter-temporal trade-off comes at a cost. Specifically, the land use systems (technologies) that are considered should not be too complex, and in particular the number of stocks ('state variables') should be kept to a minimum to ensure tractability.

Bulte *et al.* (2000) deal with the relatively simple example of pasture grazing, treating cattle as a flow ('control') variable. They use an optimal control model to analyse the behaviour of cattle ranchers in the AZ of Costa Rica where pasture production is predominantly governed by the availability of N. A key feature in their – plot level[7] - model is the size of the N stock and its development over time. How much should a rational rancher invest in maintaining the N stock, if the goal is to maximise discounted profits, assuming that the rancher is free to choose fertiliser, feed supplements, and stocking rate? In what follows, we simplify their model to retain maximum tractability (but see the original paper for details). The objective function is defined as:

$$Max \int_{0}^{\infty} \pi(t) e^{-rt} dt \tag{13.7}$$

where π(t) is profit, which is a function of (a) fattened animals bought and sold; (b) the exogenous price per animal unit; (c) production and maintenance costs (*e.g.* disease control); (d) feed supplements provided to the cattle stock and the price of these supplements; (e) N fertiliser applied and the price per unit of fertiliser; *r* is the discount rate.

Production takes place subject to a number of constraints, of which we only consider the most important one here. The main dynamic constraint is imposed by the equation–of–motion of the N pool (referred to as N in what follows). The dynamics of N are defined as:

$$\frac{dN}{dt} = R + A - G - L \tag{13.8}$$

where *R* are exogenous additions to the N pool (*e.g.* through rainfall and N fixation); A represents additions to the pool resulting from fertiliser or excreta

[7] Bulte *et al.* (2000) assume a plot with specific soil characteristics, including the initial N stock, and with a specific pasture type, managed by a single farmer; their model exhibits constant returns to scale regarding the size of the plot.

derived from feed supplements; G is net loss of N through grazing (*i.e.* uptake minus net return via faeces and urine); and L is the proportional loss of N from the system as a result of leaching, denitrification, volatilisation and other processes. The case considered is that of a typical beef-fattening farmer, growing a mixture of natural grass species (*Ischaemum ciliare*, *Axonopus compressus* and *Paspalum* spp) or naturalised species (*Cynodon nlemfuensis*) on a typical soil type (young alluvial, well-drained volcanic soil of relatively high fertility).

The link (or relation) between available mineralised N and pasture growth, based on data generated by the so-called DNDC model (Plant & Bouman, 1999), is described by a Mitscherlich production function (Bulte *et al.*, 2000).

13.3 Study Area

The explorative (SOLUS), predictive (UNA-DLV) and normative (optimal control) methodologies were developed and applied in the Atlantic Zone (AZ) of Costa Rica (Fig. 13.2). The AZ is part of the humid lowland tropics (mean daily temperature 26°C, mean annual rainfall 3500-5500 mm, elevation 0-400 m). For agricultural use, three major soil groups are distinguished, based on fertility (high and low) and drainage (good and poor) criteria (Nieuwenhuyse, 1996): SFW (fertile well-drained soils), SFP (fertile poorly drained soils) and SIW (infertile well-drained soils). Of a total area of 447,000 ha, 334,000 ha is suitable for agriculture. Of these 334,000 ha, 55,000 ha are protected for nature conservation (including 12,000 ha of national parks) or have a 'semi-protected' status (indigenous reserves, forest reserves, protected areas and wetlands), while some 28,000 ha consist of rivers, roads, urban area, *etc*. The area not available for agriculture (168,000 ha or some 38% of the total area) is under natural forest cover. Current agricultural land use is dominated by pastures for beef cattle ranching (close to 200,000 ha) and banana plantations (some 35,000 ha). Secondary crops (total about 13,000 ha) include plantain, palm heart, root and tuber crops, maize, papaya, pineapple and ornamental plants.

Tropical rain forest once covering the entire area, is now restricted to wetlands, inaccessible mountain areas and (semi-)protected areas. Colonisation of the AZ started at the end of the 19th century, but accelerated in the past thirty years (Kaimowitz, 1996). Rapid structural transformations were and are taking place in the ecological, agricultural and socio-economic conditions of the region, in part responding to structural adjustment programs (Jiménez, 1998). Even though the Costa Rican government recognises the need to base regional land use policies on an analysis of the socio-economic and environmental implications of both actual and potential technological options for land use (SEPSA, 1997), at present it has insufficient knowledge for an adequate analysis of present and possible future land use.

13.4 Results

13.4.1 Explorative SOLUS Methodology

In SOLUS, policy scenarios that reflect possibilities to combat soil degradation are compared to a base scenario in which only actual land use technologies are offered to the LP model (REALM), thus mimicking the actual situation in the AZ as closely as possible. Two scenarios were evaluated, the first of which simulates the effect of technological progress, implemented by simply offering the entire spectrum of land use technologies (as generated by the TCGs) to the LP model. The other scenario represents soil conservation, operationalised via zero N, P and K depletion restrictions.

Land use in the base scenario, while roughly in agreement with actually observed land use patterns in the AZ, results in considerable soil mining, pointing towards the unsustainability of most actual land use technologies. Combined crop and pasture land uses lead to an average N depletion of about 50 kg ha^{-1} yr^{-1} (first column in Table 13.3). Since pasture is the major land use, soil mining in beef cattle ranching is an important contributor to overall nutrient depletion. Beef cattle are mainly held on natural unfertilised pastures with high stocking rates. Combined with high rainfall, highly permeable soils and high N turnover rates (Bouman & Nieuwenhuyse, 1999), this results in, for example, an N depletion of 60-70 kg ha^{-1} yr^{-1}. This implies that the soil nutrient store becomes exhausted over time, leading to declining pasture yields; this is also the result in the optimal control model (see below).

Technological progress, essentially producing more with the same or less resources (land, labour), has important effects on economic surplus, employment, land use, and environmental indicators (column 2 in Table 13.3). Economic surplus increases by 21% between the actual technology and the improved technology scenario. Overall land productivity increases with the same percentage (the area used remains constant), while labour productivity increases by 27% (higher surplus, using less labour). As a consequence, employment decreases by 4%. The depletion of soil nutrients decreases; for N from about 50 kg ha^{-1} yr^{-1} to about 30. Thus, both environmental and economic indicators used in this study are more favourable in the *improved technology* scenario than in the *present technology* scenario. Hence, technological change results in a win-win situation, *i.e.* increased economic surplus and less environmental degradation. This general conclusion is not surprising. The alternative technologies generated in the TCGs are more productive *and* more sustainable than the actual technologies. They were generated in this way, given the agronomic possibilities, not only theoretically, but also as observed by current practices on the best farms.

Table 13.3: Impact of new technologies for combating soil degradation in the explorative SOLUS methodology.

	Base (actual technologies only)	Technological progress (all available technologies)	Zero soil nutrient depletion
Economic surplus (US $\$10^6$)	221	268	239
Employment (days 10^3)	9039	8661	8761
Land use as a % of total land area suitable for agriculture (251,000 ha)			
Pastures	69	76	12
- Natural	69	60	9
- Grass-legume	0	16	3
- Fertilised pastures	0	0	0
Forest (including unused land)	0	0	57
Crops	31	24	31
Number of animals (10^3 AU)			
Fattening system	116	138	21
Breeding system	198	254	39
Depletion of soil nutrients (kg^{-1} ha^{-1} yr^{-1})			
N	51	29	0
P	3	-0.5	-0.5
K	113	21	0

Introduction of *zero soil nutrient depletion* restrictions for N, P and K with all technologies available results in the use of both labour-intensive grass-legume mixtures and less labour-intensive fertilised pasture technologies on soils with good drainage (column 3 in Table 13.3). Overall, the pasture area diminishes considerably, while the crop area increases. The forest area increases, consisting of natural forests and Melina tree plantations. Furthermore, any land left unused by the LP model is assumed to convert to (secondary) natural forest too, and is, therefore, added to the forest area.

The economic surplus is about 11% lower than in the *technological progress* scenario. This result is in line with Cárcamo *et al.* (1994) who analysed the scope for soil conservation measures in Honduras in areas with rampant soil erosion, and found that considerable reductions in soil loss could be achieved at relatively low economic costs.

Table 13.4: Farm level results, UNA-DLV methodology.

	Actual technologies only				Actual & alternative technologies			
Farm types	Small	Medium	Hacienda	Banana	Small	Medium	Hacienda	Banana
Production structure (ha)								
Crops:	2.8	4.8	-	127.3	3.7	4.9	-	169.7
Pineapple	0.0	4.8	-	-	0.6	4.9	-	-
Banana	-	-	-	127.3	-	-	-	169.7
Plantain	2.8	0.0	-	-	1.8	0.0	-	-
Palm heart	0.0	0.0	-	-	1.3	0.0	-	-
Pastures:	0.8	26.5	190.3	-	0.9	20.8	211.9	-
Natural	0.8	26.5	190.3	-	0.7	9.6	99.8	-
Fertilised	0.0	0.0	0.0	-	0.2	11.2	112.1	-
Fallow	5.3	7.8	-	99.0	4.3	13.5	-	56.6
Livestock (AU)	1.0	75.2	304.3	-	1.1	58.7	341.7	-
Sustainability:								
Depletion of N (kg ha^{-1} yr^{-1})	160	31	50	140	94	46	49	0
Resource use intensity:								
Labour intensity (days ha^{-1} yr^{-1})	121	20	5	114	80	28	5	204

13.4.2 Predictive UNA-DLV Methodology

Model results under actual production techniques (Table 13.4) show that small farms specialise in plantain production, while beef and milk production are important for household consumption. Medium size farms specialise in local and export quality pineapple production and beef production, based on natural pastures in combination with supplementation. Both farm types face capital constraints, that limit cash crop production and explain the extensive fallow areas. The *hacienda* farm type specialises in beef cattle production on natural pastures, with an average stocking rate of 1.6 AU ha^{-1}. Further expansion of the pasture area and cattle stock is limited by net returns from beef production. Under current technological options, the banana plantation uses just over 55% of the available area, as less suitable soils cannot profitably be cultivated. Soil nitrogen depletion levels are especially high in the cultivation of pineapple, plantain and banana, while pasture-based beef cattle production leads to significantly lower levels of nitrogen depletion, especially in combination with supplementary feeding.

Introduction of technological change increases cash crop production on medium and small size farm types, as production systems become available that produce more output using less inputs, effectively relaxing the capital constraint. Small farms now also produce pineapple and palm heart, while medium size farms fully focus on export-quality pineapple production at the expense of beef production. Alternative, more productive, fertilised pastures are selected on fertile soil types by the *hacienda*, although fertiliser application levels remain low. The pasture area is extended even more, at the expense of frontier forest areas, as net returns from beef production increase, thanks to increased pasture production at similar input levels. Alternative banana production systems allow increased use of less suitable soils. The partial (small, medium and *hacienda* farms) or even complete (banana plantations) shift to alternative and thus sustainable production systems leads to lower soil nitrogen depletion levels. However, the increased importance of pineapple production relative to pasture-based beef cattle production on medium size farms leads to higher average soil nitrogen losses.

Model results for the whole region (Table 13.5) show that pastures dominate land use in the AZ, while the forest area exceeds the total crop area. Introduction of alternative production systems results in expansion of both the crop and pasture areas, at the expense of the agrarian frontier forest area. The economic surplus increases by more than 25%, while total agricultural labour demand remains within the available regional agricultural labour force. Alternative, more productive and sustainable production systems result in reduced soil nitrogen depletion, *i.e.* from more than 100 kg ha^{-1} to less than 50.

Table 13.5: Regional results, UNA-DLV methodology.

	Unit	Actual technologies		Actual & alternative technologies	
		#	%	#	%
Economic surplus	US $ 10^6	172		219	
Cultivated area					
Pastures:	ha 10^3	203	60	211	62
Natural	ha 10^3	203	60	99	29
Fertilised	ha 10^3	0	0	112	33
Grass-legume	ha 10^3	0	0	0	0
Crops	ha 10^3	55	16	69	21
Forest	ha 10^3	82	24	59	17
Number of animals	AU 10^3	285		280	
Labour use	d/month 10^3	856	86	885	89
Sustainability indicator:					
Depletion of N (crop & pasture areas only)	kg ha^{-1}	127		46	

13.4.3 Normative Optimal Control Methodology

The results of the optimal control model indicate that the maximum attainable dry matter yield in the study region is 9 t ha^{-1} yr^{-1} when using natural grasses, assuming N is non-limiting. No yield is feasible for soil N stock levels below 3,600 kg ha^{-1} (or about 2.8% soil organic matter). Solving the dynamic model formulated in Equations (13.7) and (13.8), the estimated Mitscherlich production function and some additional constraints (Bulte *et al.*, 2000) yields values for the optimal N stocks in the steady state for various combinations of the price of beef and the discount rate (Table 13.6). In the steady state, outflow of N, in crop products and unavoidable losses, is exactly compensated by the inflow (natural or otherwise), and the shadow price of N is constant over time. The numerical results for the range of prices and discount rates indicate that nutrient stocks are increasing in beef prices and decreasing in discount rates, but for most combinations at a fairly low level. Indeed, the figures indicate that, for the combination of positive discount rates and past prices (¢ 160-240 kg^{-1} liveweight for the period 1985–1997; ¢ is *Colones*, the currency unit of Costa Rica), it is economically efficient to deplete the nutrient stock. Recall that a steady state N stock of 3,600 kg ha^{-1} corresponds to zero production in our Mitscherlich production function, hence this level represents economic depletion of the soil. For most price levels, sustainable livestock keeping (in the sense that the soil N stock is not fully depleted and that a positive stocking rate exists in the steady state) is only consistent with economic efficiency for discount rates below 4%. For a zero discount rate, optimal N stocks are close to, or even higher than, the initial stock of 4,800 kg ha^{-1}.

Table 13.6: Calculated optimal steady state N stocks (in kg ha^{-1}) in 0-15 cm topsoil in humid Costa Rica for beef fattening, at different beef prices and discount rates.

	Price of beef in ¢ kg^{-1}				
Discount rate (%)	140	180	250	350	500
0	4404	4592	4811	5016	5221
2	3600	3782	4001	4206	4411
4	3600	3600	3705	3910	4115
6	3600	3600	3600	3725	3930
8	3600	3600	3600	3600	3795

If ranchers, in practice, apply a discount rate of 7% (consistent with the value applied in the SOLUS methodology) for future revenues and costs, the results in Table 13.6 imply that, in the long run, grazing land will be mined and abandoned, which is consistent with current pasture degradation in the study region, and with work by Cárcamo *et al.* (1994) for soil erosion in Honduras. They report that high levels of erosion are optimal from a private decision maker's perspective, even taking into account high future productivity losses caused by soil erosion. Note, however, that these results do not imply that soil mining is also *socially* optimal. While the opportunity costs of capital in a country like Costa Rica may well be around or even higher than 7% (which supports such a claim), the social rate of time preference is more likely to be around the long-term per capita economic growth rate of about 1.5 to 3%[8]. Furthermore, there might be external effects involved in soil mining that could dominate the private gains. This has not been considered in our models so far, and is left for future work.

The model not only considers steady states, but also provides the dynamics of depletion. These show that landowners should mine the soil as fast as possible, and invest the proceeds elsewhere. This implies that a maximum stocking rate should be chosen, and farmers should not invest in N inputs (although obviously replenishment from natural sources occurs). Assume that the initial stock of N is 4,800 kg ha^{-1} and that the optimal steady state stock is 3,600 kg ha^{-1}. Then, given that natural pastures deplete the soil N stock by about 60-70 kg ha^{-1} yr^{-1} (Bouman *et al.*, 1999b), the pasture is fully degraded over a period of about 15 years.

[8] The average annual growth rate of GNP per capita in constant prices equalled 2.6% for the period 1985-1997 and 1.5% for the period 1960-1997, based on World Bank, 1999.

It is interesting to again consider the efficiency-sustainability trade-off. Above, in the scenarios studied with the SOLUS methodology, zero soil nutrient depletion, including N, was imposed to study the impact of such a constraint on the economic surplus. This implied an overall reduction in N depletion of about 50 kg ha^{-1} yr^{-1} (Table 13.3). If we impose a comparable additional restriction in the optimal control model such that ranchers should supplement the N that is lost from the system during grazing to maintain a zero nutrient balance (or restrict grazing to the level that losses do not exceed natural enrichment), what would be the consequences for ranchers? In other words, how much profit would ranchers give up if we adopt a strict sustainability requirement? It was found that the discounted value of profits from a pasture plot, as specified for the optimal control model, is reduced by about 25%. The trade-off between efficiency and sustainability thus appears to exist for this case of soil management in Costa Rica.

As explained in Section 13.2.3, taking the inter-temporal dimension into account implies that we have to work with fairly simple models to maintain analytical tractability. A convenient way to incorporate technological change in the optimal control model outlined above is as follows. Recall that a Mitscherlich production was used to describe the relation between available N and pasture yield. For the base situation, natural grasses were being considered. Technical change may be envisaged as a shift towards 'improved' higher yielding species such as *Brachiaria brizantha*. If ranchers choose to switch to improved grasses, a given N stock may result in higher yields per hectare. Available evidence suggests that the maximum attainable dry matter yield could be increased from 9 t ha^{-1} to 12 if high yielding grasses are introduced. Of course, that involves additional costs, such as the costs of ploughing and seeding. Calibrating the model accordingly and accounting for these costs, we find that the effects on nutrient stock in the steady state are small (Table 13.7). The steady state stocks for improved pasture are somewhat larger than the corresponding stocks in Table 13.6 (evaluated at a beef price of ¢ 180 per kg, corresponding to roughly US $ 1.00 at the average 1994-1996 exchange rate) for natural pasture. Using improved grasses implies that investing in soil quality becomes more attractive, and for that reason ranchers might be willing to use more fertilisers to support the nutrient stock. However, depletion is still optimal for discount rates higher than 2%. In a similar fashion, the effect of changing the prices of inputs and outputs may be analysed.

Table 13.7: Optimal steady state N stocks (in kg ha^{-1}) in 0-15 cm topsoil in humid Costa Rica for beef fattening for an improved pasture system.

	Discount rate (%)				
	0	2	4	6	8
N stock (kg ha^{-1})	4737	3927	3631	3600	3600

The beef price is set at the benchmark value of ¢ 180 kg^{-1}.

13.5 Conclusions

The process of opening-up of the Costa Rican economy has been ongoing for the past two decades and has led to significant changes across the country, the AZ included. For example, changes in land use patterns in the AZ include both attempts at diversification and concentration on a more limited number of crops, though generally with an increasing share of so-called non-traditional export products such as palm heart, roots and tubers, ornamental plants, *etc.*, and largely at the expense of basic food crops, such as maize and beans. However, this should not distract from the fact that many production technologies currently employed in the AZ, deplete soil nutrient resources and are therefore unsustainable in the long run.

This chapter has provided a description, as well as a comparison, of a number of interdisciplinary methodologies for regional land use analysis. As their character ranges from exploratory via predictive to normative, each methodology has its own role to play in the *ex-ante* analysis of the potential of agricultural innovations aimed at combating soil mining, and of the scope for policy interventions.

The *explorative* SOLUS methodology can provide a number of visions (scenarios) about future production possibilities and their environmental consequences in the AZ, given actual and alternative technologies, taking into account both domestic and export market limitations, as well as competition for labour with other economic sectors in Costa Rica. Its major strengths are the careful simulation of relevant markets, based on estimated demand elasticities and resulting in endogenous output prices, and the empirical foundation of input-output coefficients – including environmental indicators - of several thousands of actual and alternative technologies. In addition, the SOLUS methodology simulates the labour market, thus effectively endogenising the most important input market. A weakness of the SOLUS approach is the abstraction from differences in farm types, each with different objectives. That point is taken up by the *predictive* UNA/DLV methodology, which makes it more suitable for evaluation of the effects of different policies in the short-to-medium term. The aggregate version of the UNA/DLV methodology also incorporates endogenous output prices, although wages are assumed constant. Furthermore, no account is taken of the possible exchange of land between different farm types, *e.g.* the expansion of *haciendas* or banana plantations at the expense of small and medium farms, or *vice versa*. Another shortcoming of both the SOLUS and the UNA-DLV methodologies is that both are based on single period optimisation models, although future costs and benefits of perennials are discounted and made comparable via annuities. The *normative* optimal control model is a truly inter-temporal model, but this comes at a cost. To retain a tractable and solvable model, one can only analyse fairly simple plot-level production situations with one specific technology; in the present chapter, the dynamic effects of a changing N stock for a particular soil type with a particular pasture type were simulated.

A general conclusion of the evaluation of the different scenarios in the present chapter might be that technological progress of the right kind is able to generate complementarity between income and sustainability objectives, whereas policies to reduce negative environmental consequences – other than policies to stimulate appropriate technological development – often come at the expense of income. To some extent this may be acceptable, as both income and environmental quality are part of a wider concept of (social) welfare. Furthermore, the size of the trade-off between income and sustainability outcomes can be minimised by careful policies. Analysing such policies in an *ex-ante* way could help in the search for the right policies at the right place and time. The three methodologies illustrated in this chapter are valuable tools for policy studies related to land use issues, and as such are able to better inform policy makers.

13.6 References

Barbier, B. (1995): Policy implications of land degradation in West Africa. Paper of the International Food Policy Research Institute (IFPRI) presented at the workshop on 'Land degradation in the developing world: implications for food, agriculture and environment to the year 2000', April 4-6, 1995, Annapolis, Maryland

Blaikie, P. & H. Brookfield (Eds.) (1987): Land degradation and society. Methuen, London

Bosworth, D., P. Dawkins & T. Stromback (1996): The economics of the labour market. Longman, Harlow, UK

Bouman, B.A.M. & A. Nieuwenhuyse (1999): Exploring options for sustainable beef cattle ranching in the humid tropics; a case study for the Atlantic Zone of Costa Rica. Agricultural Systems 59, 145-161

Bouman, B.A.M., A. Nieuwenhuyse & H. Hengsdijk (1998a): PASTOR: a technical coefficient generator for pasture and livestock systems in the humid tropics; version 2.0. A users guide. Quantitative Approaches in Systems Analysis, no. 18. AB-DLO-PE, Wageningen, The Netherlands.

Bouman, B.A.M., R.A. Schipper, A. Nieuwenhuyse, H. Hengsdijk & H.G.P. Jansen (1998b): Quantifying economic and biophysical sustainability trade-offs in land use exploration at the regional level: a case study for the Northern Atlantic Zone of Costa Rica. Ecological Modelling 114, 95-109

Bouman, B.A.M., H.G.P. Jansen, R.A. Schipper, A. Nieuwenhuyse, H. Hengsdijk and J. Bouma (1999a): A framework for integrated biophysical and economic land use analysis at different scales. Agriculture, Ecosystems and Environment 75, 55-73

Bouman, B.A.M., A. Nieuwenhuyse & M. Ibrahim (1999b): Pasture degradation and restoration by legumes in humid tropical Costa Rica. Tropical Grasslands 33, 98-110

Bouman, B.A.M, H.G.P. Jansen, R.A. Schipper, H. Hengsdijk & A. Nieuwenhuyse (Eds.) (2000): Tools for land use analysis on different scales. With case studies for Costa Rica. Kluwer Academic Publishers, Dordrecht

Brooke, A., D. Kendrick & A. Meerhaus (1992): GAMS release 2.25: a user's guide. The Scientific Press, South San Francisco, 289 pp

Bulte, E., B.A.M. Bouman, A. Nieuwenhuyse & H.G.P. Jansen (2000): The economics of soil nutrient stocks and beef production in the humid tropics: optimal pasture degradation in Costa Rica. European Review of Agricultural Economics (in press)

Bunce, A.C. (1942): The economics of soil conservation. University of Nebraska Press, Lincoln

Cárcamo, J.A., J. Alwang & G.W. Norton (1994): On-site economic evaluation of soil conservation practices in Honduras. Agricultural Economics 11, 257-269

DGEC [Dirección General de Estadística y Censos] (1987): Censo agropecuario 1984. Ministerio de Economía, Industria y Comercio, San José, Costa Rica

DGEC [Dirección General de Estadística y Censos] (1990): Avance de resultados. Encuesta Nacional de Ingresos y Gastos de los Hogares, Informe No.1. Ministerio de Economia, Industria y Commercio, San Jose, Costa Rica

Geurts, J.A.M.M., H.G.P. Jansen & A. van Tilburg (1997): Domestic demand for food in Costa Rica: a double hurdle analysis. Serie técnica. Informe técnico No. 286, CATIE, Turrialba, Costa Rica, 100 pp.

Hazell, P.B. & R.D. Norton (1986): Mathematical programming for economic analysis in agriculture. Collier MacMillan Publishers, London, 400 pp.

Hengsdijk, H., A. Nieuwenhuyse & B.A.M. Bouman (1998): LUCTOR 2.0. Land Use Crop Technical coefficient generatOR. A model to quantify crop systems in the Atlantic Zone of Costa Rica. Quantitative Approaches in Systems Analysis 17. AB-DLO-PE, Wageningen

Hijfte, P.A. van (1989): La ganadería de carne en el Norte de la Zona Atlántica de Costa Rica. Report 31, Atlantic Zone Program (CATIE/MAG/WAU), Turrialba, Costa Rica, 52 pp., plus appendices

Jansen, H.G.P. & J.J. Stoorvogel (1998): Quantification of aggregation bias in regional agricultural land use models: application to Guácimo county, Costa Rica. Agricultural Systems 58, 417-439

Jansen, D.M., J.J. Stoorvogel & R.A. Schipper (1995): Using sustainability indicators in agricultural land use analysis: an example from Costa Rica. Netherlands Journal of Agricultural Science 43, 61-82

Jiménez, R. (Ed.) (1998): Estabilidad y desarrollo económico en Costa Rica: las reformas pendientes. Academia de Centroamerica, San José, Costa Rica

Kaimowitz, D. (1996): Livestock and deforestation in Central America in the 1980s and 1990s: a policy perspective. Center for International Forestry Research, Jakarta, 88 pp.

Kruseman, G. & H. van Keulen (2000): Soil degradation and agricultural production: economic and biophysical approaches. This volume

Kruseman, G., H. Hengsdijk, R. Ruben, P. Roebeling & J. Bade (1997): Farm household modelling system for the analysis of sustainable land use and food security: theoretical and mathematical description. DLV Report No.7. AB-DLO/WAU, Wageningen, the Netherlands. 56 pp.

Kuiper, M. (1997): The impact of land degradation on agricultural productivity: a multi-period economic land use model. A case study of the Neguev settlement, Costa Rica. Report 124, REPOSA. CATIE/MAG/WAU, Turrialba, Costa Rica

Kutcher, G.P. (1972): Agricultural planning at the regional level: a programming model of Mexico's Pacific Northwest. Ph.D. thesis, University of Maryland

Kutcher, G.P. (1983): A regional agricultural programming model for Mexico's Pacific Northwest. In: R.D. Norton & L. Solis (Eds.) The book of CHAC: programming studies for Mexican agriculture, John Hopkins University Press, Baltimore, 317-351

Leonard, D. & N. van Long (1992): Optimal control theory and static optimisation in economics. Cambridge University Press, Cambridge

Mamingi, N. (1997): The impact of prices and macroeconomic policies on agricultural supply: a synthesis of available results. Agricultural Economics 16, 17-34

McConnell, K.E. (1983): An economic model of soil conservation. American Journal of Agricultural Economics 64, 83-89

Nieuwenhuyse, A. (1996): Soils, geology, and soil-related sustainability aspects of the perhumid tropical Limón basin, Costa Rica. CATIE Serie Técnica, Informe Técnico no. 272, Turrialba, Costa Rica

Plant, R.A.J. & B.A.M. Bouman (1999): Modelling nitrogen oxide emissions from current and alternative pastures in Costa Rica. Journal of Environmental Quality 28, 866-872

Roebeling, P.C., R. Ruben & F. Sáenz (1998): Políticas agrarias para la intensificación sostenible del sector ganadero: una aplicación en la Zona Atlántica de Costa Rica. In: E. Castro & R. Ruben (Eds.) Políticas agrarias para el uso sostenible de la tierra y la seguridad alimentaria en Costa Rica. UNA-CINPE / WAU-DLV, Costa Rica, 156-174

Roebeling, P.C., H.G.P. Jansen, R.A. Schipper, F. Sáenz, E. Castro, R. Ruben, H. Hengsdijk & B.A.M. Bouman (2000): Farm modelling for policy analysis on the farm and regional level. In: B.A.M. Bouman, H.G.P. Jansen, R.A. Schipper, H. Hengsdijk & A. Nieuwenhuyse (Eds.) Tools for land use analysis on different scales. With case studies for Costa Rica. Kluwer Academic Publishers, Dordrecht

Sadoulet E. & A. de Janvry (1995): Quantitative development policy analysis. The Johns Hopkins University Press, Baltimore

Schipper, R.A., B.A.M. Bouman, H.G.P. Jansen, H. Hengsdijk & A. Nieuwenhuyse (2000): Integrated biophysical and socioeconomic analysis of regional land use. In: B.A.M. Bouman, H.G.P. Jansen, R.A. Schipper, H. Hengsdijk & A. Nieuwenhuyse (Eds.) Tools for land use analysis on different scales. With case studies for Costa Rica. Kluwer Academic Publishers, Dordrecht

SEPSA (Secretaría Ejecutiva de Planificación Sectoral Agropecuaria) (1997): Políticas del sector agropecuario (revisión y ajuste). Ministry of Agriculture and Livestock (MAG), San José, Costa Rica

Singh, I., L. Squire & J. Strauss (1986): Agricultural household models: extensions, applications and policy. Johns Hopkins Press for the World Bank, Baltimore

Stoorvogel, J.J. & G.P. Eppink (1995) Atlas de la Zona Atlántica Norte de Costa Rica. Programa Zona Atlántica (CATIE-UAW-MAG), Guápiles, Costa Rica, 84 pp

Takayama, T. & G. Judge (1971. Spatial and temporal price and allocation models. North Holland, Amsterdam.

Thampapillai, D.J. & J.R. Anderson (1994): A review of the socio-economic analysis of soil degradation problems for developed and developing countries. Review of Marketing and Agricultural Economics 62, 291-315

Valk, A.L. van der (1999): Elasticities ¡ Claro que sí !: an analysis of the domestic Costa Rican food demand. M.Sc. thesis. Department of Marketing and Marketing Research, Wageningen Agricultural University, Wageningen

Walker, D.J. (1982): A damage function to evaluate erosion control economics. American Journal of Agricultural Economics 64, 145-55

World Bank (1999): World development indicators 1999 on CD-ROM. World Bank, Washington

Zuñiga, N. (1996): Estimación de la tasa de interés promedio. El Financiero, 22-28 January 1996

14 Effects of Economic Policies on Farmers, Consumers and Soil Degradation: a Recursively Dynamic Sector Model with an Application for Burkina Faso

Daniel Deybe

Economics, Policies and Markets Programme (ECOPOL), Advanced Methods for Innovation in Science Department (AMIS), Centre de Coopération Internationale en Recherche Agricole pour le Développement (CIRAD), Nogent sur Marne, France

Abstract

In this paper, a bio-economic model is described to support the process of policy formulation. The model incorporates the results of crop growth simulation models (with detailed information on the process of soil erosion) within the framework of an economic model designed to represent farmers', herders' and consumers' behaviour at the agricultural sector level. It consists of three interacting and recursive modules: ***production****, a regional optimisation model that simulates the individual decisions about crop allocation based on expected prices;* ***marketing/consumption****, a partial equilibrium model where 'real' prices are estimated as a result of the confrontation between marketed production and urban demand; and* ***macro-economic context****, in which the general economic variables affecting farmers and consumers are defined. The production choices and their effects on soil degradation influence the future choices of farmers within the model, allowing to take into account explicitly the dynamic recursive impact. The model gives indications on the way different actors might probably respond to the simulated policies and the influence of these responses on production, income, and consumption. Also, part of the positive and negative effects on soil quality is estimated. The use of these variables allows to compare the results of the simulation of different policies and facilitates the choice for the policy maker.*

14.1 Introduction

A strong dichotomy exists between long-term sustainability of physical, economic and social systems, and short-term objectives of society in general and of politicians in particular. Understandably, much attention is paid to urgent problems such as poverty or local development, but sometimes environmental considerations are rather neglected, in particular in low-income countries. However, it would be erroneous to forget the latter, because it might jeopardise the future productivity of the agricultural sector and thus transfer the problem to the next generation. Therefore, there is a need to incorporate future effects in the analysis of alternative present policies in order to avoid these consequences.

The process of policy formulation, specially when considering the impact on the natural environment, has to be dealt with very carefully. One way to provide elements for a judicious choice among policies is to simulate their impact. However, formalisation of the decision making processes of both the policy makers and the actors has not always produced successful results. Three factors may explain this lack of accuracy: the difficulty to represent the implicit contradiction between the policy makers' goals and actors' objectives, the use of unrealistic hypotheses, and the difficulty in evaluating environmental externalities. The agricultural sector presents additional problems: the number of farmers and herders is large, they are dispersed in space and most of their actions will have an impact (positive or negative) on the environment. Besides, in agriculture at least two levels of prices will have a strong impact on the sector outcome: the one corresponding to farmers expectations when they decide to produce a commodity, and the one they receive at the moment of marketing their production, which depends on total output and specially on consumers' demand. This requires adequate tools to evaluate *ex-ante* the effects of policies on farmers and on their environment.

Forward-looking studies, therefore, include in the analysis the present value of future losses (or lack of production) due to current use of the natural resources. This approach can be helpful to estimate how much the present generation could or should agree to pay (or forego in income), if the well-being of next generations is not to be jeopardised. However, Taylor (1996) mentions two constraints that limit the extension of the analysis: first, the lack of societal consensus on environmental or economic issues and second, the intrinsic 'risk' characteristics of any decision related to the environment. Besides, sometimes these analyses neglect to indicate how to turn the potential transfer effective and therefore, how to achieve future generation's well-being[1].

Solow (1993) indicates that part of the solution to the problem of sustainable development should reside in investments for the future, but without neglecting the needs and requirements of the current population. Even if in any human

[1] The results of these analyses depend strongly on the actualisation rate used to do the estimation, and its value is more related to a political or societal issue than to a subject of research. Is there any moral basis to assume that all societies should put in the same effort to preserve the environment for the future (or for the present)? Should developed societies invest more for the future? All these aspects classify most of the forward-looking studies in the field of ideology rather than in the field of 'objective' research.

decision there is always a part of altruism for future generations, the short-term consideration always predominates. This dichotomy can not be neglected, if the answers that will be provided by the analysis are to be realistic.

The soil, one of the principal factors for agricultural production, is a good example of what human action can produce and of the trade-off between short- and long-term objectives. In some cases, when the soil is badly managed, it can be rapidly depleted (often generating high short-term returns). In some other cases, good management can provide for sustainable use of this resource, even if in the short run the economic returns might be lower (Deybe, 1994).

But analysing the impact on the soil (as well as on most environmental resources) of human activities is also the addition of numerous decision making processes. To analyse each of them in an isolated way, without considering the whole context, might induce errors of estimation. Dent *et al.* (1995) indicate the utility of an ex-ante evaluation of the effects and of environmental externalities of any policy. Boussard (1994) advocates mathematical formalisation to avoid misinterpretations and approximate reasoning. It should be remembered that when formulating a policy, there is a difference between the process of policy decision making (by a limited number of individuals) and the influence of these policies on agricultural production, as a result of the actions of a large number of individuals (farmers, herders, *etc.*) (Candler *et al.*, 1981).

In this paper, we present a sector model that tries to address both problems: it considers the future impact on the soil, and therefore on productivity, of present choices in the decision making process, and it is based on the aggregation of individual decisions within a global context. Firstly, the principal hypotheses and the structure of the model are presented, secondly, some results of alternative policy scenarios in Burkina Faso after the devaluation are discussed, to demonstrate the relevance of this approach.

14.2 The Multilevel Analysis Tool for the Agricultural Sector

The Multilevel Analysis Tool for the Agricultural sector (MATA) (Gérard *et al.*, 1995) is a multiregional model developed to try to represent farmers' and consumers' decision making processes to provide insight in their reactions to different economic policies.

The model is based on the assumption that there is a gap between the moment the farmers decide to produce a certain good, basing their decision on expected prices and yields, and the moment they sell their production, when they do not have control over prices, which depend on other farmers' production, as well as on consumers' preferences and budgets (Gérard, 1991). A second premise is that some resources have to be dealt with within a global framework (regional and/or national) to avoid imposing individual constraints that might be artificial. A third assumption is that farmers make up their mind considering both short and long-term consequences in the process, but with different weights in the decision. This process also implies that farmers can modify their choices at any moment, according to the history of the farm, what they experience at each moment, and their expectations. This process can then be associated with a recursive and dynamic decision making process, which is particularly well adapted to capture the impact of management on soil processes. The latter take place slowly through time, and thus the impact of short-run decisions will have an unavoidable effect in the long-run.

To represent these processes, the model is solved recursively. First, the production aspects are simulated for each of the regions, based on expected prices and yields. Second, the total surpluses of all regions, based on 'corrected' yields[2], considering random weather variability, are aggregated in order to estimate the total supply arriving on the 'national' market. Third, the marketing and consumption module is solved to estimate the prices at a national 'urban' level, based on consumers' preferences and budgets. Fourth, the transaction costs are deduced from the national price for each of the regions and the resulting prices are used to estimate farmers' net income. Finally, a portion of the resulting revenue is transferred to the following period to reinitialise the availability of cash flow. The prices simulated are also used to modify price expectations. The status of the resources, *i.e.* the soil, is reinitialised after each simulation (Fig. 14.1).

Therefore, the model is built up of three modules: *a production module*, in which the mechanisms that determine agricultural supply are represented; *a marketing-consumption module*, in which the confrontation of total supply with aggregated individual demand is simulated; and a *macroeconomic module* in which the global context is described. Each of these modules operates differently: the first one is based on optimisation of the aggregated individual objective functions, the second one is a partial equilibrium model, the third one is a descriptive module. Each module will be described in detail in the following sections.

14.3 The Production Module

Two objectives are pursued in this module:

1. to represent farmer's individual decision making process based on an optimisation procedure, subject to individual and global constraints, and
2. to identify effects of the policies simulated on each type of farmer.

A typology is used to describe the main characteristics of each group of farmers, the interactions between the farmers of the different groups, and the representativeness of each of the groups. For each group, a farm model, detailing

[2] The average yield modified randomly but according to the standard deviation of each type of crop management.

its own resources, the constraints and the requirements of external resources, is included in this module.

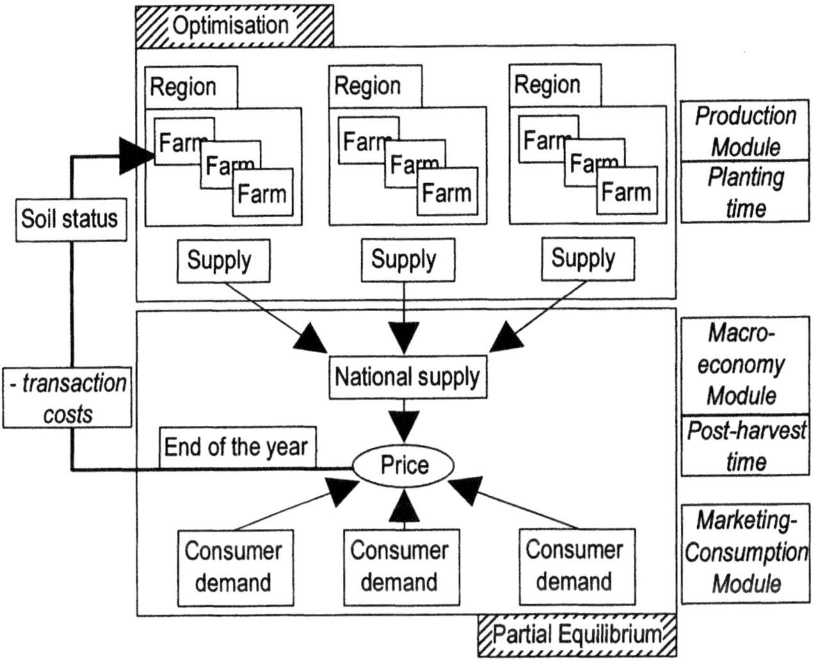

Figure 14.1: The recursivity in the model.

The activities considered in the production module include different crops with different technologies, livestock, fallow and deforestation. For each farm type:

$$\sum_j a_j x_{j,i} \leq b_i \qquad (14.1)$$

where a_j are the technical coefficients of each activity j (which thus are assumed identical for the various farm types), $x_{j,i}$ the levels of each activity j in farm i and b_i the available resources that can not be hired in or out.

For resources (bb) that can be hired in from (or out to) other farmers, like labour, oxen traction, equipment, *etc.*, for each type of farm i holds:

$$\sum_j a_j x_{j,i} \pm \sum_{j,k} aa_{j,k} xx_{j,i,k} \leq bb_{j,i} \qquad (14.2)$$

where k are the other types of farms, aa the technical coefficient of the activity and xx the levels of use of the resource of farm k by farm i or of farm i by farm k.

Then, equilibrium between the supply and the demand of the resource is reached by:

$$\sum_{j,i,k} aa_{j,k} xx_{j,i,k} n_i = \sum_{j,k,i} aa_{j,i} xx_{j,k,i} n_k \tag{14.3}$$

where n is the number of farms within each farm group. The transfers among farmers respect thus the 'actual' availability of the resources.

In some cases, some of the production factors, like credit can be dealt with in an aggregated way (*Credtot* is the total availability of credit in the region), the optimal allocation ($Cred_{j,i}$) is determined by the model, or their use can be exogenously determined (*Credtot* then acquires the farm index i).

$$\sum_{j,i} x_{j,i} Cred_{j,i} n_i \leq Credtot \tag{14.4}$$

To define the technical coefficients, the yields and the variability in yields of the different crops (the latter for risk consideration) that will be considered in the model, crop growth simulators can be applied (Fig. 14.2) (Deybe, 1994). These models are interesting for several reasons:

1. They use the same characteristics of soil and weather as starting point for the simulation of all techniques and crops, which allows objective comparison of performances.
2. They can be used to estimate the performance of potential activities not yet practised in the area under analysis.
3. They provide indications on the variability in yields of each of the alternatives in the local environment.
4. They help in estimating deterioration and/or improvement in the soil characteristics and status and their impact on future production.

This last point is very important when considering the impact on soil degradation for two reasons:

a. the magnitude of soil losses can be used to compare the impact of different policies;
b. the growth of a 'reference' crop can be simulated to measure the long-term impact of the short-run nutrient depletion generated by each activity.

The second procedure allows estimation of an economic value of all activities (crops and fallow) that allows their comparison. The procedure followed is:

i. to simulate the growth of a reference crop under actual soil conditions for a long period of time: the results serve as a baseline for comparison;

ii. after simulating the growth of any crop/management combination, the final status of the soil is used to simulate again the growth of the reference crop, so that a new set of yields is estimated;

iii. the differences in yields of the reference crop in the baseline simulation and in the other simulations can then be assigned an economic value using the reference crop average actualised price.

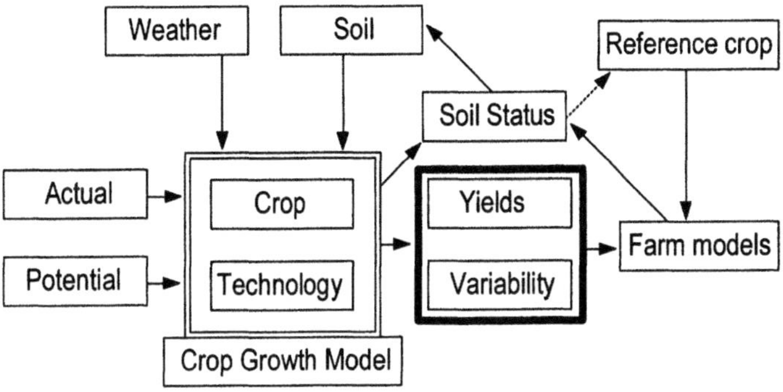

Figure 14.2: Integration of Crop Growth Models in the farm models.

The results are added to or deduced from the economic results of each of the crop/management combinations. Thus, the economic result of each activity considered in the model not only reflects the short term output, but also its influence in the long term. The soil may reach a degree of depletion where it is more beneficial to implement activities that help in regenerating this resource. Even if the results are not accurate, they represent more accurately farmers' behaviour, as mentioned by the farmers during the survey in 1995 (Kafando, 1995). Antle and Stoorvogel (this volume) propose a similar approach, but consider site-specific soil characteristics, which provides more precise results on the impact on soil depletion.

Cattle activities (sedentary and transhumant) are also included in the module, because they are a source of income and manure and a reserve of cash. The animals are differentiated by type (cattle, sheep, donkey, goat), and the decision to keep the animals in the farm is taken on the basis of farmers' expectations on fodder availability, on the value of the animals at the end of the year, on their productivity and on cash requirements. Animals will be sold within the year when the profitability of selling is higher than the potential profitability of keeping them in the farm. However, in some cases farmers can also sell the animals to finance unexpected expenditures, or in case of severe lack of cash to assure consumption or to continue carrying out cropping activities, even if it is not economically profitable. The animals consume fodder from two sources: open access land and fallow, whose fodder production is estimated for different periods within the year, and crop residues, that are estimated with the crop growth model as a fraction of the biomass, and consumed after harvest. Deforestation activities are only linked to the necessity of expanding the crop area, and require labour.

Each of the farm models considers also some other constraints related to farmers' characteristics: consumption and risk related attitude. For consumption, two options can be implemented: either a minimum requirement is considered or it is a function of farmer's wealth. In the first case, Equation (14.5), a minimum level of consumption (*CalMin*) is introduced as a constraint, with additional constraints that reflect consumption preferences of the household, that can be overcome either through own production or through purchases (*Purch*), both expressed in their caloric (*cal*) value. For the second option, Equation (14.6), two components of consumption are considered: a minimum consumption level that has to be reached and a 'surplus' consumption defined as a decreasing function (α) of the level of wealth (*Wealth*). Whatever the alternative chosen, different scenarios of weather conditions are included, affecting production of each alternative (y is the yield of crop j for the weather observation ye) in order to consider farmers' behaviour with respect to consumption 'risk', specially in the situation when own production satisfies most consumption needs, and thus will be extremely dependent on weather conditions.

$$cal_j x_{j,i} y_{j,i,ye} + Purch_{j,i} cal_j \geq CalMin_{i,ye} \tag{14.5}$$

$$CalMin_{j,i,ye} x_{j,i} + \alpha_i Wealth_i Cal_j = Cons_{j,i} \tag{14.6}$$

For the attitude towards economic risk, the variability in prices of the different products is also considered. Two methods are applied in the model, depending on the situation analysed. The first (Equations 14.7 & 14.8) is an application of the Target MOTAD methodology (Tauer, 1983), which introduces a set of minimum income (*MinInc*) constraints allowing the possibility of deviation (*Dev*) from these minimum values to quantify farmers' attitude (*Devtot*). In the second one, Equation (14.9), (Gérard *et al.*, 1999) it is considered in the utility function (*U*) that increases in value with increasing expected revenue (*E(Wealth)*), but decreases with revenue variability (σ^2) and the risk aversion coefficient (A), which depends also on wealth level (Markowitz, 1959; Gérard *et al.*, 1999)

$$\sum_j y_{j,i,ye} x_{j,i} p_{j,ye} + Dev_{i,ye} \geq MinInc_{i,ye} \tag{14.7}$$

$$\sum_{ye} Dev_{i,ye} \leq Devtot_i \tag{14.8}$$

$$MaxU(Wealth) = \sum_{i} (E(Wealth_i) - 0.5`\sigma^2 A_{wealth_i})\, n_i \tag{14.9}$$

The objective function tries to reproduce the goals of farmers within a framework representing individual and regional maximisation objectives. It aggregates at the regional level the maximisation of net individual income for each type of farmers, Equation (14.10), (revenue minus costs *co*, multiplied by the number of farmers *n* in each category *i*) or of the individual wealth, Equation (14.9), plus the optimisation of the use (and share) of common resources (*i.e.* common lands are a source of fodder that the livestock activities can consume and thus produce animals that can be sold, computed in the level of the livestock activities), and of transfers among farmers (*i.e.* the labour transfers generate an income/cost that is computed as such in the objective function). The share of use of common lands or the intensity of transfer of labour between type of farmers are determined endogenously by the model.

$$MaxZ = \sum_{i} \sum_{j} [p_j(y_{j,i}x_{j,i} - cons_{j,i}) - co_{j,i}]\, n_i \tag{14.10}$$

The final output of this module is the marketed supply of production at the regional level. The regional supplies are aggregated in order to estimate total supply at the urban market. The next step is to determine the prices of the different agricultural products.

14.4 The Marketing/Consumption Module

This module represents the confrontation between production and food demand from the urban consumers that determines the prices (Fig. 14.3).

National surplus, estimated with price expectations in the previous module, is confronted with the demand of the different urban consumers under restricted food budgets and preferences for food types. A typology is necessary to identify the different categories of consumers, generally based on income level, and to determine the different coefficients of the demand functions. Therefore, a system of demand functions with convenient behaviour is calibrated and its first order condition model is derived and introduced in the module.

It is possible then to estimate the prices for a given level of production of each product and for all products combined, based on the preferences of consumers and on the endogenously estimated prices. Thus, the adjustment of supply to demand is not simultaneous, but takes place in the recursive part of the model.

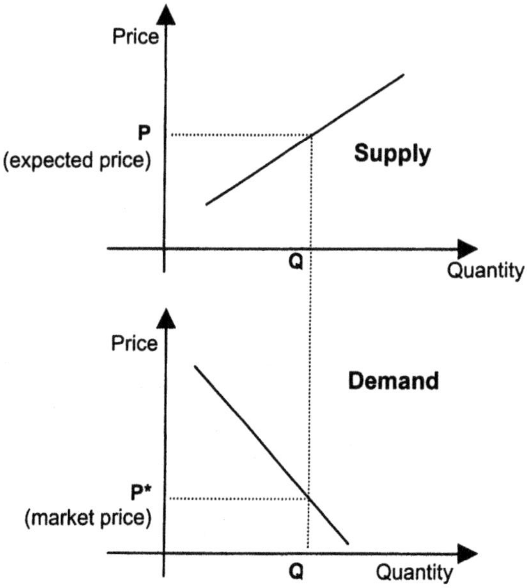

Figure 14.3: The process of decision making based on expected prices and the determination of market prices in the model.

For the Constant Elasticity of Substitution (CES) function, the first order condition can be represented by:

$$x_{i,m} = \frac{R_m (\alpha_{i,m})^{\sigma}}{P_i^{*\,\sigma}(\sum_i \alpha_{i,m}^{\sigma} P_i^{*\,1-\sigma})} \tag{14.11}$$

where $x_{i,m}$ is the quantity consumed of product i by household m, R_m is the food budget of household m, P_i* is the price of product i, $\alpha_{i,m}$ is the coefficient of the CES function specified per product and household type and σ is the elasticity of substitution between products.

The resulting price estimates, corrected for transaction costs, are used to calculate the revenue of farmers in the different regions. Part of the revenue will constitute the cash available for the subsequent season, part will be used to reimburse the debts, part can constitute savings and the rest is 'set aside' to provide for consumption and other unspecified expenditures.

The production and marketing-consumption modules generate the information required for the analysis. However, farmers and consumers are affected by the economic context in which they operate. Its characteristics are described in the remaining module.

14.5 The Macro-Economic Context

In MATA only some aspects of the macroeconomic context that are relevant to understand the functioning of the agricultural sector, are described: the rate of interest, credit availability (for the agricultural sector or for specific activities within the sector), the demand of labour by the rest of the economy, the level of salaries, *etc*. The choice of the components that will be used in the model will depend on the situation analysed. They will be selected in order to represent the main aspects that will have an influence on the choice of alternatives by the farmers and thus are necessary to represent actual behaviour. The reference value of these variables is determined exogenously. Some can be modified to design scenarios at the macro-economic level, *i.e.* the amount of credit available, a specific line of credit, a change in interest rate, economic growth (which can be translated in better salaries), *etc*. The model represents only the reactions of the agricultural sector to these incentives (positive or negative), but it does not indicate the effects (endogenous) of these reactions on the other sectors. In this portion of the model, the contribution of the sector at some levels is summarised, like its impact on the trade balance, the employment level of the economy, *etc*. To demonstrate the capabilities of the model, some results of simulations of economic scenarios are presented in the following section.

14.6 Results of Policy Simulations for Burkina Faso

As mentioned, the model allows to determine different types of impacts. Both general (*i.e.* changes in interest rate, adjustments in general salaries, *etc*.) and agricultural macro-economic (changes in agricultural and input prices) scenarios can be defined. As an illustration, some results of scenarios of policies for the case of Burkina Faso are chosen (Deybe & Castella, 1999): the base run represents continuation of the pre-1994 policies for a period of 5 years; the devaluation of the FCFA by 100% in 1994[3], reflecting the impacts observed between 1993 and 1995 (changes observed in cotton, rice, wheat and fertiliser prices); a liberalisation of all agricultural prices (including fertilisers) after the devaluation; and a decrease in the availability of fertilisers (with the same parameters observed after the devaluation) as a consequence of the application of the Rio conference recommendations. The policies undertaken after 1995 are not included in the model, even if the prices observed in reality followed the trends simulated by the

[3] The same exercise was carried out for the case of Benin (Senahoun *et al*., 1999).

model in the devaluation scenario, but of higher magnitude in the case of cotton and maize.

Three levels of results are presented: farmers, consumers and soil degradation, to show the possibilities of comparison between policies and to rank them according to established priorities.

At the farm level it can be observed in Fig. 14.4 that for the Sahelian region, the average farmer might not have benefited from any of the alternatives, given the lack of flexibility of the production systems in this area. On the other hand all three alternatives are extremely beneficial for farmers in the West and Center regions, due mainly to the rich natural endowment of these regions, as well as to the possibility to grow cash crops, specially cotton, which benefits strongly from the devaluation and the liberalisation.

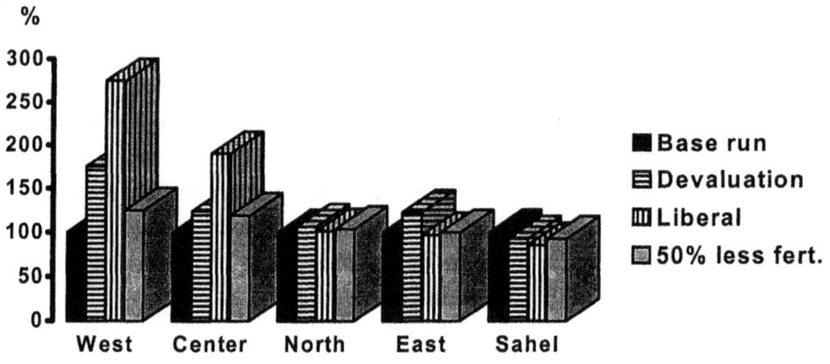

Figure 14.4: Changes (%) in farmers' revenues as a consequence of different policies.

At the consumers' level (Fig. 14.5) the liberal alternative negatively affects caloric intake of the urban consumer, due to the increase in food prices and the constant urban salaries. The other two have a positive impact, because they lead to increased local cereal production, even under a reduction in the availability of fertilisers, as the cultivated area of sorghum-millet is expanded.

Finally, with respect to the impact on soil degradation (Fig. 14.6) it is found that the devaluation, inducing intensification of production, will generate lower erosion as a whole. The liberal option leads to both expansion of the cultivated land and intensification of production, increasing total erosion, and the reduction in the availability of fertilisers leads to expansion of the cultivated area and thus, to higher erosion.

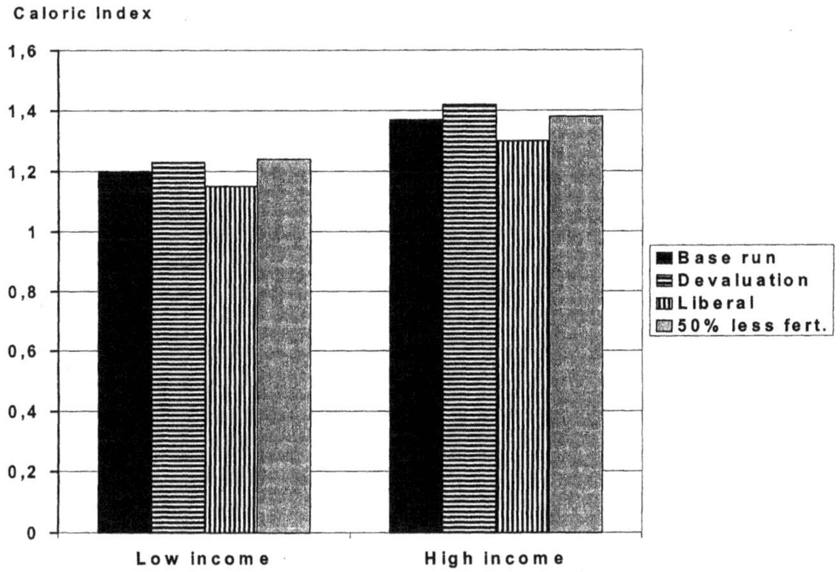

Figure 14.5: Changes in caloric intake as a consequence of different policies (1=2300 Kcal).

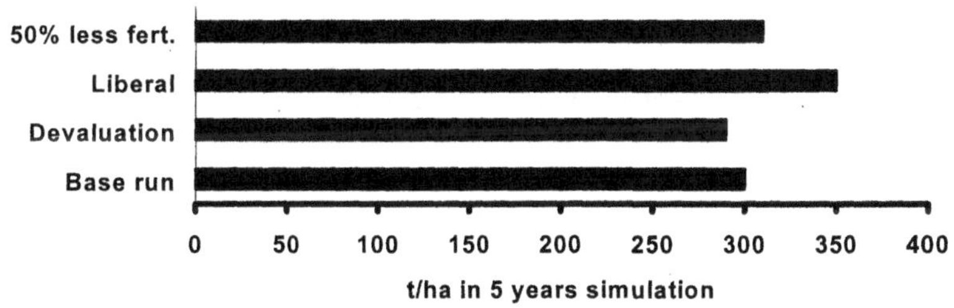

Figure 14.6: Soil erosion as a consequence of the different policies (t/ha during the 5 years simulation period).

It follows that the type of policy that should be adopted will depend on the objectives of the policy maker and the orientations that he wants to stress. Should the long-term predominate in the decision, jeopardising the well-being of present generations, with the risk to loose the elections? Should the short-term be the only worry, knowing that there will be a future negative impact ? The model can be a useful tool for a more judicious decision-making process, whatever the decision made.

14.7 Conclusions

The model described in this paper can provide useful insights in the impact of agricultural and economic policies on different farm types, consumers and on soil degradation.

The technique used combines crop growth models with more traditional economic models at different levels. This combination allows to reproduce aggregate farmers' behaviour and to include the long-term effects on soil quality in the analysis, using a common basis for the comparison of trade-off between actual and future profitability of agriculture activities.

This representation still has some limitations that have to be alleviated. First, the low degree of accuracy of the results on erosion resulting from the consideration of global soil types and the associated underestimates due to the aggregation process. The procedure proposed by Antle and Stoorvogel (this volume) may improve the accuracy of the results, even though the linkages among farmers do not appear explicitly in their framework, which in some cases might be an important factor defining the use of soils and thus the impact on erosion. Secondly, even though it is possible to include in the model the use of common resources, the processes of overuse of these resources and their associated degradation, often observed in reality, have not yet been incorporated. Thirdly, the model does not consider the structural changes that will occur whatever the policy simulated. Finally, it does not take into account the notion of risk sharing among farmers, very important in times of catastrophe. These last two factors would require to include in the model not only the aggregation of individual decision making but also some alternative methods, like the use of cellular automatas, to represent the process of negotiation between farmers.

14.8 References

Antle, J.M. & J. Stoorvogel (2000): Integrating site-specific biophysical and economic models to assess tradeoffs in sustainable land use and soil quality. This volume

Boussard, J.-M. (1994): Splendeur et misère de l'économétrie. Economie Rurale 222, 50-51

Candler, W., J. Fortuny-Amat & B. McCarl (1981): The potential role of multilevel programming in agricultural economics. American Journal of Agricultural Economics 63, 521-531

Dent, J.B., G. Edwards-Jones & M.J. McGregor (1995): Simulation of ecological, social and economic factors in agricultural systems. Agricultural Systems 49, 337-351

Deybe, D. (1994): Vers une agriculture durable. Un modèle bioéconomique. Cirad, Paris

Deybe, D. & P. Castella (1999): Is trade devaluation the right policy after a devaluation? The case of Burkina Faso. In: A.K. Dragun & C. Tisdell (Eds.) Sustainable agriculture and environment. Globalisation and the impact of trade liberalisation. Edwar Elgar, Cheltenham, 287-296

Gérard, F., 1991. Instabilité économique des prix des produits agricoles et influence de l'incertitude sur les comportements économiques: essai sur les problèmes associés à la régulation de l'offre. Thèse de Doctorat, Université de Paris I Panthéon-Sorbonne, Paris, 567 p.

Gérard, F., J.-M. Boussard & D. Deybe (1995): MATA: un outil d'analyse multi-niveaux pour le secteur agricole. In: Comportement micro-économique et réformes macro-économiques dans les pays en voie de développement. Collection Actualité Scientifique. Aupelf-Uref, Paris, 347-364

Gérard, F., Erwidodo & I. Marty (1999): Evaluation of the impact of trade liberalization on food crop production and farm income in lowland Java, Indonesia. In: G.H. Peters & J. von Braun (Eds.) Food security, diversification and resource management: refocusing the role of agriculture? Proceedings Twenty-third International Conference of Agricultural Economist. Oxford: Ashgate, 491-501

Kafando, B.E. (1995): Analyse de l'impact de la dévaluation du Franc CFA sur la production agricole et la sécurité alimentaire et propositions d'action. Systèmes de production. Document de travail. République du Burkina Faso - FAO, TCP/BKF/3453. 124 p. + An.

Markowitz, H.M. (1959): Portfolio selection: efficient diversification of investments. Wiley, New York

Senahoun, J., F. Heidhues & D. Deybe (1999): Structural adjustment programs and soil erosion. Quarterly Journal of International Agriculture 38, 245-262

Solow, R. (1993): The economics of resources or the resources of economics. In: R. Dorfman & N.S. Dorfman (Ed.) Economics of the environment, 3rd Ed. W.W. Norton & Co., New York, 162-178

Tauer, L.W. (1983): Target MOTAD. American Journal of Agricultural Economics 65, 606-610

Taylor, L. (1996): Sustainable development: an introduction. World Development 24, 215-225

15 Soil Degradation in Macro CGE Models

Solveig Glomsrød

Statistics Norway, Oslo, Norway

Abstract

Soil degradation is a widespread and serious threat to poor developing countries. The large contribution of agriculture to the national economies of these countries makes partial analysis of soil degradation insufficient. Soil productivity loss affects the urban economy through food prices, demand for farm inputs and consumer goods. Also, general economic policies create incentives that influence farm practices and the pressure on soil resources. To focus these interactions, soil degradation has been incorporated in macro CGE models. Two approaches are presented in this paper. One incorporates soil productivity with nitrogen as limiting factor in a CGE model for Tanzania, forecasting inputs of fertiliser, land and output by eleven crops. The use of fertiliser and amount of recycled crop residues determine the rate of soil productivity loss. Deforestation for subsistence agriculture has been modelled in a CGE for Nicaragua. Migration to the agricultural frontier is encouraged by the imputed income from producing their own diet, and serves as an alternative to rural/urban employment or urban unemployment. Policy simulations show that economic reforms like devaluation have significant impacts on land use and nutrient mining. Deforestation for subsistence agriculture is sensitive to income distribution and changes in food prices. High economic growth does not guarantee forest conservation.

15.1 Introduction

Agriculture is a dominant activity in most developing countries. Soil and labour are the main input factors. Although soil plays a major role in the production process, the dependence of the national economy upon the quality of this resource has hardly been studied in quantitative terms. Relevant questions are: does current soil degradation affect production and consumption decisions, and how is sector and aggregate income affected when farmers and consumers have modified their practices and habits according to new costs and constraints? And further: how do economic policies affect soil degradation by creating incentives for agriculture? Recent economic reforms in developing countries contain elements that influence farm practices significantly by rearranging prices of inputs and output. The consequences of such policies should preferably be assessed within a macro framework, since the extent of restructuring may be appreciable in aggregate terms.

Here I will describe two computable general equilibrium (CGE) model approaches that take soil productivity into account. One is a CGE model of Tanzania, incorporating a specific soil module that translates mining of nutrients into loss of soil productivity in various crops. The soil module draws upon Aune & Lal (1995), whose model framework has been used for analyses for Tanzania (Aune *et al.*, 1997; Grepperud *et al.*, 1999) and for Ghana (Alfsen *et al.*, 1997).

The other approach, a CGE for Nicaragua is more rudimentary in dealing with soil (Glomsrød *et al.*, 1998). It does not consider the problem of declining soil productivity in the established and market integrated agriculture, but focuses on the dynamics behind migration to forest reserves and deforestation for subsistence farming that rapidly degrades the fragile forest soil. The model incorporates a rural-urban migration mechanism in the tradition of Harris-Todaro (1970) where migration is encouraged by inequality between rural wage and expected urban wage, taking the probability of unemployment into account.

Both types of models have been used for studies of structural adjustment policies. This chapter focuses on the main features of the two model approaches.

15.2 Environment and Productivity

Economic activity constitutes a permanent pressure on the environment, which looses capabilities important to our production systems. In addition, the feedback from the environment to the society has pure welfare effects (health, well-being, aesthetic value). In our context, only feedback to the production system is included, not the cost of loosing pristine nature, *etc.* Fig. 15.1 illustrates linkages between the conventional market economy and the more realistic arena where the environment has a role to play.

The upper part of the figure defines the resources available to the economy, like the stocks of labour, capital, technology and soil (representing natural resources in general). The efficient use of resources depends on boundary conditions like taxes, and regulations, world market prices and the level of public expenditures. Each sector i adapts to the situation by optimising input of labour (L_i), capital (C_i), fertiliser (F_i) and materials (M_i).

The factor input combination and level of production activities affect the environment in various ways, as do the level and mix of household consumption. Among material inputs, fossil fuel consumption pollutes the air, which affects health and the supply of labour. Air pollution also reduces the lifetime of constructions and machinery, thus diminishing the capital stock. Fertiliser use may pollute to various degrees, but also supplies nutrients to the soil thus maintaining productivity. Feedback from the environment to the economy can be positive or negative. In our context, we deal with feedback from agriculture to soil productivity, hence, we need a module that describes the soil productivity implications of farm practices.

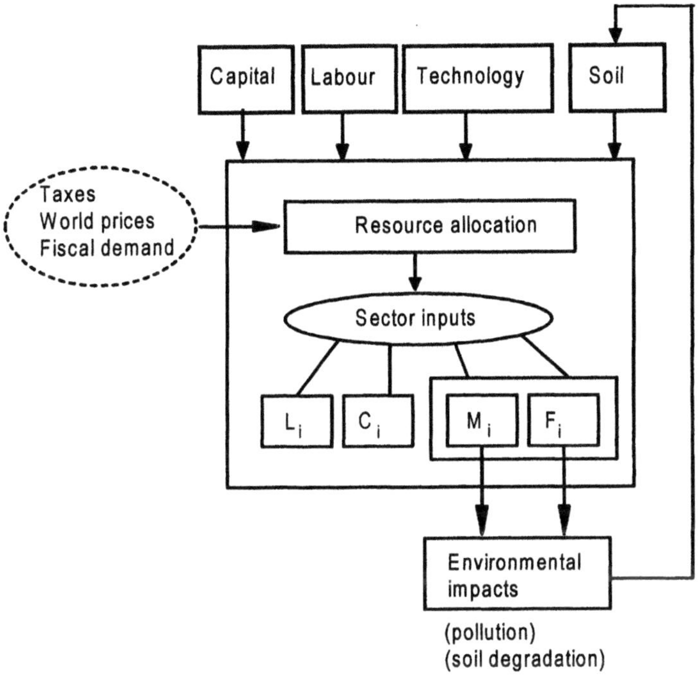

Figure 15.1: Integrating environment in economic models.

Environmental problems relate to the flows of goods through the economy, and the technologies used for their processing or consumption. Some environmental problems are mainly local and do not really interact with economic decisions, although they may be serious and costly. Others are deeply rooted in behaviour of producers and consumers, such as air pollution associated with fossil fuel combustion. In that case, it makes sense to incorporate the environmental aspects in macro-economic models. Soil mining in a predominantly agricultural economy is closely related to the economic incentives surrounding most farmers and consumers.

To model environmental effects, the material flows and technologies must be adequately described. In the model presented below, agriculture has been described as a multi-sector activity, and the inputs of materials like fertiliser and pesticides are specified in model versions where soil productivity is the main topic. The degree of detail in a macroeconomic approach can never be perfectly satisfactory, however, the approach may reveal the strength of policy-induced effects on soil productivity and the role of agriculture.

15.3 A CGE Model with Soil Nutrient Mining

15.3.1 Economic Core Model

The economic models in this chapter are fairly standard (quasi static) computable equilibrium models (Dervis *et al.*, 1982). Concerning the static character of the model, it tracks year by year development by assuming a constant saving rate and a residually determined public saving, thus generating capital accumulation. No optimisation over time is performed in the model. Producers and consumers behave as maximisers of profits and utility of consumption, respectively. The models contain no capital market, because credit facilities are considered inadequate so that profitability of investments and capacity to pay interest do not yet correspond to the actual investment criteria. A complete version of the model can be found in Grepperud *et al.* (1999).

There are some 20 sectors, each producing a single commodity, of which roughly half represent agricultural activities. Commodities are produced by labour, capital and intermediates, that are generally used in fixed proportions to the production level, while use of labour is optimised. In agriculture, land capital, fertiliser and pesticides are additional factors that farmers adjust in their efforts for best achievements. Production functions are Cobb-Douglas.

The wage rate is assumed to be constant and demand for labour is passively met by supply from the unemployed labour reserve.

Private savings equal total income less taxes and total private expenditure. Public saving is a residual of public revenues minus exogenous public expenditure.

Investments in real capital are determined by total savings. Foreign savings are assumed to be exogenous, *i.e.* domestic savings essentially determine economic growth potential. Each sector has an exogenous initial capital stock and receives a fixed share of total nominal investments. Sector capital stock changes by the balance of depreciation and investment. Allocation of investment to sectors is determined by fixed (base year) coefficients. The sectoral pattern of investment is translated into demand for investment goods by sector of origin.

Domestically produced goods compete with foreign goods, both on the world market and at home, assuming non-perfect substitution (the Armington assumption). The higher the price of the domestically produced commodity relative to the price of the imported variety, the higher the import share. On the production side, a constant elasticity of transformation (CET) function indicates that domestic output cannot be shifted costlessly between the domestic market and the world market. The higher the price of exports relative to domestic price, the higher the export share.

World market prices are exogenous, *i.e.* the model typically represents a small country unable to affect external terms of trade. The nominal exchange rate is set by the government. Prices of imported and exported goods are determined as world market prices multiplied by exchange rate and adjusted for tariffs.

In the model for Tanzania, land capital is assumed to be variable in crop production, except for coffee and tobacco. For coffee, this reflects that suitable land in the main production region Kilimanjaro, already is scarce. For tobacco, the assumption reflects that farmers are strongly dependent upon credit for its intensive cultivation, and credit is assumed to be rationed over the model time horizon.

15.3.2 The Agro-Ecological Model

The agro-ecological model describes soil productivity over time, and the effects of agricultural technology on the resource base. The soil module builds upon the Tropical Soil Productivity Calculator (TSPC) developed by Aune & Lal (1995) to deal with the variety of crops and limiting factors in tropical agriculture. Further, it allows analysis over several years time horizon.

Output in agriculture is described as a product of potential yield and indices ranging from zero to one for various growth factors, such as nitrogen and phosphorus. Potential growth is based on growth periods, water supply, radiation and temperature. Actual yield is also adjusted for soil acidity and application of pesticides.

In African agriculture, the main nitrogen source is soil organic matter, while supply from fertiliser is about 20 kg ha^{-1}. Hence, yields mainly represent resource exhaustion or soil mining. In the models for Tanzania, soil nutrient mining with respect to nitrogen, representing the limiting factor to yield was the focal point, and the nitrogen cycle (Fig. 15.2) was incorporated in the model framework.

Soil nitrogen is removed in crops [8] and lost through leaching and volatilisation [9]. Part of the residues is returned to the soil [7] with nitrogen partly in mineral [2] and partly in organic [3] form. Mineral nitrogen enters as chemical fertiliser and atmospheric deposition. The relevant equations are:

$$NR_t = r \cdot NS_t + \frac{\beta}{3}\sum_{k=2}^{4} NRR_{t-k} + na_t \,. \tag{15.1}$$

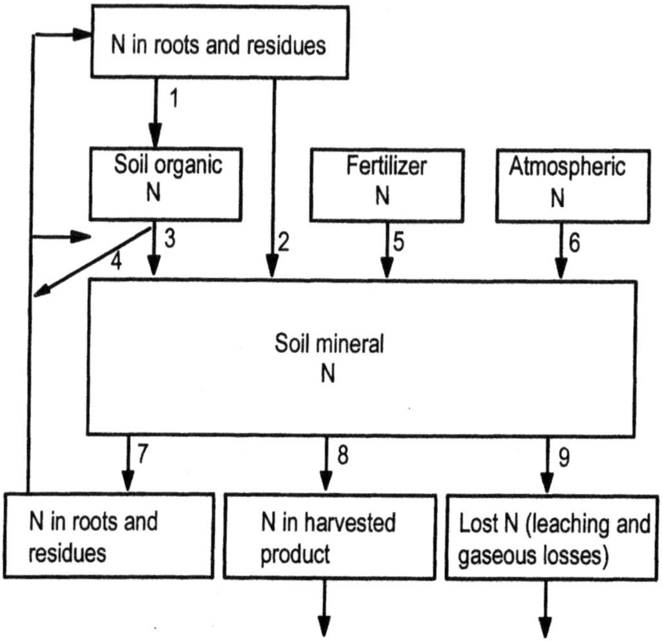

Figure 15.2: Nitrogen (N) cycle.

Mineral nitrogen (*NR*) is released from the stock of soil organic nitrogen (*NS*) in soil organic matter. Nitrogen in residues (*NRR*) is mineralised over a 3-year period with a lag due to immobilisation by micro-organisms. β is the proportion of nitrogen in residues available for mineralisation. na_t is deposition from the atmosphere [6].

$$NS_t = (1-r)NS_{t-1} + (1-\beta)NRR_{t-1} - NE_{t-1} \tag{15.2}$$

Soil organic nitrogen is mineralised at an annual rate r, receives direct contributions from residues (*NRR*, [7]) and is lost through soil erosion (*NE*).

$$NRR_t = \frac{X_t}{KL_t} \cdot (hp) \tag{15.3}$$

hp represents crop harvest parameters like stover to grain ratio, proportion of stover retained on land and the nitrogen content of stover. Together with yield (output *X* per unit land area *KL*)) they determine the level of nitrogen in residues (*NRR*).

$$NE_t = a \cdot \frac{NS_t}{D_t}\left(b - c \cdot \frac{X_t}{KL_t}\right) \tag{15.4}$$

Nitrogen loss in soil erosion (*NE*) is dependent upon the density of soil organic nitrogen (*D* is soil depth) adjusted for the modifying impact of high yields. Soil loss entails loss of soil organic matter, and the nitrogen content of eroded material is twice that of the bulk soil (*a*), as erosion removes the finest and most nitrogen-rich particles from the soil. Soil loss is calculated by means of the universal soil loss equation (USLE), that includes precipitation, soil properties, plot length and slope and vegetation cover (FAO, 1979). High yields are associated with high vegetation cover which provides a significant contribution to soil conservation. Hence, this is an important link between the economic and ecological realms. Erosion is considerably lower under forest and grassland than under annual crops such as cereals. The model adjusts soil erosion for level of yields. Finally, soil depth is updated:

$$D_t = D_{t-1} - d\left(b - c\frac{X_t}{KL_t}\right). \tag{15.5}$$

Equation 15.5 updates soil depth according to USLE, adjusting for the current vegetation cover effect. Factor *d* basically represents the USLE impact on the bulk of soil transformed to soil depth in correspondence with soil density.

15.3.3 Links Between the Two Modules

To link the two modules, description of the agricultural activities in the CGE has to be fairly detailed. In the model for Tanzania, 12 crop producing activities are distinguished. In their production functions, fertiliser, pesticides and land capital are included.

The CGE model generates sector output and use of land and fertiliser. Yield is then transferred to the soil module for calculation of nutrient loss and vegetation cover factor. The associated decline in soil productivity is then returned to the economic model. Use of fertiliser increases yield, and contributes to the soil nitrogen store. In the Tanzania version of the TSPC the vegetation cover factor hardly affects the rate of erosion, as agriculture primarily takes place in flat areas.

Feedback is modelled in two ways. In Aune *et al.* (1997), soil productivity is treated as a factor neutral technical change parameter that is adjusted annually according to last year's practice and yield. This is consistent with a view that the farmer is not aware of the extent of nitrogen loss and therefore does not adjust next year's farm practice. In Grepperud *et al.* (1999), nitrogen from both chemical fertiliser and natural sources is treated as a joint input factor, implying that the farmer will optimise factor use while taking into account the development with respect to soil quality. The latter version has a tendency to increase fertiliser use to compensate the loss of soil organic nitrogen.

When linking the modules, it was taken into account that the calibrated crop production parameters of the CGE model reflect farm practice, while the agro-ecological production module refers to field level. The experimental field productivity of fertiliser is very high at low input levels, but the CGE productivity parameter (the cost share), calibrated from base year data, gives the impression of low productivity. The difference relates to imperfections in the credit and fertiliser markets that constrain farm input. When linking the economic and the agro-ecological module, the challenge is to conserve the realistic demand incentives, while incorporating the productivity capacity of nitrogen. In the approach of Grepperud *et al.* (1999), the value of nitrogen from natural resources is assumed equal to the market price of nitrogen from chemical fertiliser, when calibrating the productivity parameters. The cost share of total nitrogen is then calculated and the productivity of nitrogen set accordingly. The productivity of natural nitrogen is thus somewhat dampened compared with the estimates from field studies, while the productivity of commercial nitrogen is more close to what it would be without the market imperfections. The effect will be to bias farmer behaviour by ignoring the institutional obstacles present in the base year, but it reflects a more realistic productivity of nitrogen as such. The type of analysis carried out should be considered when deciding on the approach.

15.3.4 Policy Simulations

When studying impacts of economic reforms (Aune *et al.*, 1997), it was found that a devaluation that increased GDP by 25% over 10 years resulted in expansion of the cultivated area by 17%. Devaluation makes fertiliser more expensive (all is imported). To meet the rising demand for output in the process of economic growth, new land compensates for the more expensive fertiliser. This result refers to the version with a Hicks neutral soil productivity index, *i.e.* a generally negative trend in output that cannot be related to any particular input. The approach taken by Grepperud *et al.* (op. cit.) implies that farmers are fully aware of their productivity loss and regard chemical fertiliser as increasingly productive when natural soil fertility declines. This creates a tendency for rising fertiliser demand at constant fertiliser prices, and will affect the response to higher prices of fertiliser due to devaluation. A crucial question is then to what extent farmers are aware of the negative trend in productivity against significant variation in output due to climate variability.

15.4 A CGE Model with Deforestation, Soil Degradation and Migration

15.4.1 Model Specification

The model approach discussed in Section 3 deals with the steady decline in soil productivity in a predominantly agricultural economy. Here, the focus is on deforestation for subsistence agriculture, causing rapid deterioration of tropical forest soils. The heart of this model approach for Nicaragua is the dynamics of migration to and from the agricultural frontier.

The core model contains totally 24 sectors, of which 11 are market-integrated agricultural sectors, but adds a frontier sector of subsistence farming, covering two distinct areas that are the dominant settlement centres, where rapid expansion of cultivation takes place in virgin forest. The migrators are settling with a cultivation pattern and a technology as described below, based on surveys of the frontier activity carried out by the Ministry of the Environment (GTZ/MARENA, 1992) and CIPRES (1992).

The frontier labour force (LF^t) produces zl_{mz}, zl_{fr} and zl_{ar} units of maize, beans, and rice respectively, per man-year. The corresponding frontier production level for a single crop b is:

$$XF_b^t = zl_b LF_b^t,\ b \in (mz, fr, ar) \tag{15.6}$$

where b represents maize (mz), beans (fr), and rice (ar). This production level generates an imputed income WF^t per man-year:

$$WF^t = \sum_b PF_b^t zl_b,\ b \in (mz, fr, ar) \tag{15.7}$$

at the frontier. Whether this income level attracts migrators, depends on alternative income opportunities in the economy. The product price obtained at the frontier (PF_b^t) is set equal to the price facing the market-integrated producers of basic grains (maize, beans, and rice) in the economy:

$$PF_b^t = P_b^t,\ b \in (mz, fr, ar). \tag{15.8}$$

Production at the frontier is for own consumption and as such, grain has a basic value independent of market prices. However, for accounting purposes it is reasonable to calculate an imputed income based on general market prices. The important assumption here, is that a change in market prices signals a change in the opportunity cost of remaining within the market economy and not getting access to the frontier diet. If frontier production would be commercialised, farm gate prices would in general be lower than in central agricultural areas with better infrastructure and lower transportation cost.

Alternative employment and income is available in the established rural economy, and in urban areas, both with a probability less than one. Equilibrium is obtained when there is equality between the expected income options. The rural wage level WR^t, an index for wages in established or market integrated rural production sectors, is equal to the urban wage level WU_0, adjusted for the probability of being employed in urban sectors:

$$WR^t = \left(1 - UR^t\right)WU_0 \tag{15.9}$$

where UR^t is the unemployment rate in the urban economy. Farming at the frontier is primarily an alternative to rural farming. We assume that landless labourers in rural farming leave for the frontier, if income at the frontier is large relative to income in rural farming.

In considering how to model this relationship, an extreme alternative would assume that people leave for the frontier if income at the frontier exceeds that in rural farming, making the income levels equal in equilibrium. A less extreme alternative, which we chose, is to assume a constant elasticity of transformation (CET) macro-relationship between working in rural farming and working at the frontier. Hence, some landless labourers leave for the frontier immediately when income at the frontier exceeds income in rural agriculture. Others need a larger income incentive to move, for reasons of transport costs, non-pecuniary preferences or risk aversion. Among the non-pecuniary costs of leaving the established communities in rural areas or the cities is the loss of access to public schools and health services. At equilibrium, a non-zero income differential is associated with a finite stock of frontier farmers.

To model the CET relationship, we use the dual and specify the macro-index (WX^t) of income in rural agriculture as function of the imputed frontier income and the rural wage level:

$$WX^t = \left[\eta_1\left(WF^t\right)^\rho + \eta_2\left(WR^t\right)^\rho\right]^{\frac{1}{\rho}}. \tag{15.10}$$

The elasticity of substitution (ρ) is set to -0.6, which suggests significant costs of resettling. We explain the stock of frontier farmers as a function of income relative to the wage index (which, since the wage index consists of frontier income and rural income, is equivalent to defining it as a function of frontier income relative to rural income):

$$LF^t = \eta_1 LS^1\left[WF^t/WX^t\right]^{\rho-1}. \tag{15.11}$$

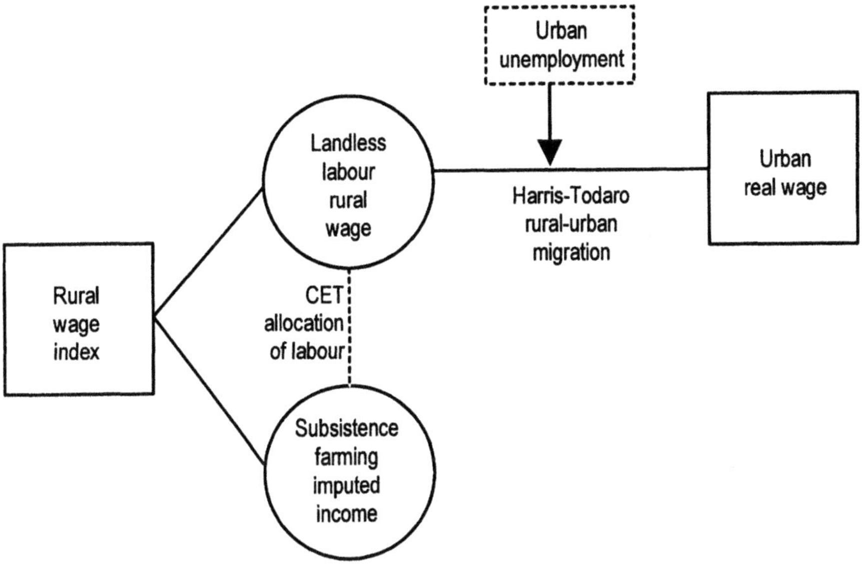

Figure 15.3: Income dynamics behind migration flow.

To fully describe the labour market, it remains to determine the stock of landless rural labourers. The model, therefore, allocates potential labourers between urban unemployed and rural labour, such that the urban unemployed equal the number necessary for the expected urban wage to equal rural income for the landless labourer.

Total labour supply (LS^t) in the economy grows at an annual rate g:

$$LS^t = LS_0(1+g)^t \tag{15.12}$$

where LS_0 is the base year labour supply.

Deforestation consists of two elements. One is land cleared by new colonisers, and the other is land cleared by the population that has settled earlier, but has to clear new land due to the rapid soil degradation. The deforested area in year t (DF^t) is:

$$DF^t = A\left[LF^t - LF^{t-1} + 1/3\, LF^{t-1}\right] \tag{15.13}$$

where A is the average area cleared per man-year. We assume that one-third of formerly settled frontier farmers have to clear new land each year. The formulation ignores the fact that the frontier population is composed of waves of colonisers of various sizes over the years, but this is a minor omission.

15.4.2 Policy Simulations

The subsistence farmers receive an imputed income determined by the market prices of basic grains. If food prices are low, it is more attractive to be employed in the market-integrated sectors for a given real wage. Low food prices enable urban and rural workers to consume beyond subsistence and buy "luxury" goods and services, not available in the remote areas. This mechanism makes the rate of deforestation sensitive to distributional change and demand-induced changes in food prices. Studies of economic reform policies illustrate that increased as well as reduced deforestation may be poverty-driven. Further, a high GDP growth rate does not guarantee less pressure on forest reserves. Introduction of a flexible urban real wage will increase economic growth significantly, but concurrently increase food prices, making subsistence farming more attractive relative to both urban and rural employment. Consequently, the forests may keep burning in spite of good macroeconomic performance.

Particularly, improving the fiscal balance by cutting public expenditures or by means of a sales tax reform, both enhance economic growth and conservation of forests. The former policy is efficient in stimulating growth, while the latter is efficient in conservation. While rural income is conserved under reduced public expenditures, it is being eroded by a sales tax increase, through a demand-induced decline in prices of basic grains. With low prices for basic grains, migration to the cities appears a better choice than subsistence agriculture at the frontier. Thus, forest conservation associated with a higher sales tax is driven by a general rural poverty increase.

15.5 Some Final Comments

To start from the macro level to study soil erosion and deforestation is based on the assumption that economic framework conditions have a systematic aggregate effect on land productivity. Analyses based on a computable general equilibrium model can be characterised as policy-oriented by including the important economic policy tools and limiting the scope of research to overall trends in land use and soil quality. Data limitations prevent an economic policy approach at a very detailed regional level, since data on regional interactions are scarce. However, to select one or a few regions of special concern might be useful as well as feasible with respect to input data. In particular, this might be the case with a region of fairly homogenous agricultural and ecological characteristics. Few dimensions of economic interactions makes the approach more feasible, but less relevant if the existing link is not of high economic significance. The Nicaragua model is an example of a one-dimensional link (wage income) that is deeply rooted in economic policies and thus induces policy effects on land use at the otherwise remote agricultural frontier. A satisfactory description of individual plot destiny cannot be obtained in a policy study. However, the micro and macro approaches co-develop in a fruitful way in other areas of environmental-economic research, such as air pollution and greenhouse gases where economic measures are considered relevant tools. Possibly they are so in the case of land use and soil quality as well.

15.6 References

Alfsen, K.H., T. Bye, S. Glomsrød & H. Wiig (1997): Soil degradation and economic development in Ghana. Environment and Development Economics 2, 119-143.

Aune, J.B. & R. Lal (1995): The tropical soil productivity calculator – a model for assessing effects of soil management on productivity. In: R. Lal & B.A. Stewart (Eds.) Soil management: Experimental basis for sustainability and environmental quality, III series: Advances in soil science. CRC Press, Boca Raton, Florida

Aune, J, S. Glomsrød, V. Iversen & H. Wiig (1997): Structural adjustment and soil degradation in Tanzania: A CGE-model with endogenous soil productivity. Discussion Papers No. 189, Statistics Norway

CIPRES (Centro para la Investigacion, la Promocion y el Desarollo Rural y Social) (1992): El campesinado en la zona de amortiguamiento de la reserva biologica. Cuadernos CIPRES/13

Dervis, K., J. de Melo & S. Robinson (1982): General Equilibrium Models for Development Policy. Cambridge University Press, Cambridge

FAO (1979): A provisional methodology for soil degradation assessment. Rome

Glomsrød, S., M.D. Monge & H. Vennemo (1998): Structural Adjustment and Deforestation in Nicaragua. Environment and Development Economics 4, 19-43

Grepperud, S., H. Wiig & F.R Aune (1999): Maize Trade Liberalization vs. Fertilizer Subsidies in Tanzania. A CGE Model Analysis with Endogenous Soil Fertility. Discussion Papers No. 249, Statistics Norway

GTZ/MARENA (1992): Estudio Básico. Protección de Recursos Naturales y Desarollo Rural Sustenido en la Zona Río Waspute-Bonanza-Siuna Región Autónomia Altántico Norte (RAAN) de Nicaragua, Managua

Harris, J. & M.P. Todaro (1970): Migration, unemployment and development: a two sector analysis. American Economic Review 60, 126-142

16 The Effect of Soil Degradation on Agricultural Productivity in Ethiopia: A Non-Parametric Regression Analysis

Michiel A. Keyzer & Ben G.J.S. Sonneveld[*§]

Centre for World Food Studies of the Vrije Universiteit (SOW-VU), Amsterdam, the Netherlands

Abstract

The paper estimates the effect of soil degradation on crop yields for dominant cereals in Ethiopia at a nation-wide level and analyses its relation with population density and fertiliser use. A soil degradation index is derived from an ordered qualitative classification on the degree of soil degradation and the area extension. Biophysical variability is incorporated by using, as dependent variable, the yield ratio (actual/potential yield) to correct for agro-climatic and crop genetic differences, and by including soil fertility as explanatory variable. The data set is cross sectional and obtained from gridded overlays on soil degradation, climate, soil, land form, population (and cattle) density. The relationships are estimated via non-parametric (kernel density) regression and the estimation results are depicted in 3-D graphs. It appears that the relationship between yield ratio, land degradation and soil fertility is not very strong. Yet, three stylised facts can be identified. First, land degradation has its major impact on soils of lower fertility, where population levels are low. Secondly, on fertile soils, land degradation is largely compensated by fertiliser application. Finally, most people can be found on the slope facing a deep and dangerous precipice. A spatial representation of the elasticity of crop productivity with respect to soil degradation indicates that most vulnerable areas are located in the northern part of the country.

[*] The comments by participants of the workshop Economic Policy Reforms and Sustainable Land Use in LDCs, Wageningen, the Netherlands, and the contribution by our colleague R.L.Voortman in calculating the yield potentials are gratefully acknowledged.

[§] Colour versions of Figures that appear in this chapter are available at http://www.sow.econ.vu.nl/output.htm

16.1 Introduction

In the highlands of Ethiopia, overgrazing and mounting demographic pressures may cause soil losses that annul within one decade the creation of a millennium. This by itself is not sufficient cause for alarm. The land lost will often deposit as fresh sediments downstream and it is good to recall that the most fertile lands in the world often are of this depositional origin. However, the loss becomes serious if the topsoil layer is shallow already, and if the settlement pattern is dense. These are typically the conditions prevailing on the Ethiopian highlands, and consequently, soil degradation is generally seen as a major threat to the long run food security of the country, that calls for immediate action. To the extent that low cost soil conservation measures are adequate to counter the process, it is safe to advocate such intervention without reservation but if the measures to be taken are capital intensive or involve bans on the cultivation of currently occupied soils, the issue becomes more subtle and calls for an assessment of the trade-offs in agriculture between short term losses and long-term gains. Such an assessment should be effectuated at the regional or national scale rather than for selected, possibly most affected fields.

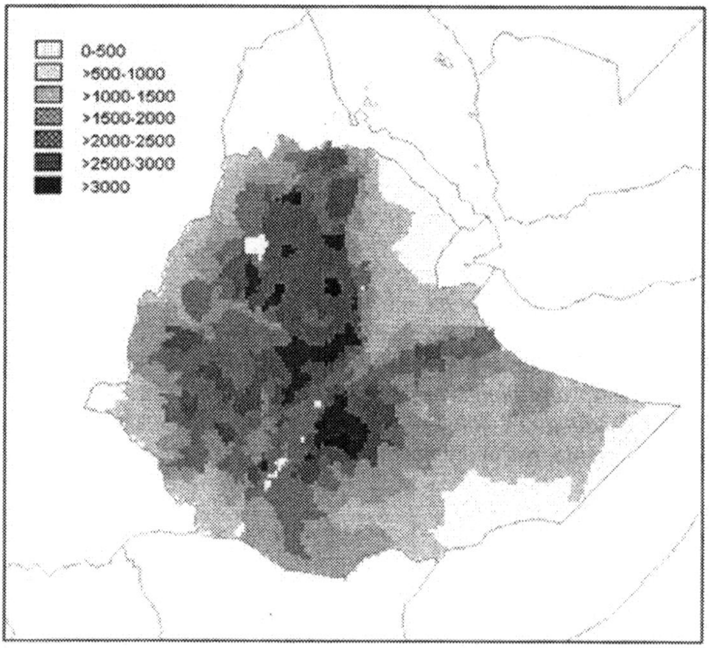

Figure 16.1: Location and altitude map (in m) for Ethiopia.

In Ethiopia, agricultural production largely takes place on fertile highlands (above 1500 m; see Fig. 16.1), which constitute 43% of the country and account for 95% of the cultivated area. The favourable agro-ecological conditions enable them to sustain remarkably high population and cattle densities, as they carry 88% out of a total population of 60 million people (see Fig. 16.2) and 75% of the cattle population of 20 million head.

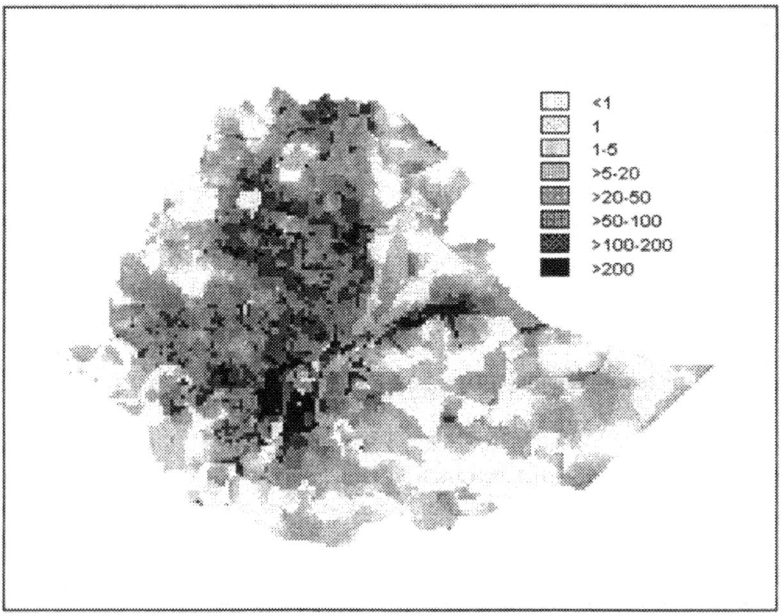

Figure 16.2: Population density (persons/km^2).

Thanks to the altitude, temperature is moderate, and tropical diseases are rare. Moreover, the physiographic abruptness of the high altitude Ethiopian land mass has a major influence on prevailing winds and results in a substantially higher rainfall than in the neighbouring, low-lying countries (Voortman *et al.*, 2000). However, under these topographic and climatic conditions the high population and cattle densities constitute a major threat to long term food security. Poor cultivation practices cause soil losses to reach alarming levels of up to 200-300 Mt per hectare per year (Hurni, 1993; Herweg & Stillhardt, 1999), already affecting 50% of the agricultural areas (UNEP/GRID, 1992). In addition, the high population growth rate (2.2% annually; World Bank, 1998) and cattle stock expansion steadily increase the pressure on land. As one of the poorest countries in the world, with an average income of less than one dollar per capita per day (World Bank, 1999), Ethiopia will in the foreseeable future have little opportunity to restore soil productivity through capital intensive rehabilitation programs, or to compensate for productivity losses through intensified fertiliser application.

In this study we assess at national level the effect of soil degradation on crop and cattle productivity so as to identify the areas where the need for soil conservation measures is most pressing. Ideally, one would like to base such a nation-wide assessment on a time series analysis of data on changes in soil quality at various spots. However, such information is only available for a small number of plots, and far from representative. Consequently, the assessment of the influence of soil degradation must rely on cross-sectional, spatial information. This raises various difficulties, basically because local agro-ecological conditions, farming practices and cropping patterns also affect crop yields. For example, yields might be low at sites with both poor soil fertility and a low degree of soil degradation that are hardly cultivated, whereas they could be high at sites with good soils that are intensively cultivated but count with high levels of degradation. Hence, when analysing the effect of soil degradation on yields, it is necessary to control for climatic conditions as well as soil fertility. In this study the climatic conditions will be accounted for by expressing productivity as the ratio of actual to potential yield for the dominant crops, while soil fertility appears explicitly.

16.1.1 Constraints on Incorporating Soil Degradation into Economic Models

The relationship between soil degradation and crop yield is critical for the assessment of the negative impact by the degradation process and, conversely, for the evaluation of the benefits from soil conservation measures. Then, it may seem surprising that research has been rather unsuccessful in elaborating such a relationship (Foster, 1999; de Roo, 1993). Statistically estimated reduced form models like the RUSLE (Renard *et al.*, 1997) and SLEMSA (Elwell & Stocking, 1975) do not perform much better than the calibrated process based models like EUROSEM (Rickson, 1994) and WEPP (Lane & Nearing, 1989) that simulate the soil erosion processes themselves, as both produce poor correlations between observed data and model calculations of soil loss, runoff and crop yields (*e.g.* Bjorneberg *et al.*, 1999; Reyes *et al.*, 1999; Klik *et al.*, 1997; Quinton, 1997; Littleboy *et al.*, 1996). Moreover, these models are very demanding in terms of data and often require following all the basic steps in the erosion process of detachment, transport and deposition, while the effect on crop yields is partly quantified with rule-based procedures that rely to a large extent on discrete variables such as texture and soil type (Kassam *et al.*, 1991). In short, current knowledge is insufficient to warrant the representation of the relationship between crop yields and soil degradation by means of a theoretically based functional form. This makes it very hard to incorporate soil degradation within an economic model.

Within an economic model soil degradation might be represented as a transition of land area from a given quality class to a lower one with a lower yield for given land management practices. Indeed, at the conceptual level it may be very useful to design and analyse this type of model, to highlight the specificity of soil degradation as compared to other environmental problems. However, at the empirical level the problems in implementing such an approach seem very serious. One could assign a yield to every land class but the problem is to specify a reliable function that relates prevailing agricultural practice to land transition. One would also have to ensure that this function is endowed with the convexity properties necessary for inclusion within an optimisation model. Furthermore, the rates of degradation are often so low compared to the prevailing discount rate that the conservation measures resulting from such a decision model are unlikely to arrest the degradation process. Alternatively, it would also be difficult to include the relationship into a more descriptive, non-optimising economic model as very little is known about the attitudes of economic agents and institutions with respect to soil degradation.

We conclude that an approach is called for that characterises a basic physical relationship between soil degradation and crop yield without attempting to account for economic behaviour. Hence, when specifying the set of controlling variables, it is advisable to avoid the inclusion of variables such as population density or fertiliser intensity as explanatory factors as these depend themselves on prevailing yields and soil degradation conditions. Rather than invoking these human related factors to explain differences in crop productivity, we will merely study the nature of their association to soil degradation. Finally, although livestock is of pre-eminent importance in Ethiopia, we focus on crops as georeferenced production data are not available at a nation-wide scale for livestock[1].

16.1.2 A Non-Parametric Approach

To address the above concerns, we opt for a non-parametric model, to be estimated by kernel density regression (*e.g.* Bierens, 1987). This approach is very flexible, which is an important advantage when little robust a priori information is available on the functional form to be adopted. In addition, kernel density regression allows to generate measures of statistical reliability at each point rather than for the full sample only. This eases the task of detecting the influence of outliers and for identifying weak and strong data domains. We apply the Mollifier program (Keyzer & Sonneveld, 1998) to generate statistical information on the estimates and to visualise the kernel density regressions as 3D-graphs that map the dependent variable against the independent variable(s) for fixed values of other exogenous variables. Information on associated statistics is shown in colours or shading of the surface plot and the ground plane.

The chapter is organised as follows. Section 16.2 describes the data sources including a brief overview of the soil degradation processes in Ethiopia. Section

[1] Crop residues count for approximately one third of the animal feed (de Leeuw, 1997).

16.3 briefly introduces the non-parametric regression technique, as a more detailed treatment is given in the Annex. The results of the non-parametric analysis are discussed in Section 16.4 and Section 16.5 concludes.

16.2 Data

16.2.1 Sources

The data base for this study consists of several gridded overlays with a grid size of approximately $10*10$ km^2. All data are georeferenced according to (the central points of) 460 polygons of the Crop Production System Zones (CPSZ; Fig. 16.3) derived from FAO (1998).

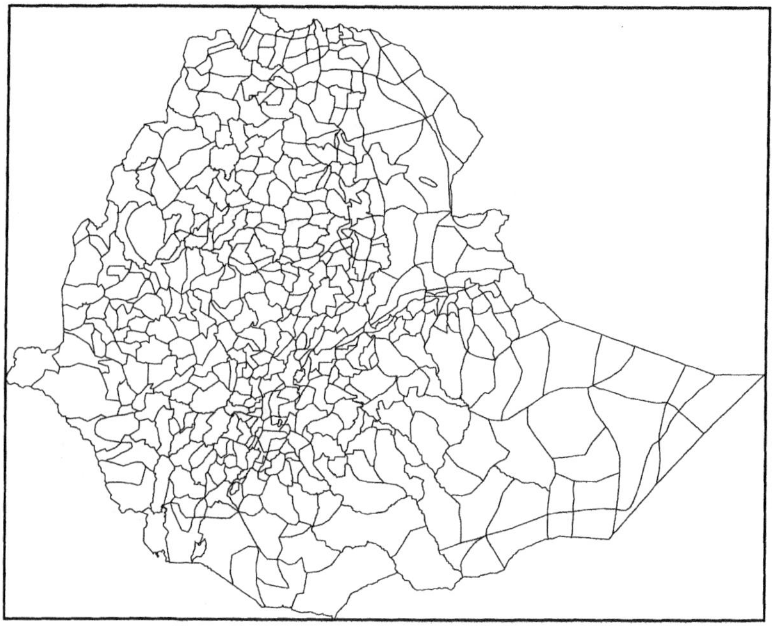

Figure 16.3: Crop Production Systems Zones.

These map entities correspond to administrative units (Auraja's), or subdivisions thereof whenever steep ecological gradients occur. The CPSZ data that were used in this study comprise qualitative classification on soil fertility, continuous data on mean monthly rainfall, slope gradients, altitude, temperature (monthly mean, minimum and maximum), length of growing period and crop yields of dominant cereal crops. Data on radiation were obtained from the world wide agroclimatic data base of FAO (1994). Further, overlays were performed with the African population density map (Deichman, 1997) and the African cattle density map (Kruska *et al.*, 1995). Finally, the data were overlaid with a map on soil degradation derived from an African continental inventory (UNEP/GRID, 1992)[2]. This inventory is based on assessments by experts who gave for physiographic units in Ethiopia a qualitative classification on the degree of soil degradation and its areal extent.

Fig. 16.4 shows the degree of the dominant degradation processes and makes clear that water erosion related processes play the major role everywhere, except in the extreme south-eastern part where wind erosion prevails. The detachment of top soil is the most widespread soil degradation characteristic, followed by mass movements that mainly occur on the steep soils of the highlands. Main causes of soil degradation are, in order of importance, overgrazing, deforestation and agricultural activities. Table 16.1 shows the dominant degradation processes in those areas where cereals are cultivated and where we will concentrate our analysis.

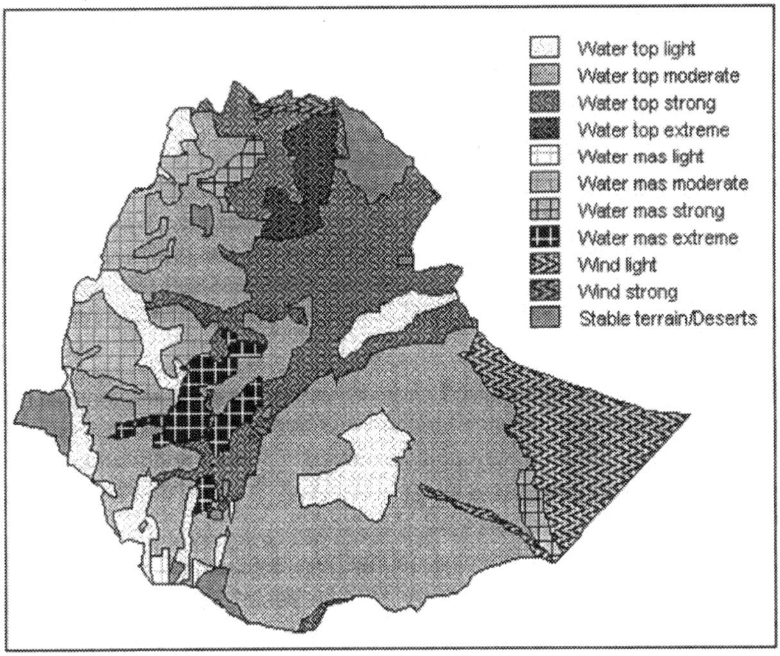

Figure 16.4: Dominant degradation processes ('Water top' is top soil loss; 'Water mass' is mass movements, both caused by water erosion).

---

[2] The study had terms of reference that were similar to those of the global assessment study (Oldeman *et al.*, 1991), although the inventory was at a spatially more detailed level.

Table 16.1: Degradation type and percentage area of cereal zone.

Degradation type	% area
Water erosion top soil	57%
Water erosion mass movements	29%
Wind erosion	8%
Other	3%

Source: UNEP/GRID, 1992.

16.2.2 Selection of Variables

Climatic information was used to calculate *potential yields* of the dominant cereal crops. The potential yield calculations are based on radiation, temperature, length of growing period and crop specific phenological and physiological characteristics (Kassam *et al.*, 1991). The Agricultural Planning Toolkit (Voortman & Buurke, 1995) was used to perform these calculations. The CPSZ data base gave actual yield statistics for the cultivated crops. The *yield ratio* (actual/potential) is used as an indicator to determine the influence of soil degradation on crop production.

Overlaid grid maps give the areal extent and degree of soil degradation in five qualitative classes (nil, low, moderate, severe, extreme) for each CPSZ.

The *soil fertility* variable characterises the influence of intrinsic soil properties on crop production. The information at hand did not allow a crop specific soil suitability assessment. Its ruling effect over soil productivity was therefore expressed in a general soil suitability rating, analogous to the FAO (1978) AEZ methodology. The observations on soil degradation (1992) and soil fertility (1995) demarcate a relatively small time span and we assume that soil fertility remained constant during that period.

The influence of *population density* and *cattle density* represents on one side the pressure on the land through intensive land use and overgrazing and on the other side input availability like labour, animal draught power and dung. Since population and cattle density are closely correlated it was decided to use population density only.

The Shoa, Gojam and Arsi regions account for over 75% of the total fertiliser consumed (FAO, 1995). To express these differences in *fertiliser use* and its possible masking effect on soil degradation, we introduced a variable in the data base that equals 1 for the regions Shoa, Gojam and Arsi and 0 elsewhere.

16.3 The Mollifier Program: 3D-Visualisation of Kernel Density Regressions

This section provides some background on the non-parametric analysis by kernel density regression. A more detailed specification is given in the Annex.

16.3.1 Mollifier Mapping

The mollifier mapping is defined as the following stochastic model:

$$y = E(R(x+\varepsilon)) \tag{16.1}$$

where y is the observed yield ratio, x a vector of explanatory variables and ε denotes measurement errors in x. The function $R(x+\varepsilon)$ is the unknown function, and the mollifier mapping is the expected value of this function. For an infinite sample of observations, spread evenly over the domain of x, it would be possible to evaluate this expected value. However, for a finite sample of size S, the value of y can only be estimated using the Nadaraya-Watson kernel density estimator:

$$\tilde{y}(\mathrm{x}) = \sum_{\mathrm{s}} \mathrm{y}^{\mathrm{s}} P_{s}(\mathrm{x}^{\mathrm{s}}) \tag{16.2}$$

where y^s and x^s denote observations. Thus, the estimate $\tilde{y}(\mathrm{x})$ is a probability-weighted sample mean. The probabilities are computed on the basis of the distance of x^s to point x, attributing higher weight to nearby points, on the basis of a postulated density function (the kernel) for ε whose spread is controlled by the window size parameter θ. We suppose that all the elements of ε are independent and normally distributed. For small samples, a misspecification of this density will affect the estimate, but this effect disappears as the sample size becomes larger.

16.3.2 Mollifier Program

The mollifier program offers the possibility to exhibit the estimated $\tilde{y}(x)$ in 3-D graphs as a surface plot or blanket against two independent variables on, for example, a 50x50 km grid, while controlling for other explanatory variables by setting them at a pre-defined value, *c.f.* their sample mean. In the default mode the program generates a colour shift or shading in the surface plot to reflect the likelihood ratio of the observation density which measures the number of observations on which the function evaluation is based at that point. The colours or shading in a ground plane below the surface plot show the probability of the actual y falling within a prescribed interval around the mollifier mapping, whose upper and lower bounds are specified as a percentage (default = 10) of the sample mean $\bar{y}$.

The mollifier program also calculates the partial derivative of the regression curve with respect to the explanatory variable x_k^t at data point x^t , as well as a measure of reliability for it. For this, it evaluates at every data point:

$$\frac{\partial \tilde{y}(x^t)}{\partial x_k^t} = \sum_s \frac{\partial P_s(x^t)}{\partial x_k^t} y^s \ . \tag{16.3}$$

The mollifier program uses the band (or window) width as a control variable to specify the neighbourhood of x whose points affect the prediction of $\tilde{y}$. The user can vary the window size relative to a benchmark (optimum) level defined by:

$$\theta = \left(\frac{4}{n(d+2)} \right)^{\frac{1}{d+4}} \tag{16.4}$$

with n the number of observations and d the number of exogenous variables (Silverman, 1986). If the averaging should emphasise nearby points, the window size should be small. The larger the window size, the tighter the blanket and the less it will follow the profile of observations. We will keep the window size at its benchmark level.

16.4 Results of the Non-Parametric Analysis

16.4.1 The Qualitative Assessment of Soil Degradation and its Relation to Crop Yield

In our data set, the impact of the degradation process is only expressed on an ordinal scale (such as 'moderate', and 'severe'), based on the perception of the evaluators only. Therefore, to obtain a first indication of the objective meaning of these concepts used, we regress the yield ratio (actual over potential yield) on the area under various classes of degradation.

Table 16.2: Regression with soil degradation classes.

Variable	Prob. of error in sign 1st derivative
% low degradation	0.52606
% moderate degradation	0.44950
% severe degradation	0.43413
% extreme degradation	0.49311

Table 16.2 shows the average probability of an error in sign of the first derivatives of this regression curve, as a (non-parametric) measure of significance of the associated variable. A value of 0.5 indicates that on average the slope information is uninformative, above 0.5 that it has the wrong sign, and the more below 0.5, the more reliable the average slope. We notice that the 'moderate' and 'severe' degradation classes exhibit the lowest values and have relatively more reliable derivatives.

The mollifier picture in Fig. 16.5 takes a closer look at this relationship. The horizontal axes indicate the area percentages under the 'severe' and 'extreme' degradation classes. The area percentages of 'low' and 'moderate', are shown as grey shifts, in surface curve and ground plane, respectively, where the corresponding boundary values and frequency distribution are given in the upper right and lower left legend. The vertical axis measures the yield ratio as dependent variable. Note that the graph is turned 180 degrees from its point of origin.

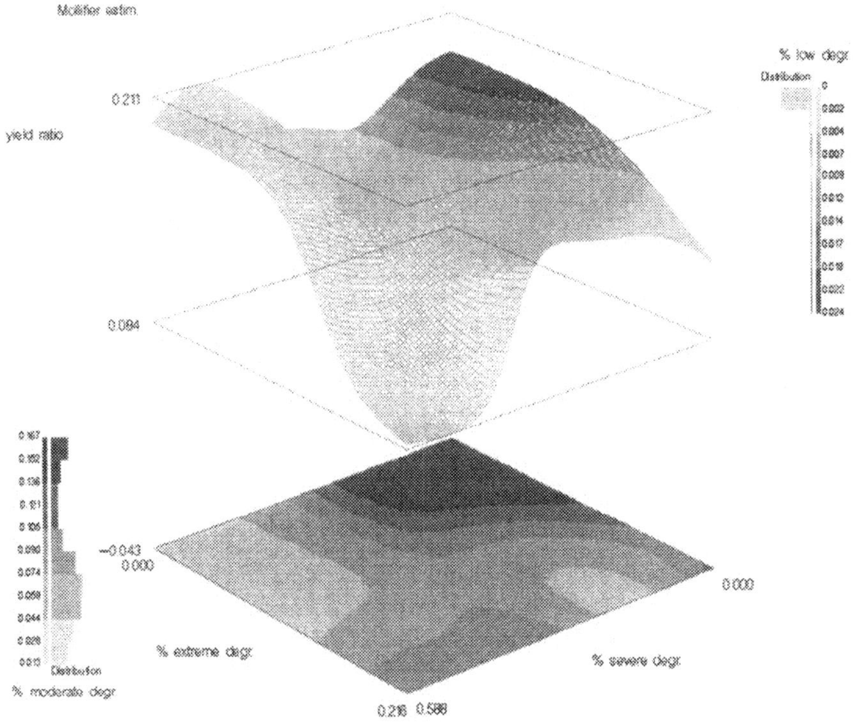

Figure 16.5: Yield ratio against area share under severe and extreme degradation (covariates: area share of low and moderate degradation).

The graph indicates that yield ratio is negatively correlated to the area shares of 'severe' and 'extreme' degradation and positively to 'low' and 'moderate'. It also appears that the impact on yield reduction of the 'severe' class corresponds to area percentages that are two up to three times larger than those of the 'extreme' class. Consequently, it seems possible to develop an aggregate index of both degradation types that attributes twice the weight to the area percentage of the extreme degradation.

16.4.2 Reliability of Relationships Between Yield Ratio and Common Explanatory Variables

Next, we turn to our main exercise, and study the shape and reliability of the relationship between crop productivity and degree of soil degradation for different levels of soil fertility. The relationship between *e.g.* soil loss and productivity has been well documented for specific crops at field level (*e.g.* Follet & Stewart, 1985), but here our aim is to test its validity in a nation-wide cross sectional analysis.

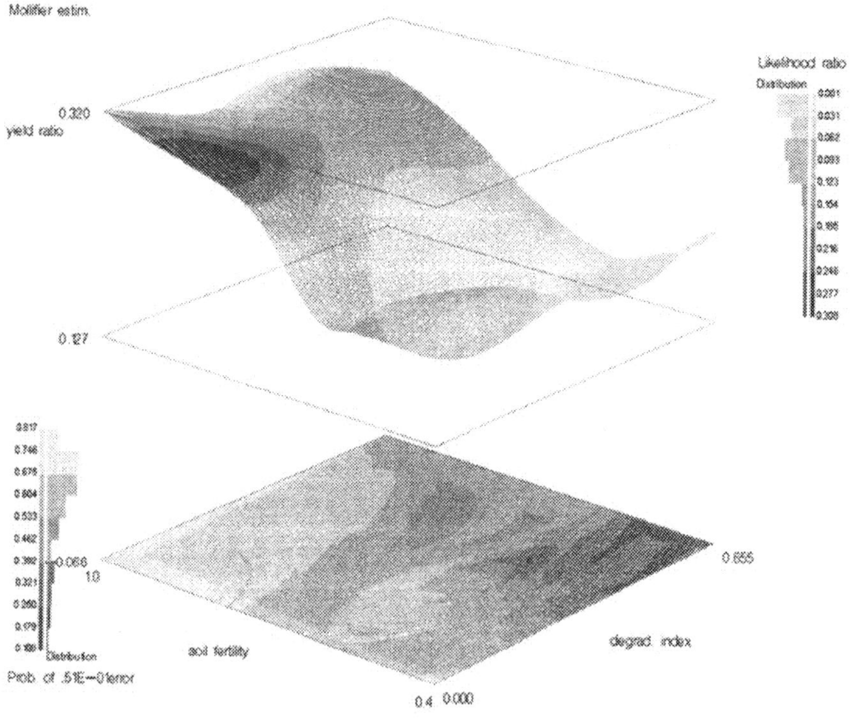

Figure 16.6: Yield ratio against land degradation index and soil fertility (covariates: likelihood ration and probability of error).

To assess the quality of the non-parametric regression at a given point x we measure (i) whether there are many observations x^s in the neighbourhood of that point (likelihood), (ii) the fit (probability of a given percentage error from calculated regression value), and (iii) the reliability of the slope (probability of a wrong sign for the derivative)[3]. Fig. 16.6 shows the yield ratio on the vertical axis depicted against soil degradation and soil fertility on the horizontal axes[4]. The grey shift in the surface curve measures the observation density, whereas the probability of a wrong estimate appears in the ground plane.

The likelihood density measure is low for the soils with low fertility and higher degradation. The probability of error for a 10% deviation from the estimated value is reasonably low for the major part of the graph. Also we observe a 'dip' in the blanket where it approaches lower soil fertility and soil degradation indices. We postpone further discussion of this and pursue our assessment of the reliability of individual variables.

Table 16.3: Regression with biophysical variables.

Variable	Prob. of error in sign 1st derivative
Degradation Index	0.44588
Soil fertility	0.47662

Table 16.3 shows the average reliability of slope at data points. On average, the derivative is more reliable for soil degradation than for soil fertility, but neither is convincing[5]. The low correlation with soil fertility is partly due to the fact that the soil map used to generate the soil fertility data depicts associations of soils whose location within the map unit is unknown. These associations are represented in the database through a single number, the general soil fertility rating which is a relatively crude measure since farmers tend to select the better soils within a unit. Note the unexpected increase in the yield ratio in the lower soil fertility range, that could be due to the prevalence of crops on these soils that are less demanding in terms of soil fertility.

[3] It must be emphasised that these measures all take the assumed normal density to be the correct kernel density, without relying on any application of the Central Limit Theorem.

[4] The extreme values of the axes were adjusted to avoid areas with low observation densities where the relationship is unreliable.

[5] Since unlike maximum likelihood estimation, kernel density regression has no standardized test-statistics such as a chi-square, it does not yield a formal procedure for rejection or acceptance.

The probabilities of error in the sign of the first derivative shown in Table 16.3 are mere averages. Fig. 16.7 depicts their distribution as covariates in the surface curve and plane, respectively.

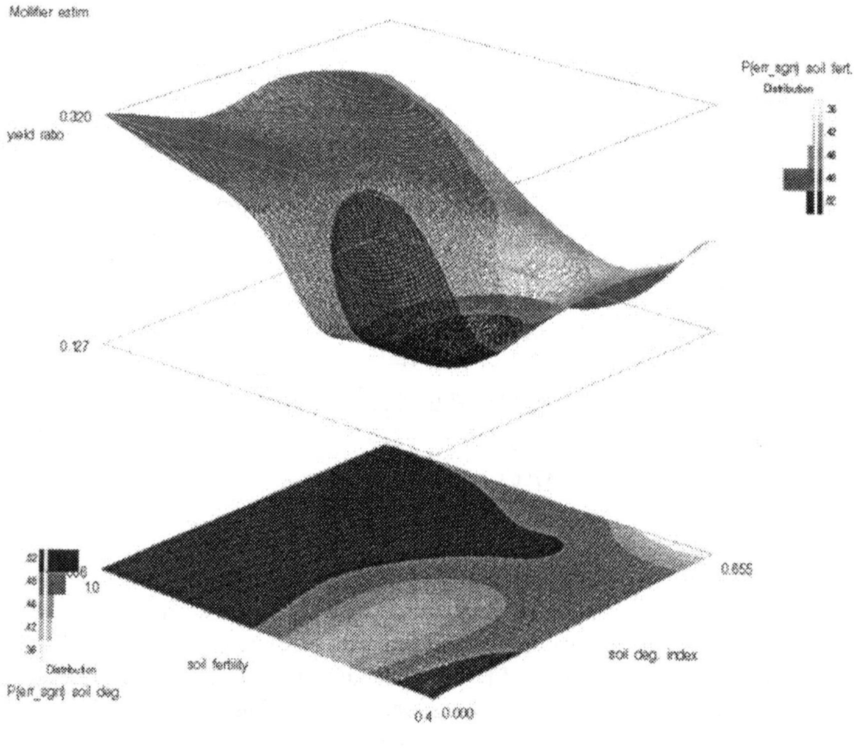

Figure 16.7: Yield ratio against land degradation index and soil fertility (covariates: P(error sign of first derivative) soil degradation and soil fertility).

Comparing the values shown in the horizontal histograms at the sides of the graph suggests that also for the distributed error values, the relationship is more reliable for the soil degradation index. Through a grey shift in the surface curve and plane we can readily notice that the highest error values for soil fertility are concentrated in the 'dip' coinciding with a low observation density. The lower values concur with the highest values of the degradation index. The probabilities of error for the degradation index are highest at the more fertile soils while the relationship is more reliable when it approaches the lower values.

16.4.3 Spatial Correlation

The values of the error term of the regression in the latitudinal and longitudinal direction (Table 16.4) are close to 0.5, which is an indication that it is devoid of spatial correlation and thus that no correction was necessary. Figure 16.8 shows the spatial distribution of the error term and confirms the absence of a clear spatial pattern.

Table 16.4: Regression of error term with latitude and longitude.

Variable	Prob. of error in sign 1st derivative
Longitude	0.44166
Latitude	0.45069

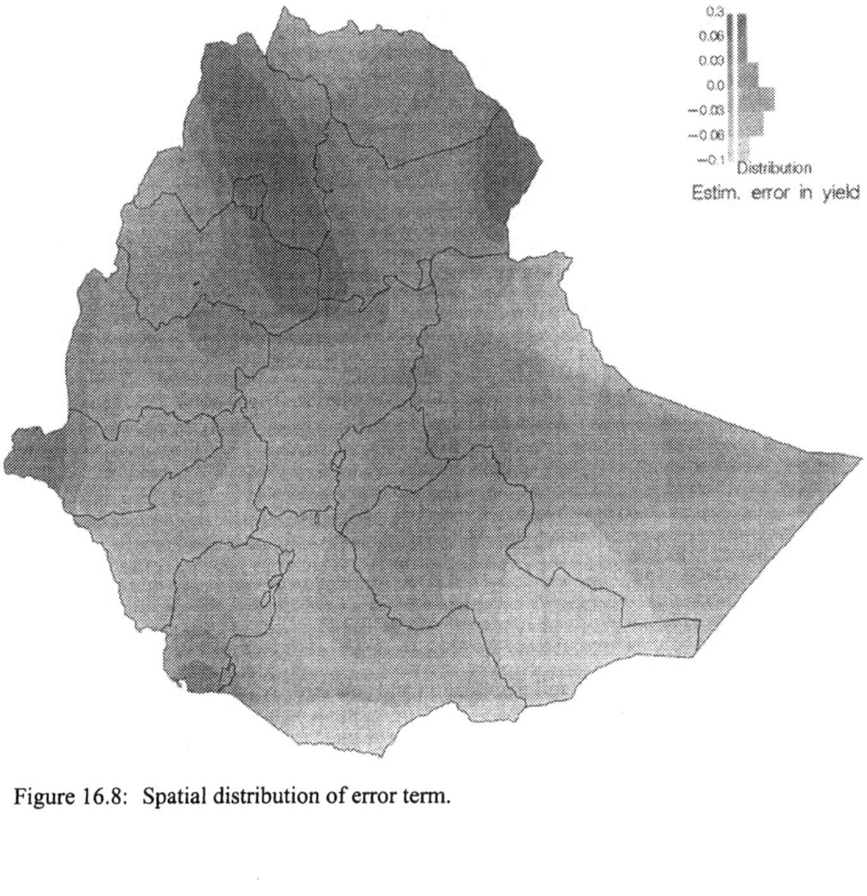

Figure 16.8: Spatial distribution of error term.

16.4.4 The Relation with Population Levels and Fertiliser Use

The surface planes in Figs 16.6 and 16.7 can be thought of as representing an abstract landscape that summarises the situation in Ethiopia. The next step is to investigate where human settlement takes place. From a food security perspective one would favour settlements on the higher parts of the curve, far away from the precipice and the barren 'lowlands' of the relief. Hence, we plot in Fig. 16.9 the population levels (population density in each CPSZ times the area) and fertiliser use as covariates in the surface and ground plane, respectively in the same relationship depicted in the Figs 16.6 and 16.7.

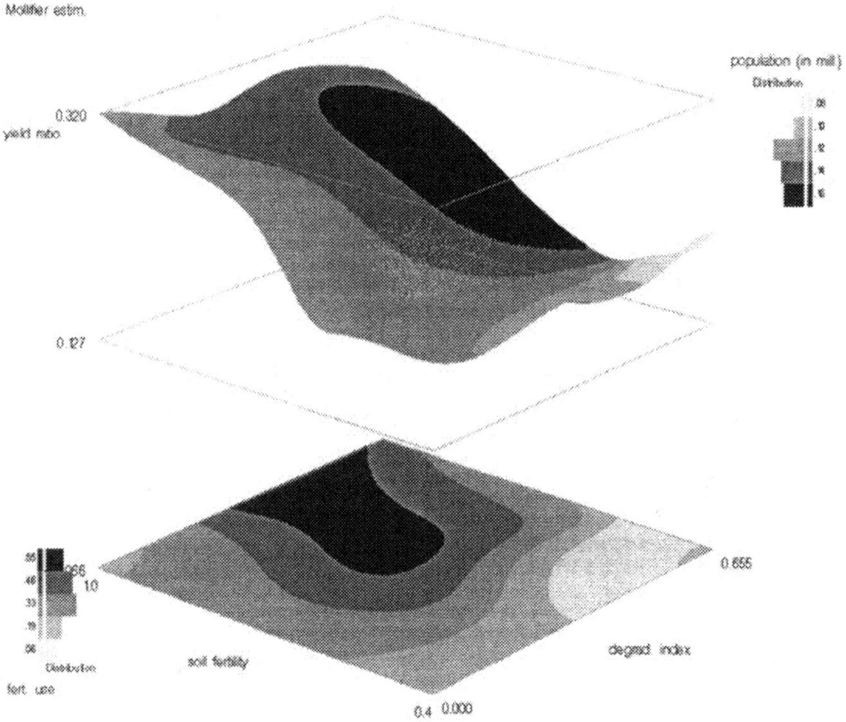

Figure 16.9: Yield ratio against soil degradation and soil fertility (covariates: population level and fertiliser use).

Our exploration enables us to identify three 'stylised facts':

1. Soil degradation has its major impact on soils of lower fertility and at low population levels. Due to their vulnerability and limited carrying capacity, these soils are restrictive in crop selection and offer little prospect for accommodating future expansion of the population.
2. On fertile soils, soil degradation is largely compensated by fertiliser application.
3. The effect of soil degradation differs according to soil fertility. On better soils, the negative effect is gradually compensated by fertiliser use. The highest population (and cattle) levels are sliding down into the precipice. Most revealing is that the high population levels are to be found on an uncomfortable, downward slope. Continued soil degradation would cause a further decline in soil fertility and result in severe yield reductions. While it should be possible in principle to compensate for such losses through application of external inputs such as fertilisers, this would seem highly unrealistic under the prevailing economic conditions. Therefore, it is important to work along the other axis and prevent further soil degradation.

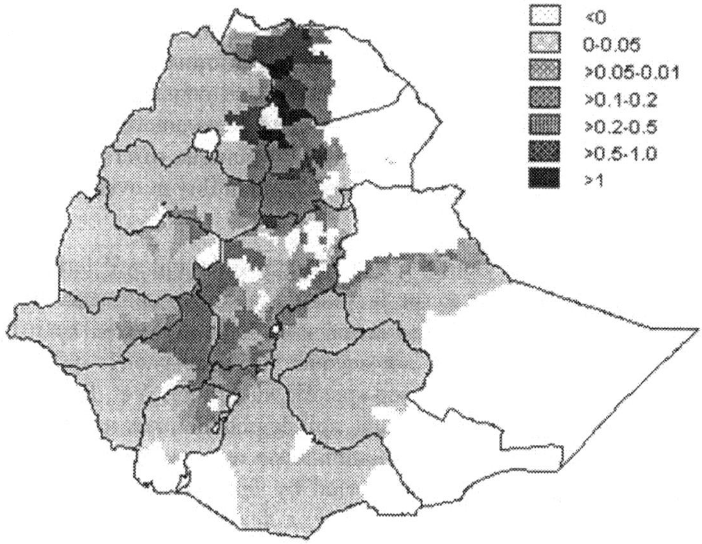

Figure 16.10: Crop productivity elasticity of soil production.

16.4.5 Back to GIS

Finally, at the end of the exercise, we leave our virtual landscape and return to the geographic map of Ethiopia where we apply the regression to identify the areas most sensitive to soil degradation. For this, we evaluate the elasticity, *i.e.* the percentage reduction in crop yield resulting from a one percent increase in the soil degradation index.

Fig. 16.10 shows that most of the land has a low elasticity and a few spots even have a small negative response. Higher elasticities are found in the northern Kefat and eastern Shewa provinces and along a line that follows part of the Rift Valley and then goes up to the Northern provinces of Welo, Gondar and Tigray. The latter provinces contain the real 'hot spots' that suggest themselves as priority areas for intervention. Whether soil conservation measures or increased fertiliser application is the answer, or a combination of both, will depend on the costs involved.

16.5 Summary and Conclusions

We have applied a non-parametric (kernel density) regression to analyse the effect of soil degradation on crop productivity at a nation-wide scale, and to examine its relation to population (and cattle) density and fertiliser use.

The non-parametric technique uses a flexible curve fitting method to characterise the multivariate relationships without imposing a fixed structure on the data. We adopted this approach because there is little *a priori* information available on the functional form that describes the relation between soil degradation, soil fertility and crop productivity at a national level. Compared to parametric methods, such as spline regression or variogram estimation, the technique offers the additional advantage that it generates a measure of statistical reliability at every point, rather than for the full sample only.

The relationship between the yield ratio, soil degradation and soil fertility is not very strong. This is possibly due to the fact that (1) the map units chosen are too heterogeneous to reflect accurately the actual use of the cultivated land; (2) the generalised soil fertility rating does not account for crop-specific demands and therefore (slightly) misspecifies the crop-specific suitability. Yet, the regression analysis identified three stylised facts. First, soil degradation has its major impact on soils of lower fertility and where population levels are low. Secondly, on fertile soils, soil degradation is largely compensated by fertiliser application. Finally, most people can be found on the slope facing a deep and dangerous precipice. Tumbling into it could be avoided either by compensating the fertility loss through intensified use of external inputs, or by preventing further degradation through soil conservation measures.

A spatial representation of the elasticity of crop productivity with respect to soil degradation indicates that the most vulnerable areas are located in the northern part of Ethiopia. Further research is required to assign costs and returns to soil conservation measures and fertiliser application so as to identify the most suitable policy for these areas.

16.6 References

Bierens, H.J. (1987): Kernel density estimations of regression functions. Advances in Econometrics 6, Cambridge University Press

Bjorneberg D.L., T.J. Trout, R.E. Sojka & J.K. Aase (1999): Evaluating WEPP predicted infiltration and soil erosion for furrow irrigation. Paper presented at the 10th International Soil Conservation Organization conference, May 23-27, Purdue University, Lafayette, Indiana

Deichman, U (1997): African population database documentation. National Center for Geographic Information and Analysis, University of California. Through GRID: http://grid2.cr.usgs.gov/data/

Elwell, H.A. & M.A. Stocking (1982): Developing a simple yet practical method of soil-loss estimation. Tropical Agriculture (Trinidad) 59, 43-48.

FAO (1978-1981): Reports of the Agro-Ecological Zones project. World Soil Resources reports Vol. 1-4. FAO. Rome

FAO (1994): World wide agroclimatic data. Agro-meteorology group. FAO. Rome

FAO (1995): Country Information Brief, FAO June 1995: Agricultural production and diversification programmes. Food and Cash Crops. http://www.sas.upenn.edu/African_Studies/eue_web/docs/fao_agr.htm

FAO (1998): Soil and terrain database for north-eastern Africa/ Crop Production System Zones of the InterGovernmental Authority on Development (IGAD). Land and Water Digital Media Series 2. Rome

Follet, R.F. & B.A. Stewart (Eds.) (1985): Soil erosion and crop productivity. American Society of Agronomy, Crop Science Society of America, Soil Science Society of America. Madison, Wisconsin, USA

Foster, G. (1999): Soil erosion modelling. Paper presented at the 10th International Soil Conservation Organization conference, May 23-27, Purdue University, Lafayette, Indiana

Herweg, K. & B. Stillhardt (1999): The variability of soil erosion in the Highlands of Ethiopia and Eritrea. Research Report 42. Centre for Development and Environment. University of Berne

Hurni, H. (1993): Land degradation, famines and resource scenarios in Ethiopia. In: World Soil Erosion and Conservation, D. Pimentel (Ed.) Cambridge University Press, Cambridge, 27-62

Kassam, A.H., H.T. van Velthuizen, A.J.B. Mitchell, G.W. Fischer & M.M. Shah (1991): Agroecological assessment for agricultural development planning. Main report FAO/IIASA. Rome

Keyzer, M.A. & B.G.J.S. Sonneveld (1998): Using the mollifier method to characterize datasets and models: the case of the Universal Soil Loss Equation. ITC Journal 1997-3/4, 263-272

Klik A., B. Hebel, A. Zartl & J. Rosner (1997): Measured vs. WEPP simulated runoff and erosion from differently tilled plots. Annual International Meeting, Minneapolis, Minnesota, USA, 10-14 August, 1997. American Society of Agricultural Engineers., No. 972120, ASAE, 10 pp

Kruska, R.L., B.D. Perry & R.S. Reid (1995): Recent progress in the development of decision support systems for improved animal health. In: Proceedings of the Africa GIS '95 meeting, 'Integrated Geographic Information Systems Useful for a Sustainable Management of Natural Resources in Africa', March 6-9, Abidjan

Lane. L.J. & M.A. Nearing (Eds.) (1989): USDA Water erosion prediction project: hillslope profile version. NSERL Report No 2. USDA-ARS National Soil Erosion Research Laboratory, West Lafayette, Indiana, 272 pp

Leeuw, de P.N. (1997): Crop residues in tropical Africa. In. Crop residues in sustainable mixed crop/livestock farming systems. (C. Renard, ed). CAB International, 41-78

Littleboy, M., A.L. Cogle, G.D. Smith, D.F. Yule & K.P.C. Rao (1996): Soil management and production of Alfisols in the semi-arid tropics. I. Modelling the effects of soil management on runoff and erosion. Australian Journal of Soil Research 34, 91-102

Oldeman L.R., R.T.A. Hakkeling & W.G. Sombroek (1991): World map of the status of human induced soil degradation. ISRIC/UNEP. Wageningen

Quinton, J.N. (1997): Reducing predictive uncertainty in model simulations: a comparison of two methods using the European Soil Erosion Model (EUROSEM). Catena 30, 101-117

Renard, K.G., G.R. Foster, G.A. Weesies, D.K. McCool & D.C. Yoder (1997): Predicting soil erosion by water: A guide to conservation planning with the Revised Universal Soil Loss Equation (RUSLE). Agriculture Handbook number 703. USDA. ARS

Reyes, M.R., K.D. Cecil, C.W. Raczowski, G.A. Gayle & G.B. Reddy (1999): Comparing, GLEAMS, RUSLE, EPIC and WEPP soil loss predictions with observed data from different tillage systems. Paper presented at the 10th International Soil Conservation Organization conference, May 23-27, Purdue University, Lafayette, Indiana

Rickson R.J. (1994): EUROSEM: preliminary validation on non-agricultural soils. Conserving soil resources: European perspectives. Selected papers from the First International Congress of the European Society for Soil Conservation

Roo, A.P.J. de (1993): Modelling surface runoff and soil erosion in catchments using Geographical Information Systems. Netherlands Geographical Studies

Silverman, B.W. (1986): Density estimation for statistics and data analysis. Chapman & Hall

UNEP/GRID (1992): World Atlas of Desertification. E. Arnold: A division of Hodder & Stoughton, London, 38-39

Voortman R.L. & B.J. Buurke (1995): Climatic data analysis and biomass/crop yield potential assessment. FAO/SOW-VU version 1.0. FAO/Centre for World Food Studies, Rome/Amsterdam

Voortman R.L., B.G.J.S. Sonneveld & M.A. Keyzer (2000): African land ecology: opportunities and contraints for agricultural development. Centre for International Development at Harvard University, Harvard

World Bank (1998): African Development Indicators 1998/1999. World Bank. Washington D.C.

World Bank (1999): World Development Indicators 1999. World Bank, Washington D.C.

Annex Chapter 16

Further Background on the Mollifier

Let us start the explanation of the mollifier method by considering a given data set S of real-valued observations indexed s, and partition it into a vector of n (bounded) endogenous variables y^s and a vector of m exogenous variables x^s from the bounded set X. The mollifier calculates a value $y(x)$ at intermediate points x, thus creating a blanket that fills the gaps between the observations. The mollifier uses for its estimation a weighting function $w^s(x)$ which equals the probability P^s of y^s being the correct value of $y(x)$. This means that errors have to be accounted for and relaxes the requirement of conventional interpolation methods to let the curve pass through the observations. The resulting specification will be:

$$\tilde{y}(x) = \sum_s y^s P_s(x) \qquad \text{(A1)}$$

This defines a non-parametric regression function, whose shape will depend on the postulated form of the probability function. For example, if y^s is a scalar and x^s a two-dimensional vector of ground co-ordinates, every observation s can be viewed as a pole of height y^s located at point x^s. The regression curve lays a 'soft blanket' on these poles that absorbs the peaks of the highest poles (upward outliers) and remains above the lowest poles. The analytical form of the probability function $P^s(x)$ of this model can be obtained in various ways. Here we will apply the mollifier approach.

For a finite sample of size S, the value of this mollifier function (A1) can be estimated by a Nadaraya-Watson estimate, *i.e.* a weighted sample mean with window size θ as parameter:

$$\tilde{y}(\mathrm{x}) = \sum_{\mathrm{s}} \mathrm{y}^{\mathrm{s}} P_s(\mathrm{x}) \qquad \text{(A2a)}$$

for

$$P_s(\mathrm{x}) = \psi((\mathrm{x}^{\mathrm{s}} - x)/\theta) / \Psi^S(x) \text{ if } \Psi^S(x) > 0 \text{ and } 0 \text{ otherwise} \qquad \text{(A2b)}$$

where

$$\Psi^S(x) = \sum_{s=1}^{S} \psi((\mathrm{x}^{\mathrm{s}} - x)/\theta), \qquad \text{(A2c)}$$

and where the density function $\psi(\varepsilon;\theta)$ has its mode at $\varepsilon = 0$ and is such that for θ going to zero its support goes to zero.

In this approach, expression $\psi(x^s - x)$ in (A2c) can be interpreted as the likelihood of x being associated to the observation s and $\Psi^S(x)$ the likelihood of x being associated to any of the observations in the sample. Hence, probability $P_s(x)$ is the probability of x being associated to observation s, conditional on its association to at least one observation in the sample, and $\tilde{y}^s(x)$ is the expectation of the y^s-values associated with the sample. We also define the likelihood ratio:

$$\Lambda(x) = \sum_{s=1}^{S} \psi((x^s - x)/\theta) \Big/ \sum_{s=1}^{S} \psi(0) \tag{A3}$$

as well as the probability $Q(x;a)$ of y falling outside a given range $a = \alpha\bar{y}$ around $\Psi^S(x)$, where $\bar{y}$ is the sample average:

$$Q(x;a) = \sum_{s \in S(x;a)} P_s(x)\text{, for } S(x;a) = \left\{ s \middle| \left| y^s - \tilde{y}(x) \right| \geq a \right\} \tag{A4}$$

This probability serves as a measure of fit.

The mollifier program also assesses the partial derivative of the regression curve as well as a measure of its reliability. For this, it calculates the first partial derivative to x_k at point x^t, where k represents an explanatory variable, at all data points:

$$\frac{\partial\, \tilde{y}(x^t)}{\partial\, x_k} = \sum_s \frac{\partial\, P_s(x^t)}{\partial\, x_k} y^s \tag{A5}$$

Since $\sum_s \frac{\partial\, P_s(x^t)}{\partial\, x_k} = 0$ we can write:

$$\frac{\partial\, \tilde{y}(x^t)}{\partial\, x_k} = \sum_s \frac{\partial\, P_s(x^t)}{\partial\, x_k} (y^s - y^t) \tag{A6}$$

where y^t refers to the tth observation. As by definition,

$\frac{\partial\, P_s(x^t)}{\partial\, x_k} = P_s \frac{\partial \ln P_s(x^t)}{\partial\, x_k}$, it follows that:

$$\frac{\partial\, \tilde{y}(x^t)}{\partial\, x_k} = \sum_s P_s(x^t) \left[\frac{\partial \ln P_s(x^t)}{\partial\, x_k} (y^s - y^t) \right] \tag{A7}$$

Let us now rewrite and interpret the term in square brackets:

$$\frac{\partial \ln P_s(x^t)}{\partial x_k} = \frac{\partial \ln \psi_s(x^t)}{\partial x_k} - \sum_{h=1}^{s} P_h(x^t) \frac{\partial \ln \psi_h(x^t)}{\partial x_k} \tag{A8}$$

Now, for a density $\psi_s(x^t) = \psi \frac{(x^s - x^t)}{\theta}$ where ψ is a normal joint density with diagonal variance matrix and variance σ_k^2 around x^t it follows that:

$$\frac{\partial \ln \psi_s(x^t)}{\partial x_k} = \frac{x_k^s - x_k^t}{\sigma_k^2}. \tag{A9}$$

Hence the term in square brackets can be rewritten as:

$$\frac{\partial \tilde{y}(x^t)}{\partial x_k} = \sum_s P_s(x^t)\left[\xi_k^s \delta_k^s\right], \tag{A10}$$

where $\xi_k^s = \frac{x_k^s - x_k^t}{\sigma_k^2} - \sum_h P_h(x^t)\frac{(x_k^h - x_k^t)}{\sigma_k^2}$ and $\delta_k^s = (y^s - y^t)$.

In other words, the term in square brackets is the contribution of observation s to the slope.

For given x^t this enables us to define the probability of a positive sign for the slope as:

$$P_k^+(x^t) = \sum_s P_s((x^t)\left|\xi_k^s \delta_k^s \geq 0\right)$$

Hence the probability of a wrong sign can be calculated as:

$$P_k^\#(x^t) = P_k^+(x^t)\ ,\ \text{if } \frac{\partial \tilde{y}(x^t)}{\partial x_k} < 0, \text{ and}$$

$$1 - P_k^+(x^t)\ ,\ \text{if } \frac{\partial \tilde{y}(x^t)}{\partial x_k} \geq 0.$$

17 Between Free Riders and Free Raiders: Property Rights and Soil Degradation in Context

Franz von Benda-Beckmann

Law and Governance Group, Department of Social Sciences, Wageningen University and Research centre, Wageningen, the Netherlands

Abstract

This paper aims at a reorientation of the assumptions and perspectives with which the significance of property rights for more or less sustainable land use is viewed. The conventional four property categories do not offer a useful guidance into the complexity of property relationships. Generalisations over relations between categorical types of property rights and environmental consequences are dangerous and do not seem to be warranted given the empirical evidence. Rather, characteristics of property rights seem to play a dominant role for land use, relatively independent of the legal status of the land, such as the time horizon of land users, their dependence on the resource and the kind of dependence, and the spatial proximity/distance to the resource. In particular, actual constellations of property rights, in other words the distribution of wealth, over people and resources are important characteristics. Institutions should be approached from a post-institutionalist perspective which problematises the relationships between the normative-legal institutional framework, the social relationships that develop in their context and the social practices these rules purport to regulate, and which maintain or change the relationships and the normative institutional framework.

17.1 Introduction

'There is a tendency for all knowledge, like all ignorance, to deviate from truth in an opportunistic fashion. The fact that conceptions of reality, and ideologies and theories, are influenced by the interests as commonly perceived by the dominant groups in the society where they are formed, and that they so come to deviate from truth in a direction opportune to these interests, is easily seen and, in fact, taken for granted when we look back at an earlier period of history. But in our own intellectual endeavours we ordinarily preserve a naive non-awareness about such influences working on our minds - as, indeed, people have done in every earlier epoch of history. We believe — as they did and with equal firmness — that we are simply factual, basing ourselves on observation of reality when we think, argue and conclude. A first precondition when trying to unfetter our minds from biases in order to reach a truer perception of reality is to see clearly the opportunistic interests affecting our search for truth and to understand how they operate. In this attempt to overcome naivety, a backward look becomes helpful' (Myrdal, 1970, 21, 22).

Myrdal's words are quite appropriate when one looks at the question I want to address in my paper: the interrelationships between institutions, property in particular, resource use and its ecological effects. Property has become an important issue in more or less all academic specialisms and disciplines concerned with resource management and its social, economic and biological-ecological consequences[1]. This is not surprising, given the multifunctional and contested character of property rights in all societies. Natural resource property rights serve to legitimate control over the means of production, whether production is for the market or for subsistence. They can be an important material basis for the social continuity of groups. They usually have important political functions for states and non-state social organisations, and tend to be a source of individual power and prestige. Natural resource property often also has considerable cultural-religious meanings. Moreover, the distribution of property is a crucial issue of social-philosophical and political debates over equity and social justice. Last and certainly not least an important function is the significance for the sustainability of resources and the preservation of environmental values. Given this political, economic and ideological importance, legal property regulations and rights therefore constitute crucial social resources in people's strategies, negotiations and struggles over natural resources. These struggles concern the legitimate power to control, manage, and use resources and profit from their yields. They also concern the relative significance of the economic, political, social, cultural and ecological functions property should have. It is therefore not surprising that analytical frameworks, social philosophies, ideological justifications of past, present or future property regimes have become closely intertwined, and that unbiased perceptions of past and present are frequently sacrificed to partisan and opportunistic ideas of how property relations do and should work[2].

This was and is true with respect to developed states, the relationship between free market and communist plan economies, and it certainly is true with respect to developing countries. The early confrontations between European powers and the societies colonised by them were dominated by evolutionist assumptions (Schott,

[1] See McCay & Acheson (1987); Berkes (1989); Ostrom (1990); Schlager & Ostrom (1992); Brouwer (1992); Rose (1994); Spiertz & Wiber (1996); Balland & Platteau (1999); Hanna *et al.* (1996); Berkes & Folke (1998).

[2] Most theories and policies concerning property regimes are characterised by strong ideological and political overtones in the justification of current and future property regimes (see Reeve, 1991, 111).

1982). Western property, epitomised in legal and ideological notions of individual private ownership, was regarded as the outcome of a long process of evolution and civilisation. Moreover, it had been the basis for the economic development in Europe and the US. This contrasted dramatically with the various communal property regimes which were seen to exist in those archaic and economically undeveloped societies. In the modernisation-driven economic policies, the assumption consequently was that legal reform creating marketable individual private ownership rights would significantly contribute to economic growth, by creating greater legal security, freeing individuals from communal constraints, and providing collateral to obtain productive credit. These assumptions have been a great impetus for the land rights reforms in many developing countries (von Benda-Beckmann, 1989; Bruce *et al.*, 1994).

In the recent past, ecological and environmental concerns have become prominent (see Lele, 1991). The issue of sustainability has rekindled earlier debates, pressing the question of how concerns for sustainable resource use can be balanced with the concerns for optimal productivity of property resources and for a just distribution of resources, and not only in the present, but taking into account the needs of future generations. Especially following Hardin's essay on the *Tragedy of the commons* (1968), discussions of property rights have shifted towards the merits of different types of property rights for sustainable resource management. Common property, Hardin argued, induced rational actors to ever increase resource use, because they would gain the full benefits of each additional labour or capital input while sharing the costs of resource use (degradation) - the externalities - with all other users. Without internalisation of such environmental costs or coercive action by government, the tragedy of overexploitation could not be avoided. Hardin's construction has been used to advocate an increasing government role in dealing with property and the environment; at the same time it has been used for propagation of the division of common property regimes into private individual ownership.

It is now generally accepted that Hardin failed to distinguish open access as a theoretical condition in which there are no relevant institutions from common property as a social institution (the commons) (McCay & Acheson, 1987)[3]. Moreover, most resource complexes in developing countries, such as grazing lands, forests or water are common/communal property and/or have been declared state property. The new mainstream resource management literature now is mainly concerned with the value of common/communal property institutions for collective action for sustainable resource management, with a strong focus on the classical commons such as grazing lands, forests, water and fishing grounds. Communal ownership and resource management have lost their earlier scapegoat role, to be replaced by an opposite image. Local communities now seem to be the 'natural' protectors of the environment. There are many attempts to make use of

[3] Hardin's analysis is assumed to be still valid for open access property (Berkes, 1996, 94).

communal management, under the assumption that local communities are usually better equipped than other organisations to manage water or forests (IUCN, 1984; Lynch & Talbott, 1995; but see Agrawal & Gibson, 1999; von Benda-Beckmann & Von Benda-Beckmann, 1999a).

Although property rights are an important issue in discussions of resource use, they are still widely neglected in economic modelling approaches to sustainable use of natural resources, with a few exceptions[4]. In contrast, property rights are given much attention in new institutional economic approaches, and in particular in the economic property rights school. In economic approaches, there is a strong bias in favour of individual private ownership and freely transferable use rights (ideally private individual ownership). This is based on the assumption that such property rights contribute to more sustainable use of natural resources through conservation investments by providing greater security of tenure and the possibility to recover investments through transfer (see also Heerink *et al.*, this volume). For some authors, private individual property rights even are assumed to internalise costs and benefits, and thereby increase individual responsibility for the environment and rational use of its resources[5].

Common to both genres of property discussions is that they are dominated by what I call the 'big four' categories of property: open access resources (or no one's or everyone's property); private, often individual, ownership; state or public ownership; and common/communal property (see Berkes, 1996, 89). These types of property rights then are related to more or less sustainable forms of resource use based on the assumptions mentioned. Empirical evidence, however, suggests that the assumptions on the distributive and environmental effects of freely transferable private individual ownership rights seem to be naive and, in Myrdal's (*op. cit.*) sense, opportunistic indeed. On the other hand, important as the common property discussions are, the strong focus on common property as an institution detracts attention from the general question of the significance of different kinds of property rights for resource degradation. The discussions are further strongly preoccupied with institutions and 'rules of the game', but give too little attention to the problematic nature of possible interrelations between the complexity of normative institutional frameworks and social practices (North, 1990). In both genres, there is a tendency to depoliticise property rights, both held by public and private social entities, by underplaying the political and power relationships in which property is embedded in general, and those of state owned and controlled property in particular (but see Tigar & Levy, 1977; Geisler, 1995).

In this chapter, I argue that approaches in the anthropology of law, with their descriptive-analytical emphasis, offer a more promising guide to the complexities

---

4 See Larson & Bromley (1990); Holden *et al.* (1998). For an overview, see Kaimowitz & Angelsen (1998).

5 Demsetz (1967); Furobotn & Pejovic (1972). For critiques see McCay & Acheson (1987, 5); von Benda-Beckmann (1995); Platteau (1996).

of property rights than the normatively and teleologically oriented property rights discussions in the economic and resource management literature[6]. It helps us to see that the conventional four property categories are neither useful as descriptive nor as analytical categories. There is no clear correlation between these types of property and more or less sustainable resource use. Rather, other characteristics of property rights seem to play a dominant role, such as the actual distribution of property rights, the time horizon of resource users, their dependence on the resource and the kind of dependence and the spatial proximity/distance to the resource. In addressing these questions, I also shall go back to Hardin, but I shall not exit through the door of common/communal property. Rather, I shall briefly re-examine his original problem - increasing population in relation to finite resource availability - as well as his grazing land metaphor, to dwell on the problems of individual rights under all different legal statuses of natural resources. But before I come, via my discussion of Hardin, to these issues, I shall first outline some general points of departure, giving my own perspective on property rights.

17.2 Points of Departure and Assumptions

17.2.1 Empirical Generalisations

There is no clear correlation between types of property rights and more or less sustainable resource use[7]. Most degrading activities are brought about on the basis of private ownership rights or of licences and concessions of state governments given to private individuals or firms[8]. The expectation that private ownership or similarly strong economic use rights would indeed internalise negative environmental effects often proves to be wishful thinking[9]. On the contrary, all

[6] Given the voluminous literature, it is impossible to deal with all relevant conceptual and theoretical discussions. For further elaborations, see von Benda-Beckmann (1979, 1995); Moore (1986); Wiber (1993); Hann (1998); von Benda-Beckmann & Von Benda-Beckmann (1999b).

[7] I am aware that these generalisations do not sufficiently distinguish between uses that degrade through pollution, over-exploitation, specific threats to the resilience and renewability of the resources.

[8] See Gibbs's & Bromley's (1989, 30) major classes of resource degradation: The deliberate conversion of natural resources to other form of capital; the redefinition of property rights of the state and individuals; indirect side-effects of policies such as developing commercial agriculture for food self-sufficiency or commercial export, amongst others through too many chemical inputs. We can add the contamination of large resource areas through nuclear waste and radiation (Geisler, 1995; Conn, 1997) and the pollution of large areas through oil and mineral extraction.

[9] The assumption that private property rights lead to responsible and efficient management of resources, and that the market would function as guarantor of efficient use has frequently been criticised (see *e.g.* McCay, 1987, 177). The Netherlands are a good example in which agricultural soils are seriously polluted on the basis of individual

over the world governments struggle to curb private ownership rights, taking away important right-elements out of private law ownership to avoid environmental degradation. It is also common knowledge that in most developing countries, governments do not usually allocate resource rights in the pursuit of collective goods such as the environment. There is also sufficient evidence that both judicious as well as degrading use of natural resources does occur under common or communal forms of property rights, and that 'communities' of local resource users may or may not be good guardians of traditional environmental values[10]. We therefore have to look beyond these general categories of property rights.

17.2.2 Categorical and Concretised Property Rights

Property rights are legal relationships between social entities with respect to valuable objects. They connect property holders and elements of the physical environment[11] and define the rights and options holders have vis-à-vis others, the temporal duration of such rights, and the conditions under which such rights can be acquired. Ever since Maine (1861, 1883) coined the metaphor, property rights are often referred to as *bundles of rights*. This metaphor can be used as an analytical device to distinguish different aspects of socio-economic positions and activities, for instance the legitimisation of appropriation, use and exploitation of resources and the appropriation of profits of such uses; the legitimisation of different kinds of socio-political control over property, but not (necessarily) of the right to use it for own profit; and the legitimisation of management activities, to maintain or improve the object, to decide on the use that oneself or others may or should make of the resource (for other categorisations, see Schlager & Ostrom, 1992; von Benda-Beckmann *et al.*, 1997)[12]. Discussions over property as a bundle

---

private ownership, and where the 'traditional' right of ownership over land and animals is increasingly curtailed by state legislation. For the USA, see Geisler (1995). Privatisation of earlier communal resources has often led to increased overexploitation; for overgrazing in Botswana, see Hitchcock (1980); Peters (1987).

[10] The often romantic expectations of indigenous environmental knowledge and practice of local communities have been criticised accordingly, see Agrawal & Gibson (1999); Leach *et al.* (1999); von Benda-Beckmann & Von Benda-Beckmann (1998); Ellen (1986); Persoon (1991); von Benda-Beckmann *et al.* (1995); von Benda-Beckmann (1997).

[11] Elements of the environment sometimes but not necessarily become legally distinguished and defined as property objects on the basis of physical or economic characteristics. See Buck's (1989) classification of natural resources by bio-physical and functional criteria; see also Ostrom's (1990) distinction between resource stock/system and flows/units. In the literature, common pool and common property resources are often used interchangeably, see Berkes (1996, 89); Schlager & Ostrom (1992). But see Ostrom *et al.* (1999) for the need to use separate concepts.

[12] The metaphor is then also used to analyse how such aspects are bundled together in actual legal constructions of types of rights, categories of resources, or concrete

of rights, however, are often confusing because the concept of rights has several meanings that tend to be used interchangeably. It is important to make at least a difference between 'categorical' and 'concretised rights' [13]. Categorical rights are typified legal concepts that construct a general relationship of rights and options between categories of persons or groups with respect to categories of resources. Examples are legal constructions of 'ownership', 'open access resources', or 'inheritance rights'. In concretised rights, on the other hand, a right-relationship is established between actual persons or groups and an actual resource. Categorical rights only form the legally defined potentiality to establish concretised rights. The legal criteria of the rights category are, as it were, inscribed into a concrete relationship. We no longer speak of ownership as a type of right, but of Mr. A's ownership of a concrete rice field, the multinational's concession to a demarcated piece of land, forest or mineral deposit, an Indonesian descent group's rice and upland fields.

The distinction between categorical and concretised property rights is important, for it opens up the question of possible relations between categorical rights, say individual private ownership, and the distribution of concretised property rights, *i.e.* wealth. While categorical rights in legal systems tell us little about their social, economic or ecological significance, concretised rights give us a perspective on the constellation and distribution of property rights (and the wealth which they embody) over property categories, over people and over resources in the actual environmental space. Moreover, the ways in which categorical and concretised property rights are distinguished from, or interwoven with, other social and political relationships also usually vary considerably. At the categorical level, property rights may be regulated distinct from social and political relationships, while in actual social relationships property rights may be one strand in many-stranded relationships between persons, or persons and organisations. When talking of property as bundles of rights, we therefore have always to remember whether we talk about rights as categorical or concretised, as we have to expect quite different social and economic consequences of such different bundles.

---

resources. The oldest use of the metaphor (Maine, 1861) was to characterise the varied kinds of property rights held by individuals or other social entities.

[13] Von Benda-Beckmann & Von Benda-Beckmann (1999b); von Benda-Beckmann (1995). Compare Schlager & Ostrom (1992, 250) who fail to make this clear distinction. In their words, 'rights' refer to particular actions that are authorised, while rules refer to the prescriptions that create authorisations. This conceptual clarification does not bring out the essential difference between categorical rights and concretised rights, and consequently does not address the crucial significance of the social processes through which categorical rights have to be transformed into concretised rights.

17.2.3 The Conventional Four Property Categories

The four conventional property categories — open access resources, private individual ownership, common/communal property, state/public ownership — are not useful analytical guides nor sufficient descriptive categories for dealing with the complexity of bundled property rights.

First, they are selective. The property status of a natural resource, such as individual or state ownership or open access resource, only captures a limited number of rights and obligations that pertain to the resource. Natural resources, individually owned agricultural property in the Netherlands for instance, are subject to a number of legal constraints on the kinds of resource use or transferability, that are regulated in public law of the Dutch state or the European Union. Also in the cases of other property categories, whether we talk about lineage property, state owned property, communal property of villages or open access resources, the total bundle of rights is a combination of individual rights, and rights of larger social units or political organisations (Bruce, 1988; von Benda-Beckmann, 1995).

Secondly, these categories in themselves are umbrella terms that encompass a variety of quite different bundles of rights. Private law ownership of agricultural land one hundred years ago was an institution (bundled rights) quite different from what it is now. Common or communal property is probably the most varied category. Contemporary discussions are heavily dominated by specific forms of common property, the proverbial commons, grazing lands, forests, fishing grounds (the prototypical 'common pool resources' of Ostrom, 1990). But communal properties such as lineage property or inherited property of extended families or descent groups - a predominant form of property holdings of agricultural land in many parts of Asia and Africa - have a different mix of individual rights and rights of other social entities. For the relation between property rights to land and sustainable land use this is particularly important, since most agricultural land in developing countries is communal property of the latter, and not of the 'commons' kind.

The conceptual simplification is even worse if public or state property/ownership is treated as one unitary category. State property rights in developing countries are particularly relevant, because most contemporary states control and claim to own huge resource areas. That should be looked at, however, from at least four different categorical perspectives. First, states on the basis of their sovereignty, claim supreme political and regulatory rights with respect to land, water, forests and sub-soil resources, even if we see that some of the conventional sovereign powers have been transferred to, or taken over by supra-state organisations such as the United Nations. Secondly, especially in developing countries, states have assumed not only political but also proprietary rights over large tracts of natural resources termed crown land or state domain, and especially mineral resources. Thirdly, states can also be owners in private law and then are legally treated like any other citizen/owner. Fourthly, states are an ever increasing source of 'governments largess', out of which quotas, licenses or concessions are allocated by government on its own terms and held by recipients under conditions that are said to express the 'public interest' (Reich, 1964, 733). This has created a new mix of public and private rights with strong political functions (Reich, 1964, 745), for which an administratively controlled 'quasi-market' has been established (van Eldijk, 1992). These new rights, embodied in quotas and concessions, are a kind of property right. These 'new properties' have become the actual value, perhaps more important than the 'traditional' ownership right to the land surface itself (see also Wiber & Kearney, 1996).

What can be inferred from the conventional four property rights categories thus is limited. Given all these differentiations, it would be strange to expect or predict that these categories could have the same implications or significances in terms of economic or environmental parameters. They only give us an indication of the legal-institutional structure of property categories, but by themselves tell us nothing about the ways in which such categorical rights are actually embedded and distributed in the economic and political organisation of societies. Rather than using these categories as analytical categories, they - time and place specific bundles of categorical and concretised property rights - should be the object of analysis.

17.2.4 Legal-Institutional Pluralism

In developing countries, property rights are especially complex, because of conditions of legal pluralism[14]. Legal pluralism is an umbrella concept indicating the condition that more than one legal system or institution co-exist with respect to the same set of activities. The social entities that can be property holders, the elements of the environment that can be property objects, the ways in which property rights can be acquired, and the kind of rights-relationships can be, and often are defined in different and contradictory ways in the official state legal system and in traditional or newer forms of local law. The legal-institutional constellations become even more complex as a result of the increasing role played by international law and conventions that regulate environmental issues, access and exploitation rights, and due to the legal status and political and economic rights of 'indigenous peoples' [15].

[14] The literature on legal pluralism is immense. See Griffiths (1986); Merry (1988); von Benda-Beckmann (1983a, 1997). In resource management studies, there is an increasing recognition of legal pluralism, often couched as *de facto/de jure* (Schlager & Ostrom, 1992), or formal/informal rules and behavioural norms (see Leach *et al.*, 1999). For one of the few systematic studies in neo-institutional economics, see Ellickson (1991).

[15] For instance UN and ILO conventions and the Rio Declaration, see Van de Fliert (1994); Assies & Hoekema (1994); von Benda-Beckmann (1997). See also Leach *et al.* (1999).

At the level of categorical property rights, one thus usually is confronted with different constructions of property rights. The legal subsystems by no means are well-integrated wholes. At government levels, different domains of authority such as conventional property law making, agricultural development, mining, forestry, nature protection and conflict management are entrusted to different policy making and implementing institutions, all engaged in the making of laws and regulations, that often express distinct and often conflicting interests. Local forms of customary and folk legal regulations are also far from unambiguous. There are usually different, and often contradictory versions of such laws. Besides, one is increasingly confronted with conflicts between different local customary laws as a result of migration. All these legal forms provide a wide repertoire with which concretised property relationships can be defined. This complexity (the 'jungle of legal pluralism' (von Benda-Beckmann, 1983b) lies at the basis of a large extent of legal insecurity that characterises social, economic and political conditions in most rural areas, and that is part of the everyday context in which farmers, bureaucrats, and development agents live and work.

17.2.5 Towards a Post-Institutionalist Perspective

It is difficult to capture this complexity in the concept of 'institutions' as is usually done by resource management scholars and institutional economists. Institutions are usually defined at the level of categorical concepts, rules and procedures with respect to property (North, 1990; Ostrom, 1990; Feder & Feeny, 1993). They often are defined as compounds in which 'complexes of norms, rules and behaviours that serve a collective purpose' (de Janvry *et al.*, 1993) are lumped. Such compounds make it difficult to systematically address the question how actual behaviour relates to sets of rules and regulations, especially when dealing with plural and contradictory rules and regulations asserting to serve the same purpose (see von Benda-Beckmann, 1983a, 1995, 1997). The assumption that institutional rules - even if everyone would conform to them - could be easily associated with a specific economic 'substratum' (Renner, 1929), a more or less standardised set of property relationships and distribution, is also unfounded. It should be kept in mind that many rules and procedures concerning property rights, distribution, transfers or inheritance leave considerable rooms for autonomous decision making for property holders. While legally structured modalities of (trans)action, once invoked, may have the character of prescriptive rules, the decisions on whether or not to engage in such action - distribution, sales, the making of a testament - are not prescriptive. As many studies have shown, there can be considerable changes in the constellations of concretised property relationships, and, as a consequence of economic organisation, with more or less unchanged rule systems[16]. In addition,

[16] See the classical study of Renner (1929) on the relations between the institution of ownership and economic change in Western Europe. For Senegal, see Snyder (1977). For

the assumption that actual behaviour would correspond to institutional rules (state or local) cannot be maintained. How different actors establish, maintain or change concretised property rights, varies considerably and is highly context-specific. The different types of legitimisations of property rights may be clearly distinguished and one chosen over the other. Farmers, as well as bureaucrats, often deviate from or manipulate 'their own' rules. Claims based on different rule systems may be played out against each other in 'forum shopping' strategies (von Benda-Beckmann, 1981)[17]. They may also be used as material to construct ad hoc hybrid forms; or such hybrid forms may become institutionalised themselves. For a proper understanding of the significance of legally structured institutions and the interrelations between normative institutional frameworks, social relationships and behaviour, a perspective is required that does not take the normative framework of institutions as point of departure. One should not be guided by the prescriptive normative framework, but by the expectation that institutions (such as institutionalised property rights) can be multifunctional and their significance (measured in different functional parameters) varied. Institutions therefore should be approached from what I would call a post-institutionalist perspective that problematises the relationships between the normative institutional framework, the social relationships that develop in their context and the social practices these rules purport to regulate and that maintain or change the relationships and the normative institutional framework.

17.3 The Tragedy of Individual Rights under Different Property Rights Bundles

17.3.1 Revisiting Hardin and Open Access

I want to go back to Hardin as he, though talking about the tragedy of the commons, was essentially concerned with the unfettered use of *individual rights* to use resources. This comes out clearly when we look at Hardin's actual problem, the restriction of population growth. The tragedy in his view resulted from an unrestrained legally protected practice of individuals to procreate. His real 'commons' was 'the world' into which children were born. Children, in Hardin's essay, are the externalities imposed on a public good, the continued liveability of a world which has a problematic imbalance between resource availability and resource users.

Minangkabau, see von Benda-Beckmann (1979). For the Chagga in Tanzania, see Moore (1986).

[17] For approaches to the strategic uses of law as a resource in social interaction, see Turk (1978); von Benda-Beckmann (1983a, 1997). For actor-structure approaches in general social theory, see Giddens (1984).

But do such 'world-commons' represent an open access resource? Obviously not. Apart from the air children breathe after their birth, these alleged commons do not have the characteristics of an open access resource; on the contrary. Access to resources - hospitals at birth, nourishment, educational facilities, employment, or ownership of productive resources is strongly institutionalised and subject to a variety of social, economic and political relationships in which the distribution of wealth through property rights is an important factor. In his discussions of how the tragedy could be averted, Hardin also did not discuss how this institutionalised world should be restructured. He did not develop scenarios for the rights to the world's resources that children should acquire. His concern was not with private, communal or state property, but with the tension between the most fundamental human right and its coerced restriction, in the name of a totalising common good.

The metaphorical grazing commons which he used to illustrate his actual problem had a very different character. Here the model indeed is one of open access, in which individual appropriation of material benefits is legally unrestrained and where environmental impacts would be shared by all. The model, however, neglects that, in order to use and appropriate these commons and contribute to the tragedy, more property rights are required than individual appropriation rights. In order to put ever more calves on the pasture, one needs calves! It does not make sense to construct a tragedy without the herders and their calves, without the fishermen and their boats and nets, the wood cutters and their chain saws, the mining companies without their technology. The model also neglects the type of rights that herders must have in order to create the tragedy. It is implicit in Hardin's model that the rights to the calves are held in individual ownership.

This tells us that the main problem lies in a combination of an unrestrained individual right of appropriation with individual ownership rights over other means of production (animals, labour, money, technology). It therefore does not make sense to dissociate the open access status of resources from the rights over the means of appropriation and the wealth embodied in these rights. As Balland and Platteau (1999, 774) say with respect to common property resources, 'the distribution of benefits reflects the distribution of rights of access to the common property resources which is itself influenced by the distribution of wealth endowments', in other words the distribution of different kinds of concretised property rights over people[18]. This is the more so, the more sophisticated and valuable these resources are. And the more sophisticated and efficient these resources are, the more capital-intensive (in whatever meaning of capital), the greater the danger of overexploitation and degradation of the natural resources by

[18] 'Privileged access to a CPR does not necessarily take on the form of a comparative large number of harvesting units put into use, but also may manifest itself in the ability to occupy strategic positions in the common property resource' (Balland & Platteau, 1999:785). See also McCay, 1987, 207, 208.

the wealthier[19]. The tragedy of the commons seems more a problem of *free raiders* than of free riders.

17.3.2 The Tragedy and Other Property Rights

Assumptions over the relationships between property rights and resource use are not just problematic with respect to open access natural resources. They are equally problematic with respect to natural resources that are used and controlled under different property regimes. As I have already mentioned, Balland & Platteau (*op. cit.*) cogently make the point for common property resources. The conditions are similar for natural resources held in freely transferable individual ownership. Richer people obviously can buy up more rights in land than poor people. Unrestricted transaction regimes, the ideal-typical free market, open up possibilities of 'free raiding', for accumulation of property rights by the rich and thus to the redistribution of wealth. In developing countries, this process has dramatically increased through the increasing abolition of market restrictions associated with globalisation. In fact, unrestrained access to natural resources through the market has become easier than with respect to most open access resources and certainly with respect to communal property resources where there usually are much more cultural and/or legal restrictions. The same also holds with respect to state owned or controlled natural resources.

17.4 Features of Property Rights Constellations and Care for Resource Quality

The basic problem, thus, does not lie in the legal status of the resources as such. It is rather one of limiting individual rights on the (other) means of production, and this is relevant with respect to all categories of property rights to resources. Policy and law makers therefore ask for rules and regulations that constrain rights of appropriation and use of individually owned resources. When looking for elements of property rights that may be more, or less, conducive to sustainable resource use, we cannot expect much from the four general categories but have to look further.

What features of categorical property rights and distributive constellations of concretised rights then can be distilled, that would form a basis for a more

[19] As Peters (1987, 177) said with respect to the use of grazing lands in Botswana that Hardin's analysis would turn history on its head. The gradual erosion of the commons as a system of complementary and overlapping rights occurred through a long process of commercialisation in which owners of private land used the commons as free good for their large flocks of sheep to the detriment of those commoners without access to free land. See also Thompson (1993) for the analysis of the enclosure processes in England. For the baldios in Portugal, see Brouwer (1992).

responsible attitude towards resource quality? What emerges from the literature is a set of often interconnected factors. A matrix of time horizon, distribution of concretised property rights and the wealth embedded in them, the function of the property for its holders, especially their dependence on the resources for survival/livelihood, profit making, and alternative economic possibilities and the fungibility of property rights.

17.4.1 The Temporal Dimension of Property Rights and the Time Horizon of Property Holders

An important aspect of the relation between property rights and resource use is the time horizon of property holders. It is one important condition for long term planning of production and of people's attitudes towards caring for the quality of resources and the security of the right that right holders have or experience (see Ruben *et al.*, this volume). Balland & Platteau (*op. cit.*) have emphasised the importance of actors' time horizons in influencing the extent to which they are likely to care about the value and quality of resources under their control beyond the immediate present needs. They add that variations in time horizon are related to the initial distribution of wealth, either through the intervention of survival constraints or through the availability of outside economic opportunities[20].

Here again, it is important to look into the temporal dimension of both categorical and concretised property rights, for the temporal dimension is also structured by the time horizon built into categories of property rights. Let me give an example from my own research in West Sumatra (von Benda-Beckmann, 1979).

The Minangkabau people in West Sumatra are noted for their irrigated rice farming. More or less all rice fields (and also upland fields) have the legal status of inherited property held by matrilineal descent groups. The Minangkabau have a saying with respect to their inherited lineage property complexes: 'Its fruits may be eaten, its water may be drunk, but its stems remain forever' (*buahnyo bulieh dimakan, aienyo bulieh diminum, batangnyo tatap tingga*). The dominant function of these property complexes is to provide the material basis for the social continuity of the descent group and its future members. The Brundlandt report (WCED, 1987) could not have said it better. The intergenerational solidarity is solidly anchored in Minangkabau adat ideology and law. This obliged group members to increase group property and imposed severe restrictions on permanent and even temporary transfers to persons outside the group, for transactions of the lineage property stock are a fundamental threat to maintaining this intergene-

[20] Unfortunately Balland and Platteau limit their deliberation to incentives to collective action with respect to common property resources, but do not ask whether this would also hold true for individual action with respect to resources with a different property status. They also do not query whether variations in time horizons of actors are also related to time horizons embedded in different categorical constructions of property rights.

rational solidarity. 'Optimal allocation' of property in Minangkabau thus was 'non-transfer', maintenance in the group beyond the lifetime of individuals. What economists might call a market imperfection or failure, for the Minangkabau and other societies with similar structures, is the optimal allocation of property. In such societies, the property and inheritance rights structure can significantly influence the extent to which people take ecological problems of overexploitation into account in their dealings with resources. Generally, there is a likelihood that there is greater awareness of the need not only to conserve the quantity, but also the quality of the resource. Inheritance therefore must be seen as an important variable of property. Inheritance rules, and the legal capacity to direct the flow of property rights after the holder's death, are therefore frequently considered to be more important for farmers' economic and ecological motivations than the transferability of property rights through sales (see Derrett, 1965; von Benda-Beckmann, 1979, 1995; Bruce *et al.*, 1994).

Minangkabau is an example where the social continuity function of property is clearly inscribed into the most important category of property. This is in contrast with property categories, like ownership, that abstract from a specific temporal dimension and permit both long term continuity as well as frequent transfer of rights. But even then, inheritance law and people's inheritance practices are one important manifestation in which the time horizon of social actors plays an important role and that brings people to be concerned with the quantity and property beyond their own lifetime. Even if there is a market, and even if the circle of 'social continuity is narrowed down to one individual heir, the maintenance of the property complex for the next generation tends to be an important incentive for social actors' [21].

Of course, in Minangkabau as elsewhere, cultural and legal institutional rules do not determine individual action, but they do provide motivations for those who deal with property - both as holders and as future potential recipients/heirs, and they impose heavy constraints on those who want to deviate from these rules. This will remain of relevance even if a land market should emerge or be forced upon the people by the government via land reform legislation, or when group rights to group property become increasingly individualised.

17.4.2 The Functions of Property and Outside Opportunities

Balland and Platteau (1999) emphasise that variations in time horizon are related to the initial distribution of wealth, either through the intervention of *survival constraints* or through the *availability of outside economic opportunities*. Common to both factors is the element of dependence, and the actual function or significance a natural resource has for its users. This largely depends on the distribution of wealth. Poor farmers/resource users are more dependent on the land

[21] See De Haan (1994) for Dutch farmers; Marcus (1980) for American business families.

they use, and may indeed be forced to overexploit it because of survival constraints which force them to reduce their time horizon (Balland & Platteau, 1999; see also Holden *et al.*, 1998). Wealthier land owners may not need the land for agricultural production, but have alternative incomes. Depending on the function they give to the land, they may use/acquire it for leisure, living in a beautiful natural environment and be concerned with the resource quality, and easily convinced to adopt environmental improvements even at costs (see Geisler & Bedford (1996) for better-off 'inholders' in the United States). The extent to which people are likely to care about the quality of their resource is likely to vary with their time horizon. If they are not primarily concerned with optimising agricultural production, this can work both ways. They are less likely to overexploit the resource; they will also not tend to invest in conservation measures. Others may use land as an old age insurance, as object of speculation and accumulation of wealth[22]. The function of property at group and individual levels for social security and as fall-back resource (Agarwal, 1991) is easily neglected in economic analyses of property rights, yet often plays a quite important role (see Chambers & Leach, 1989; von Benda-Beckmann, 1990; van de Ven, 1994; von Benda-Beckmann & Von Benda-Beckmann, 1999b).

Outside economic opportunities also depend on the actual distribution of concretised property rights and the actual bundles of rights to different kinds of property. In societies in which farmers depend on the continued availability of their resources for themselves and for their descendants, farmers usually have no intention to sell their property anyway. If they sell, this will mainly be to escape from a miserable economic life. Whether they will then be in a bargaining position to recover any costs which they may have invested is very questionable anyway. De la Brière's research (this volume) also refutes these assumptions. She shows that the more farmers depend on their land, the poorer they are, the less they are integrated in the market, the more they adopt conservation methods. Larger richer farmers, less 'trapped in agriculture', more market integrated, are 'most resistant to adoption and hence the more significant source of soil erosion and its associated externalities'. A large agricultural land holding negatively affects adoption of conservation practices. Under these conditions, the main incentive for investments is not the possibility for recovering these investments via market transactions.

---

[22] The reforms of land rights introducing individual ownership titles in many African and Asian countries has shown that land sales do not tend to follow the normative logic of economic theory. Most land bought under the new property rights and transfer law is not used to optimise agricultural production. Neef (1999:101) found that land sales increased in Southern Benin, especially in rural areas close to towns, and that the large majority of all sales was by urban people of whom only 10% worked the land themselves. From the urban buyers' point of view, the main motive frequently is insurance against economic needs and distress, much more than the intention to valorise an agricultural plot (1999, 103). See also Fisiy (1992) on Cameroon; Werbner (1980) and Hitchcock (1980) for Botswana. For Indonesia, see von Benda-Beckmann (1979); von Benda-Beckmann & Taale (1996); van de Ven (1994).

These considerations not only apply to resources, land, held in individual ownership or as a share in communal property, but as well for state property. At household and local community level there is the observation that natural resource management, as well as new forms of co- or joint management, tends to be monopolised by village elites, or to be 'sold out' to outside interests. The household head can not be identified with the household, nor the village with its chief or the managers of a firm with its employees or shareholders. Also the state structure can not be identified with the aims and actions of those controlling and exploiting state owned natural resources, whether on the basis of state ownership or indirectly via the medium of licences and concessions. While the state structure is based on a long time horizon, those governing states and controlling the state's resources, are mortal and live through changing and unpredictable careers. Democratic decision making, job rotation schemes, and the constant pressures of the immediate present foster a short term time horizon. At the level of state politicians and bureaucracy, variations in time horizon and commitment to the common good are also likely to vary with the distribution of wealth, dependence on resources, and alternative economic opportunities. In many developing countries, high and low ranking civil servants cannot earn their livelihood without additional and conscious direct or indirect appropriation of state controlled resources. For lower ranking civil servants, this may indeed be a survival imperative (see von Benda-Beckmann & Von Benda-Beckmann, 1998). The abuse made of state power, the grabbing of the nation's and its people's resources to channel them into private pockets have become proverbial[23].

The influence of resource dependence and time horizon may be similar at the level of states as unitary organisations. Given their dependence on the revenue they can extract from natural resources, directly or indirectly, they may be forced to 'sell out' their resources to more powerful economic agents (other states, transnational corporations). They then are open to 'free raiders', the demands of predatory outside economic interests that are more concerned with shareholder profit than with the sustainability of the resources they exploit. The shorter term the exploitation licence, the greater the pressure of securing profit in this period, the greater the likelihood that they externalise environmental damage to others -- others here being the state, those affected immediately, and the world community at large.

Also the *spatial* constellation of the property right holders in relation to potentially degrading resources plays an important role here. This is also a function of their

---

[23] This is not only a problem in developing countries. For the USA, see Geisler (1995:22). 'The impact of governmental power falls unequally on different components of the private sector, so that some gain while others lose. Third, the apparatus of governmental power may be utilised by private interests in their conflicts with other interests, and thus the tools of government become private rather than public instrumentalities' (Reich, 1964, 764).

dependence on resources, and their proximity/distance to the resources and the possible degrading effects of the resource use they control. Certain forms of degradation are inflicted from a distance, where the decision makers, state officials for instance, themselves do not directly depend on the resources, do not have to live in the resource environment but may depend on the profit/revenues generated from the commercial exploitation or other of these resources. As a consequence of globalisation, we witness - despite all efforts to come to more local, community based forms of resource management - also an increasing 'deterritorialisation' (Appadurai, 1990) of property rights, and an increasing dissociation of managerial and governance organisation from the spatially grounded resources through the globalisation of environmental values, the construction of common goods at world scale.

17.5 Conclusions

Property rights, as legal-institutional frameworks and as concrete relationships between people, organisations and natural resources are an important element structuring land use and its economic, political and ecological significance. It is equally clear that property, even in this wider meaning, is only one of the social elements influencing resource use. In this chapter I have tried to contribute to a reorientation of assumptions and perspectives with which the significance of property rights for more or less sustainable land use is viewed. I have tried to show that the conventional four property categories do not offer a useful guidance into the complexity of property relationships we encounter. Generalisations over relations between categorical types of property rights and environmental consequences are dangerous and do not seem to be warranted given the empirical evidence. Rather, characteristics of property rights seem to play a dominant role for land use, relatively independent of the legal status of the land, such as the time horizon of land users, their dependence on the resource and the kind of dependence, and the spatial proximity/distance to the resource. I have further argued that in particular actual constellations of property rights, in other words the distribution of wealth, over people and resources are important characteristics. Finally, I have argued that institutions thus should be approached from a post-institutionalist perspective which problematises the relationships between the rule-aspects, the normative-legal institutional framework, the social relationships that develop in their context and the social practices these rules purport to regulate and which maintain or change the relationships and the normative institutional framework.

17.6 References

Agarwal, B. (1991): Social security and the family: coping with seasonality and calamity in rural India. In: E. Ahmad, J. Drèze, J. Hills & A. Sen (Eds.) Social security in developing countries. Clarendon Press, Oxford

Agrawal, A. & C.C. Gibson (1999): Enchantment and disenchantment: the role of community in natural resource conservation. World Development 27, 629-649

Appadurai, A. (1990): Disjuncture and difference in the global cultural economy. In: M. Featherstone (Ed.) Global culture: nationalism, globalization and modernity. Sage, London, 295-310

Assies, W.J. & A.J. Hoekema (Eds.), 1994, Indigenous peoples. Experiences with self-government. Document 76. IWGIA, Amsterdam, Copenhagen

Balland, J.M. & J.Ph. Platteau (1999): The ambiguous impact of inequality on local resource management. World Development 27, 773-788

Benda-Beckmann, F. von (1979): Property in social continuity: continuity and change in the maintenance of property relationships through time in Minangkabau, West Sumatra. M. Nijhoff, The Hague

Benda-Beckmann, K. von (1981): Forum shopping and shopping forums. Journal of Legal Pluralism 19, 117-159

Benda-Beckmann, F. von (1983a): Why law does not behave: critical and constructive reflections on the social scientific perception of the social significance of law. In: H. Finkler (Comp.): Proceedings of the Symposium on Folk Law and Legal Pluralism, XIth IUAES Congress, 1983, Vancouver, Ottawa

Benda-Beckmann, F. von (1983b): Op zoek naar het kleinere euvel in de jungle van het rechtspluralisme. Inaugural lecture. Wageningen

Benda-Beckmann, F. von (1989): Scapegoat and magic charm: law in development theory and practice. Journal of Legal Pluralism 28, 129-148

Benda-Beckmann, F. von (1990): Sago, law and food security on Ambon. In: J.I.H. Bakker (Ed.) The world food crisis: food security in comparative perspective. Canadian Scholars. Press Inc., Toronto, 157-199

Benda-Beckmann, F. von (1995): Anthropological approaches to property law and economics. European Journal of Law and Economics 2, 309-336

Benda-Beckmann, F. von (1997): Citizens, strangers and indigenous peoples: conceptual politics and legal pluralism. In: F. von Benda-Beckmann, K. von Benda-Beckmann & A. Hoekema (Eds.). Natural resources, environment and legal pluralism. Special issue of Law and Anthropology 9, 1-42

Benda-Beckmann, F. von & K. von Benda-Beckmann (1998): Where structures merge: state and off-state involvement in rural social security on Ambon, Eastern Indonesia. In: S. Pannell & F. von Benda-Beckmann (Eds.) Old worlds places, new worlds problems: exploring issues of resource management in eastern Indonesia. Centre for Resource and Environmental Studies, The Australian National University, Canberra, 143-180

Benda-Beckmann, F. von & K. von Benda-Beckmann (1999a): Community based tenurial rights: Emancipation or indirect rule? In: K. von Benda-Beckmann & H.F. Finkler (Eds.) Papers of the XIthe International Congress 'Folk Law and Legal Pluralism' Societies in Transformation. Commission on Folk Law and Legal Pluralism, Moscow August 18-22, 1997. Ottawa, 169-187

Benda-Beckmann, F. von & K. von Benda-Beckmann (1999b): A functional perspective on property rights, with special reference to Indonesia. In: T. van Meijl & F. von Benda-Beckmann (Eds.) Property rights and economic development: land and natural resources in southeast Asia and Oceania. Kegan Paul, London

Benda-Beckmann, F. von & T. Taale (1996): Land, trees and houses: changing (un)certainties in property relationships on Ambon. In: D. Mearns & D. Healey (Eds.) Remaking Maluku: social transformation in Eastern Indonesia. NT University, Darwin, 39-63

Benda-Beckmann, F., K. von Benda-Beckmann & A. Brouwer (1995): Changing indigenous environmental law in the Central Moluccas: communal regulation and privatization of Sasi. Ekonesia 2, 1-38

Benda-Beckmann, K. von, M. de Bruijn, H. van Dijk, G. Hesseling, B. van Koppen & L. Rys (1997): Rights of women to the natural resources land and water. Ministry of Development Cooperation, The Hague

Berkes, F. (Ed.) (1989): Common property resources: Ecology and community-based sustainable development. Belhaven Press, London

Berkes, F. (1996): Social systems, ecological systems, and property rights. In: S. Hanna, S.C. Folke & K.G. Maler (Eds.) Rights to Nature. Island Press, Washington, D.C., 87-107

Berkes, F. & C. Folke (1998): Linking social and ecological systems resilience and sustainability. In: F. Berkes & C. Folke (Eds.) Linking social and ecological systems: management practices and social mechanisms for building resilience. Cambridge University Press, Cambridge, 1-25

Briere, B. de la (2000): Imperfect food markets and household adoption of soil conservation practices in the Dominican Republic Highlands: household probit and duration models. This volume

Brouwer, R. (1992): The commons in Portugal: A story of static representations and dynamic social processes. In: F. von Benda-Beckmann & M. van der Velde (Eds.) Law as a Resource in agrarian struggles. Pudoc, Wageningen, 127-148

Bruce, J.W. (1988): A perspective on indigenous land tenure systems and land concentration. In: R.E. Downs & S.P. Reyna (Eds.) Land and society in contemporary Africa. University Press of New England, London, 23-53

Bruce, J.W., S. Migot-Adholla & J. Atherton (1994): The findings and their policy implications: institutional adaptation or replacement. In: J. Bruce & S. Migot-Adholla (Eds.) Searching for land tenure security in Africa. Kendall/Hunt, Dubuque, 251-265

Buck, S.J. (1989): Multi-jurisdictional resources: testing a typology for problem-structuring. In: F. Berkes (Ed.) Common property resources: ecology and community-based sustainable development. Belhaven Press, London, 127-147

Chambers, R. & M. Leach (1989): Trees as savings and security for the rural poor. World Development 17, 329-342

Conn, S. (1997): Nuclear waste and Eskimo hunters: a clash of law ways and culture. In: F. von Benda-Beckmann, K. von Benda-Beckmann & A. Hoekema (Eds.) Natural resources, environment and legal pluralism. Special issue of Law and Anthropology 9, 211-225

Demsetz, H. (1967): Toward a theory of property rights. American Economic Review 57, 347-359

Derrett, J.M.D. (Ed.) (1965): Studies in the law of succession in Nigeria. Oxford University Press, London

Eldijk, A. van (1992): Collectieve goederen en private rechten. Recht der Werkelijkheid 1, 4-7

Ellen, R. (1986): What black elk left unsaid: on the illusory images of green primitivism. Anthropology Today 2, 8-12

Ellickson, R.C. (1991): Order without law. How neighbours settle disputes. Harvard University Press, Cambridge, Mass.

Feder, G. & D. Feeny (1993): The theory of land tenure and property rights. In: K.A. Hoff, A. Braverman & J.E. Stiglitz (Eds.) The economics of rural organization: theory, ppractice, and policy. Oxford University Press, New York, 240-258

Fisiy, C. (1992): Power and privilege in the administration of law: land reform and social differentiation in Cameroon. Africa Studies Centre, Leiden

Fliert, L. van de (1994): Indigenous peoples and international organisations. Spokesman, Nottingham

Furobotn, E.G. & S. Pejovic (1972): Property rights and economic theory: a survey of recent literature. Journal of Economic Literature 10, 1137-1162

Geisler, C. (1995): Land and poverty in the United States: insights and oversights. Land Economics 71, 16-34

Geisler, C. & B. Bedford (1996): Who owns the ecosystem? LTC Paper 157, Land Tenure Centre. University of Wisconsin

Gibbs, C.J.N. & D.W. Bromley (1989): Institutional arrangements for management of rural resources: common property regimes. In: F. Berkes (Ed.) Common property resources: Ecology and community-based sustainable development. Belhaven Press, London, 22-32

Giddens, A. (1984): The constitution of society. Polity Press, Oxford

Griffiths, J. (1986): What is legal pluralism? Journal of Legal Pluralism 24, 1-50

Haan, H. de (1994): In the shadow of the tree: kinship, property and inheritance among farm families. Het Spinhuis, Amsterdam

Hann, C. (Ed.) (1998): Property relations: renewing the anthropological tradition. Cambridge University Press, Cambridge

Hanna, S., C. Folke & K.G. Maler (Eds.) (1996): Rights to nature. Island Press, Washington, D.C.

Hardin, G. (1968): The tragedy of the commons. Science 162, 1234-1248

Heerink, N., A. Kuyvenhoven & M.S. van Wijk (2000): Economic policy reforms and sustainable land use in developing countries: issues and approaches. This volume

Hitchcock, R.K. (1980): Tradition, social justice and land reform in central Botswana. Journal of African Law 24, 1-34

Holden, S.T., B. Shiferaw & M. Wik (1998): Poverty, market imperfections and time preference: of relevance for sustainable resource management? Environment and Development Economics 3, 105-130

IUCN (1984): Traditional life-styles, conservation, and rural development. Commission on Ecology Papers No.7. Gland.

Janvry, A. de, A. Sadoulet & E. Thorbecke (1993): State, market and civil organizations: new theories, new practices and their implications for rural development. World Development 21, 565-575

Kaimowitz, D. & A. Angelsen (1998): Economic models of tropical deforestation. CIFOR, Bogor

Larson, B.A. & D. Bromley (1990): Property rights, externalities and resource degradation – locating the tragedy. Journal of Development Economics 33, 235-262

Leach, M., R. Mearns & I. Scoones (1999): Environmental entitlements: dynamics and institutions in community-based natural resource management. World Development 27, 225-247

Lele, S.M. (1991): Sustainable development: a critical review. World Development 19, 607-621

Lynch, O. & Talbott (1995): Balancing acts: community-based forest management and national law in Asica and the Pacific. World Resources Institute, Washington

Maine, H.S. (1986[1861]): Ancient law. University of Arizona Press, Tucson

Maine, H. (1883): Dissertations on early law and custom. Murray, London

Marcus, G.E. (1980): Law in the development of dynastic families among American business elites: the domestication of capital and the capitalization of family. Law and Society Review 14, 859-903

McCay, B.J. (1987): The culture of the commoners: historical observations on old and new fishermen. In: B.J. McCay & J. M. Acheson (Eds.). The question of the commons. University of Arizona Press, Tucson, 195-216

McCay, B. J. & J.M. Acheson (1987): Human ecology of the commons. In: B.J. McCay and J. M. Acheson (Eds.), The question of the commons. University of Arizona Press, Tucson, 1-35

Merry, S.E. (1988): Legal pluralism. Law and Society Review 22, 869-896

Moore, S.F. (1986): Social facts and fabrications. Cambridge University Press, Cambridge

Myrdal, G. (1970): The challenge of world poverty: a world anti-poverty programme in outline. Penguin Books, Harmondsworth

Neef, A. (1999): Auswirkungen von Bodenrechtswandel auf Ressourcennutzung und wirtschaftliches Verhalten von Kleinbauern in Niger und Benin. Development Economics and Policy No. 12. P. Lang, Frankfurt am Main

North, D. (1990): Institutions, institutional change and economic performance. Cambridge University Press, Cambridge

Ostrom, E. (1990): Governing the commons. Cambridge University Press, Cambridge

Ostrom, E., J. Burger, C.B. Field, R.B. Norgaard & D. Policansky (1999): Revisiting the commons: local lessons, global challenges. Science 284, 278-282

Persoon, G. (1991): Ecological balance and innovations: cases from the forest. In: J.P.B. van den Breemer, H.Th. van der Pas & H.J. Tieleman (Eds.) The social dynamics of economic innovation/ DSWO Press, Leiden, 113-127

Peters, P. (1987): The grazing lands of Botswana and the commons debate. In: J. McCay & J.M. Acheson (Eds.) The question of the commons. Tucson: University of Arizona Press, 171-194

Platteau, J. Ph. (1996): The evolutionary theory of land rights as applied to sub-saharan Africa: a critical assessment. Development and Change 27, 29-86

Reeve, A. (1991): The theory of property: beyond private versus common property. In: D. Held (Ed.) Political theory today. Polity Press, Oxford, 91-114

Reich, C. (1964): The new property. Yale Law Review 72, 734-787

Renner, K. (1929): Die Rechtsinstitute des Privatrechts und ihre soziale Funktion. Tübingen.

Rose, C.M. (1994): The comedy of the commons: custom, commerce, and inherently public property. In: Property and persuasion: essays on the history, theory and rhetoric of ownership. Boulder: Westview Press, 105-164

Ruben, R., M. van den Berg & T. Shuhao (2000): Land rights, farmers' investment, and sustainable land use: modelling approaches and empirical evidence. This volume

Schlager, E. & E. Ostrom (1992): Property-rights regimes and natural resources: a conceptual analysis. Land Economics, Vol.68: 249-62.

Schott, R. (1982): Main trends in German ethnological jurisprudence and legal ethnology. Journal of Legal Pluralism 20, 37-68

Snyder, F. (1977): Land law and economic change in rural Senegal: Diola pledge transactions and disputes. In: I. Hamnett (Ed.) Social anthropology and law. Academic Press, London, 113-157

Spiertz, J. & M. Wiber (Eds.) (1996): The role of law in natural resource management. Vuga, The Hague

Thompson, E.P. (1993): Customs in common. The New Press, New York

Tigar, M.E. & M.R. Levy (1977): Law and the rise of capitalism. Monthly Review Press, New York and London

Turk, A. (1978): Law as a weapon in social conflict. In: Ch. Reasons and M.M. Rich (Eds.) The Sociology of law: a conflict perspective. Butterworth, Toronto, 213-232

Ven, J. van de (1994): Members only: time-sharing rice fields and food security in a Sumatran valley. In: F. von Benda-Beckmann, K. von Benda-Beckmann & H. Marks (Eds.) Coping with insecurity: an 'underall' perspective on social security in the Third World. Special Issue Focaal 22/23, 85-96

WCED (1987): Our common future: the report of the World Commission on Environment and Development. Oxford University Press, Oxford

Werbner, R.P. (1980): The quasi-judicial and the experience of the absurd: remaking land law in north-eastern Botswana. Journal of African Law 24, 131-150

Wiber, M.G. (1993): Politics, property and law in the Philippine uplands. Wilfried Laurier University Press, Waterloo

Wiber, M.G. & J.F. Kearney (1996): Stinting the commons: property, policy or power struggle? Comparing quota in the Canadian dairy and fisheries sectors. In: J. Spiertz & M.G. Wiber (Eds.) The role of law in natural resource management. VUGA, The Hague, 145-165

18 Land Rights, Farmers' Investment, and Sustainable Land Use: Modelling Approaches and Empirical Evidence

Ruerd Ruben[1], Marrit van den Berg[1] & Tan Shuhao[2]

[1] Development Economics Group, Department of Economics and Management, Wageningen University and Research Centre, Wageningen, the Netherlands

[2] College of Land Management, Nanjing Agricultural University, Nanjing, China.

Abstract

Secure land rights are generally acknowledged as a major incentive to increase agricultural yields and farm investments. In practice, however, tenure policies easily fail due to the persistence of a large number of market imperfections. This chapter provides a review of the literature on the role of land rights in farm household resource allocation decisions, and the dynamic effects of improved land property or transfer rights on labour use, input applications and capital investments. Special attention is given to the importance of land rights for enhancing (non-)attached investments in sustainable production systems and the incentives for increasing labour and material input use as components of more sustainable land management practices. Different aspects of land rights are discussed in order to provide insight in the dynamics of rural land markets under multiple market imperfections. Basic components for a generic modelling framework that addresses the impact of tenure status on access to credit and farmers' willingness to invest are presented. Empirical results derived from studies that assess the implications of land rights for fixed investment, material input use and labour use still provide rather mixed evidence. Differences in modelling approaches and empirical settings prove to be of fundamental importance for the understanding of the implications of land rights for resource management practices.

18.1 Introduction

Structural adjustment programmes for developing countries usually include measures for market liberalisation and institutional reforms. Legislation to secure land rights and effective operation of land and capital markets are considered as major devices to improve efficiency and stimulate investment among small farmers (Binswanger *et al.*, 1993). Therefore, a large number of ambitious programmes for land registration, land consolidation and land titling have been launched.

Recent theoretical and empirical studies point, however, to the complexity of land right systems and the importance of a wide range of conditioning factors that influence land use and investments. Tenure security appeared fairly well guaranteed under indigenous systems (Atwood, 1990; Place & Hazell, 1993), while legal land rights do not automatically lead to better access to credit (Gavian & Ehui, 1999). Moreover, transactions on land markets are restricted due to the high risks associated with private rights and consolidated holdings (Melmed-Sanjak & Carter, 1990; Blarel *et al.*, 1992). Without denying the importance of clearly defined and recognised property and transfer rights, welfare implications of tenure reforms strongly depend on access to factor markets, as well as on farmers' risk behaviour.

This chapter provides a review of some analytical approaches available for the appraisal of the structure and dynamics of land markets in developing countries and their impact on farm households' resource allocation decisions. Major attention is given to (i) the distinction between land property and land transfer rights, (ii) the role of land rental and rural labour markets as vehicles to provide incentives for improved efficiency, and (iii) the impact of tenancy regimes on farmers' decisions regarding labour use, input allocation and fixed investments. The latter aspect is considered especially important within the framework of the policy debate on suitable incentives for introducing sustainable land use systems and practices.

The remainder of the chapter is divided into five sections. In Section 18.2, different types of land rights are distinguished and their implications for resource management are discussed. Section 18.3 presents a brief overview of current theories regarding the impact of property rights on farm households' investment decisions. Section 18.4 provides a general modelling framework that can be used to assess the effects of credit constraints, migration and tenure insecurity on farm household decisions regarding land and labour use. In Section 18.5, results are discussed from several empirical studies that address these different implications of land rights for farm household resource use decisions. We conclude with some major challenges for rural development policies with respect to land rights reform in developing countries.

18.2 Land Rights and Resource Management

While ongoing structural adjustment programmes in developing countries devote major attention to economic reforms aiming at 'getting prices right', informal or insecure land rights and imperfect functioning of rural land markets tend to reduce the potential impact of these incentives on small farmers. Therefore, institutional reforms increasingly receive attention as a complementary policy device. Moreover, transparency of land markets and well-defined land rights are considered important instruments for environmental protection.

Land rights consist of a bundle of entitlements defining rules and duties concerning the use of natural resources (Bromley, 1991). They involve a contractual relation between heterogeneous agents (*i.e.* with unequal asset endowments) that face imperfect information and knowledge about both the behaviour of other agents and the carrying capacity of the resource base. Decisions regarding resource use are taken within an environment of missing markets or multiple market imperfections (especially on land, capital and output markets) that make agents intrinsically risk averse (Carter & Zimmerman, 1999). Therefore, mixed institutional arrangements appear to provide additional insurance (Carter, 1984).

Land rights refer to different tenure and management regimes that are effective in a certain region. They involve multiple social relationships of rights and duties between people that are guided by law and tradition. A common distinction is made between four different dimensions of land rights (Besley, 1995; Liu *et al.*, 1998):

a) *property rights*: formally registered or traditionally recognised ownership or possession rights on land acquired through legal titling, land clearing or tradition;
b) *use rights*: usufruct rights granted to an operator of a field for a certain defined period, including leasehold and sharecropping contracts;
c) *transfer rights*: rights to sell, rent, bequeath, pledge, mortgage and give land, usually exercised under lineage approval (inheritance rules);
d) *disposal rights*: rights to use the residual income derived from land use without any further obligations to community or state authorities.

These separate aspects of land rights can be combined in different ways and constitute so-called 'bundles' of rights. Specific land rights can also be assigned to different agents. These rights are sanctioned through different legal regimes or traditions. Customary and formal land tenure regimes can co-exist, giving rise to hybrid tenure systems characterised as legal pluralism (von Benda-Beckman, this volume).

The evolution of land rights shows important differences among countries and regions (Feder *et al.*, 1988; Byerlee & Valdez, 1996; Faruque & Carey, 1997). In the Asian continent, private land rights have become a common feature, established either through massive land titling programmes (Thailand) or based on long-term usufruct rights (China). Even where customary rights are still important (like in Indonesia and the Philippines), interest in individual land ownership increases when access to credit or commodity markets enables investments in attached capital (especially permanent tree crops). In most Sub-Saharan countries well-developed systems of indigenous land rights exist, that can be maintained due to low levels of agricultural intensification and scarcity of formal credit. Instead of redistributing property rights, most attention is given to the re-definition of the terms and conditions for land disposal (Platteau, 1996). In Latin American countries fairly well established private property regimes have evolved, although situations of undefined and overlapping rights still exist. Land reform programmes recognised use rights, but registration of property rights has generally been neglected.

Adjustment of land rights usually takes place in response to growing population pressure and commercialisation of agriculture. Institutional change is considered as an 'induced innovation', to take advantage of modified factor and commodity prices (Hayami & Ruttan, 1985). Provision of tradable land rights would enable the allocation of land to more productive farmers, enhancing their investment ability, and thus increasing land use intensity. In practice, however, demand for necessary credit is constrained due to high transaction costs and risks (Place & Hazell, 1993; Pickney & Kimuyu, 1994; Platteau, 1996). The direct impact of land rights on agricultural yields can be questioned, since farmers' willingness to invest depends far more on difficult to measure factors like soil quality, farmers' skills and marketing perspectives. Moreover, when land ownership is considered as an insurance against unemployment, as a pension fund, or as a source of identity, limited land sales take place (Lawrey, 1993). Customary land rights may then be preferred for insurance purposes.

The impact of land rights on sustainable land use is usually analysed through a comparison of current income flows with asset (capital and soil nutrient) stocks. Improving sustainability implies that sufficient incentives should be available for (i) capital investments in physical soil conservation measures (*e.g.* anti-erosion structures, wind shields, *etc.*), (ii) acquisition of material inputs that restore soil nutrient stocks (fertilisers, manure), and (iii) labour resources devoted to soil-quality maintenance activities. In principle, all these efforts imply a reduction in current revenues while maintaining or increasing the value of land resources that create future income streams. Their attractiveness, from the farm household viewpoint depends on its resource endowments (labour and cash availability, access to labour and capital markets), the technical possibilities for factor substitution, and inter-temporal choice variables (*i.e.* subjective time discount rate and the certainty of future income streams).

18.3 Land Rights and Investment Decisions

Land tenure reforms are usually considered as an instrument for improving the efficiency of resource allocation in agriculture. More secure land rights offer incentives for investments attached to land (*e.g.* perennial crops, physical anti-erosion measures), increase the intensity of agricultural input and labour use, and eventually improve farmers' incomes. Besley (1995) distinguishes between two primary effects derived from secure property rights:

a) *Investment Demand or Security Effect*: households perceive greater security of receiving the full benefits of long-run investment in land improvement;
b) *Credit Supply or Collateral Effect*: potential lenders consider land ownership title as a suitable collateral for credit operations and consequently, farmers could improve allocative efficiency.

A clear distinction between these two effects is, however, difficult to make. Decisions on investment can be determined either by supply or demand factors. Even when farmers' willingness to invest increases, due to improved tenure security, credit supply may still be a limiting factor. Property rights may be a necessary, but not a sufficient condition to have access to finance (Feder *et al.*, 1992; Lopez, 1996; Carter & Olinto, 1998). Besides investment in attached capital (*i.e.* land improvement), also non-attached investment (*i.e.* factor inputs and implements) as well as the use of material inputs (seed and fertilisers) and labour could be enhanced as a result of improved tenure security. The latter resources may be particularly important when attention is given to the sustainability implications of tenure reforms.

While these studies focus on the effects of property rights on investment behaviour, other approaches pay more attention to the effects of improved land transfer rights on investment in land. Carter (1999) distinguishes the following effects:

c) *Investment Regret Mitigation Effect*: greater transfer rights make households more willing to invest in land because they are less likely to regret sunk-investment in case they should exit farming;
d) *Gains from Specialisation Effect*: well-functioning land markets permit skilled farmers to specialise in input-intensive agricultural production systems;
e) *Factor Price Equalisation Effect*: increased land transfer rights facilitate the equalisation of factor rewards among farmers, permitting a shift in land use from land-abundant to land-scarce households.

The investment regret effect refers to the engagement of farm household members in the labour market and the available incentives to direct part of the wage income towards farm investment. When land is a tradable commodity, the increased land value can be made effective and investments made in land improvement continue to be attractive. Without well-defined transfer rights, ownership of land can still be important as a social security device. When rural households under these conditions become involved in off-farm employment, a net reduction in cropping area, seasonal fallowing of farmland and decreasing yields are the most likely result. In a similar vein, farmers may oppose land consolidation and tend to prefer fragmented holdings in order to reduce their income variance (Blarel *et al.*, 1992).

The gains from specialisation effect depends on the development of a land transfer or rental market. Better defined land rights or more secure tenure regimes could lead to expanding trade opportunities and thus enhance investment incentives (Besley, 1995). Assuming that working capital constraints do not bind producers behaviour, productivity will be enhanced when land-abundant households with less productive capacities, rent out land to more capable farmers. Under effective liquidity constraints, tenure reform has little or no impact on small farmers' land use.

The factor price equalisation effect refers to hiring of land by farm households that possess a surplus of family labour but cannot participate in the labour market. Wealthy households could rent-out land to labour-abundant households, profiting of the relative advantage of family labour compared to hired labour. Labour intensity decreases for renting-in households and increases for renting-out households. Consequently, the shadow price of land is equalised among households, while the economy as a whole reaches a productivity gain (Carter & Yao, 1999).

Dynamic effects of improved land rights for land use adjustments can be seriously hampered by the occurrence of market imperfections. This refers especially to land lease and rental markets that are generally considered to be most important in developing countries[1]. Rental markets offer an opportunity for getting access to land for labour-abundant and capital-constrained households[2]. The importance of rental markets depends especially on the available employment opportunities through the labour market. Carter & Yao (1999) estimate parameters that influence land rental decisions in China, concluding that increasing off-farm employment opportunities lead to renting-out of land, while farmers with constrained land resources tend to rent-in. Interim transactions (*i.e.* selling of land rights with the intention of buying back in the near future) tend to be particularly important for recent migrants (Basu, 1986). This confirms the importance of landed property as a social security device.

The effectiveness of land rights regulations also depends on access to rural markets that is likely to be spread unevenly among rural households. Major attention should be given to different causes for constrained input use. Most empirical studies do not distinguish between supply-side or demand-side causes of credit and investment constraints. This so-called fundamental identification problem asks for a detailed analysis of the relationships between tenure status, credit use and investment (Feder *et al.*, 1999). Another common problem refers to the simultaneity bias, since the relationship between property rights and

[1] This also includes sharecropping arrangements that can be considered as a variable rent contract.

[2] Gender restrictions to labour (*i.e.* women that can only work on their own farm) may limit the flexibility of rural labour supply and hence induce additional demand for land (Faruque & Carey, 1997).

investments is two-sided: land rights can provide incentives for investments, but the realisation of investments in land (*e.g.* tree planting) is also an established procedure for improving *de facto* ownership rights. In the latter situation, insecure rights may even be an incentive for investment (Otsuka *et al.*, 1997). Finally, the tenure-investment relationship can be flawed due to unobserved individual farm household characteristics that are responsible for variations in willingness to invest or access to credit. Lopez (1996) refers in this respect to the selectivity bias in most studies on land titling[3]. Farmers are likely to seek legal titles for the best land, so if one correlates tenure security with farm productivity a strong but spurious relation is generally found.

Finally, land transactions are typically related to a number of farm household characteristics. Family size and composition, age, education and gender are major factors of heterogeneity among farmers. Land transactions are part of a complex portfolio choice that involves farm household decisions regarding production, consumption and labour use under conditions of selective engagement in labour and financial/insurance markets. Willingness to invest in landed assets increases with the length of the time horizon of farmers. Rental and sharecropping contracts tend to be preferred under more risky conditions.

18.4 Modelling Framework

For the discussion on the effects of land rights for farm household decisions with respect to input use, labour allocation and fixed investment, we rely on a two-period modelling approach. First, the role of land property rights providing a collateral for lending is discussed, thus reducing borrowing constraints. Next, the willingness to invest in land improvements is considered under conditions of participation in the labour market. Finally, investment demand effects are introduced as a function of security of land transfer rights.

18.4.1 Basic Model with Credit Constraints

We consider a two-period model in which a farmer maximises the utility of consumption (C):

$$u(C_1)+u(C_2) \tag{18.1}$$

Consumption is restricted to a minimum level for survival.

[3] Studies in Latin America reveal that land titling programmes have a strong bias in favour of medium-size farmers and better endowed regions (Byerlee & Valdez, 1996; Lopez, 1996).

The farmer derives all his income from crop production, which is a concave function of land (A), capital invested in the previous period (K_{t-1}), and variable inputs (I):

$$Q_t = f\left(A, K_{t-1}, I_t\right) \tag{18.2}$$

Capital stands for land improvements, like anti-erosion structures. Production without these improvements is possible, and K_0 is zero. Land improvement increases production not only directly, but also indirectly through increasing the productivity of variable inputs ($df_I/dK>0$). The profits associated with production are:

$$\pi_t = pQ_t - p_i I_t \tag{18.3}$$

where p is the price of output and p_i is the price of inputs.

In period 1, the farmer can invest in the land and borrow at a fixed interest rate (r). Hence, his first period budget constraint is:

$$C_1 + K_1 = \pi_1 + B_1 \tag{18.4}$$

where B is the amount borrowed. In the second period, the farmer produces another crop and then liquidates his assets. The share of the asset value that he can recover (s) depends on the possibility to transfer land rights. Hence, the second-period budget constraint is:

$$C_2 + (1+r)B_1 = \pi_2 + s\left(p_A A + K_1\right) \tag{18.5}$$

where p_A is the price of land.

Total borrowing is constrained to the maximum amount that lenders are willing to provide:

$$B_1 \leq B_{max} \tag{18.6}$$

The credit limit depends not only on the farmers land endowment, but also on the transfer rights. The lender's costs in seizing land in case of default, decrease with the development of transfer rights. As a result, the lender is willing to lend more with higher transfer rights: $dB_{max}/ds>0$).

Combining the first-order conditions for borrowing and investing yields:

$$pf_{K_1} + s = 1 + r + \frac{\lambda}{u'} \tag{18.7}$$

where λ is the Lagrange multiplier for the credit constraint.

When credit is not constrained, λ is zero and the sum of the marginal productivity of capital and the recovery share is the interest rate. Otherwise, the farmer takes the maximum loan, and the total returns to capital exceed the interest rate with λ/u'. As $dB_{max}/ds>0$, λ is a decreasing function of s. This implies that credit constraints are less likely to be binding when transfer rights are better developed (*collateral effect*).

The first order condition for variable inputs is:

$$pf_I = p_i \qquad (18.8)$$

Although input use simply reflects marginal costs and returns, there is an indirect effect of the structure of property rights. As $df_I/dK>0$, the use of variable inputs is increasing in capital. In the presence of credit constraints, investment in capital increases with the development of transfer rights. Consequently, second period input use increases with transfer rights.

The above model can be adapted to include production-credit constraints similar to the investment-credit constraint. This results in a relation between input use and land tenure that is similar to Equation (18.7) for capital investment, apart from the recovery of investment which is zero for variable inputs. In this case, land rights affect production also directly through their impact on credit availability.

18.4.2 Migration

Now suppose that the same farmer has a probability P of getting a salaried job in period 2. The job earns more than farming at the non-constrained capital level:

$$Y > \pi(A, K_1) \qquad (18.9)$$

When the farmer gets the job, he exits farming and liquidates the assets at the end of the first period. This leads to the following expected utility function:

$$\begin{aligned} &u(\pi_1 + B_1 - K_1) + Pu(Y - (1+r)B + s(p_A A + K_1)) + \\ &(1-P)u\,(\pi_2 - (1+r)B + s(p_A A + K_1)) \end{aligned} \qquad (18.10)$$

The condition for investment now becomes:

$$(1+r) + \frac{\lambda}{u'} = ps + (1-P)\left(pf_{K_1} + s\right) = s + (1-P)pf_{K_1} \qquad (18.11)$$

There are two possible marginal returns to investment, depending on the outcome of the job offer: (i) the marginal productivity of capital plus the recovery share, if the farmer does not get the job; or (ii) the recovery share, if the farmer gets the job. The expected marginal return to investment is a weighted average of the two options. This is lower than the return to investment in the absence of the job offer.

In other words, the possibility of getting a non-farm job decreases farm investment. Nevertheless, even when getting the job, the farmer can recover part of the investment. Better transfer rights increase the recovery factor and make the income loss smaller. Farmers will therefore want to increase investment, in reaction to an increase in transfer rights (*regret mitigation effect*).

The presence of credit constraints determines whether this increased investment demand results in an actual increase in capital. As the presence of an employment opportunity reduces optimal investments, credit constraints are less likely to be binding for farmers with such an option.

18.4.3 Uncertainty of Land Ownership

Alternatively, suppose that the farmer is uncertain about land rights in the second period: with probability P he looses the land. This probability depends on the land rights system and is a decreasing function of s ($dP/ds<0$). When the land is lost, informal insurance mechanisms are available to the farmer, providing some minimum level of consumption with utility $\hat{u}$. In order to recover at least part of the loan, the lender forces the farmer to provide services. The services required increase with the loan taken and give the farmer a disutility of $B\ddot{u}$.

The farmer's two-period expected utility function is:

$$u(\pi_1 + B - K_1) + P(\hat{u} - B\ddot{u}) + (1-P)\,u\,(\pi_2 - (1+r)B + s(p_A A + K)) \qquad (18.12)$$

As before, total credit availability depends on the system of land rights: $dB_{max}/ds>0$. The reason is somewhat different than before: more secure land ownership gives the lender a larger probability of total loan recovery.

This results in the following investment condition:

$$P\ddot{u} + (1-P)\,(1+r)u' + \lambda = (1-P)\,u'(\pi_{2,K_1} + s) \qquad (18.13)$$

In the absence of credit constraints, the expected marginal cost of investment is equal to the expected marginal return. The farmer only gets a return when he does not loose the land, while he has costs independent of the outcome. This implies that his demand for investment is lower than in the absence of uncertainty (*investment demand effect*). The higher the level of compulsory services requested by the lender and the more uncertain land ownership, the lower the investment.

18.5 Empirical Evidence

Following the pioneering work of Feder *et al.* (1988) a large number of field studies on the impact of different tenure regimes on input use, labour allocation and investment decisions have been conducted for several regions in Africa

(Ghana, Gambia, Niger, Uganda, Ethiopia), Asia (Thailand, Indonesia) and Latin America (Paraguay, Honduras, Costa Rica). Empirical evidence is, however, rather mixed. Most studies examine (i) the probability of farmers' investment in new land use activities due to better defined land transfer rights, and (ii) the adjustment in input use and labour intensity of production originating in more secure property rights and their subsequent impact on agricultural yields. In the following we discuss a number of empirical studies regarding land rights and investment behaviour, focussing on (i) the effects for resource allocation (as analysed in Sections 18.3 and 18.4), and (ii) the derived implications for sustainable land use.

18.5.1 Effects

Table 18.1 provides a comparative overview of some recently published studies on the impact of tenure regimes on land use and investment decisions. Following the three components of the modelling framework distinguished in Section 18.4, we discuss studies that focus on improved access to credit (land as collateral for borrowing), the regret mitigation effect (certainty to recover fixed costs) and the investment demand effect.

Empirical appraisals of the *Credit Supply* effect (Besley, 1995 for Ghana; Hayes *et al.,* 1997 for Gambia; Place & Hazell, 1993 for Ghana and Rwanda) confirm the importance of property rights as collateral for borrowing. The study by Feder *et al.* (1988) found a strong correlation between land titling, access to formal credit and investment in attached capital. When land can be used as collateral, the bank interest rate might be reduced and investment in land increases. For Sub-Saharan Africa, this relationship is not generally confirmed (Bruce & Migot-Adholla, 1993). A general problem in interpreting these studies is, that it remains unclear whether farm households are credit-constrained from the demand or the supply side. Making use of switching regression approaches, the wealth differentiating effects of land titles on credit supply can be determined (Carter & Olinto, 1998). Results indicate that legal titles only have an impact on formal credit supply and fixed capital investment for medium- and large-size farmers, while for small farmers credit rationing and transaction costs continue to limit their access to rural finance.

Empirical studies on the *Investment Demand* effect are usually based on the estimation of separate yield, income and investment (land improvement) functions, distinguishing between parcels with different ownership rights. Li *et al.* (1998) examine the impact of property rights on productivity through differences in input intensity for mainland China. Similarly, Gavian & Ehui (1999) estimate total factor productivity effects for different land tenure arrangements in Ethiopia. Gavian & Fafchamps (1996) distinguish between the security effect (*i.e.* long-term effects of manuring) and the transaction effect (input applications yielding short-term returns). Most studies do not explicitly analyse farmers' willingness to invest, but rather rely on indirect measures.

Table 18.1: Empirical studies on the impact of land rights on input use, labour use and investments.

Authors	Region (type of land rights)	Method	Implications	Impact tenure security		
Hayes *et al.* (1997)	Gambia (complete, preferential, limited rights)	Probit	Fences; stone bunds	+	sign	**
			Investment in trees	+	sign	**
			Manuring		n.s	
Place & Hazell (1993); Blarel *et al.* (1992)	Ghana and Rwanda (complete, preferential, long-term use rights)	OLS	Input use		n.s	
			Land improvement	++	sign	*
Li *et al.* (1998)	North China; Fengning county (length of tenure)	SUR	Labour use		n.s	
			Nitrogen fertiliser		n.s	
			Phosphate fertiliser	+	sign	*
			Animal traction		n.s.	
			Organic manure	+	sign	*
Besley (1995)	Ghana (rights to sell, rent, mortgage, pledge, bequeath, gift)	Probit	Investments in trees	++	sign	**
			Manuring		n.s	
			Mulching	+	sign	*
Gavian & Fafchamps (1996)	Niger (own, borrowed, use)	2-3 SLS	Manuring	+	sign	**
Place & Otsuka (1997)	Uganda (customary, lease, public)	2 SLS	Tree cover	++	sign	*
Otsuka *et al.* (1997)	Indonesia; Sumatra (own, purchase, cleared)	Logit	Tree planting	--	sign	*
Gavian & Ehui (1999)	Ethiopia (rented, shared, borrowed)	OLS	Chemical fertilisers	--	sign	*
			Labour use	+	sign	*
Carter & Yao (1999)	China; Zhejang and Jiangxi province (transfer rights)	MLE	Fixed Investment	++	sign	**
			Labour use	+	n.s.	

Note: '+' refers to a positive impact of land rights, '-' refers to a negative impact of land rights ; the number of '+' or '-' signs refers to the intensity of the relationship; sign * and sign ** means a significant relationship with land rights at 1% or 5 % ; n.s. = not significant.

Carter & Yao (1999) developed a procedure for empirical assessment of the *Income Regret Mitigation* effect. Improved security of land transfer rights permits further engagement in off-farm activities, without the usual investment-depressing regret effects. Land transfers take place in a situation of an imperfect labour market, causing *factor price equalisation* effects. Panel data from China are used to identify the land and labour market conditions required to realise both effects. In essence, hiring of land or engagement in off-farm employment mitigate the usual liquidity constraints that arise from less secure transfer rights. Similarly, Otsuka *et al.* (1997) demonstrate that in Sumatra insecure land property rights could enhance long-term investments in tree planting as long as farmers maintain control over the revenues and trees are tradable independent from land.

18.5.2 Implications

Empirical studies make a difference between the implications of improved land rights on fixed investment (*i.e.* tree planting, fences, land improvement), material input use (*i.e.* fertilisers, manure, mulching) and labour use. Contradictory evidence is found with respect to the implications of more secure ownership rights on attached *investments*. In most African studies (Hayes *et al.* (1997) for Gambia; Besley (1995) for Ghana; Place & Otsuka (1997) for Uganda) a strong positive relationship between land ownership and tree planting is confirmed, whereas studies from Indonesia point to a significant negative relationship, *e.g.* more tree planting occurs with less secure property rights. The latter effect is explained by the fact that farmers try to confirm land rights through the establishment of perennial crops. The expected positive relationship between ownership rights and farm investment is probably only relevant for medium-size farmers that are most sensitive to improved access to formal credit in order to improve the input-intensity of their farming operations (Carter & Olinto, 1998). As argued before, small farms are most likely constrained from the demand side.

The analyses on the impact of tenure security on agricultural *labour intensity* are rather ambiguous: most regressions give no significant results (Li *et al.*, 1998), while sometimes positive effects are registered for farmers that can devote some more time to yield-increasing crop maintenance activities (mainly weeding labour) or are able to improve the timing of their activities (Gavian & Ehui, 1999). There is hardly any evidence that property rights directly induce farmers towards increased labour use for soil conservation. Farmers' decisions to intensify labour use for maintenance activities are probably more dependent on their subjective discount rate (which in turn is related to tenure security), and far less on the formal legal status of their land resources.

Land rights and tenure security do provide some positive incentives for higher application of *material inputs*. Increased use of organic (manure, mulch) and inorganic fertilisers is mainly observed in regions with a high population density (Hayes *et al.* (1997) for peri-urban regions of Gambia, Gavian & Fafchamps

(1996) for western Niger, Besley (1995) for Ghana and Li *et al.* (1998) for North China), where farmers are likely to intensify their resource use through land-saving technologies. Compared to fixed capital investment, non-attached resources can be used with equal ease on less secure parcels. This is especially the case with chemical fertilisers that provide macro-nutrients for the current production cycle, while manuring and mulching practices, that have a long-term effect on soil organic matter balances, are usually restricted to parcels with more secure property rights. In the Ethiopian highlands more chemical fertilisers are applied on less secure land, because rented fields have a lower natural fertility and fertilisers are used also to compensate for non-optimal timing of labour activities (Gavian & Ehui, 1999).

The positive and significant relationships found in only some studies ask for some further comment. Final results depend basically on two features: (i) the modelling framework and statistical procedures used in the research, and (ii) the market and institutional environment in which research took place. Models that focus on credit supply are usually far more optimistic about the effects of land rights on attached investment, while investment demand models that take into account farmers' willingness to invest, tend to register mainly positive effects for material input use. Since land rights also influence farmers' relative risk aversion (Shively, 1997), high cross-correlation can be found between the likelihood of adoption of different soil conservation measures and resource endowments (farm size, land quality, access to credit, family labour). OLS techniques or probability analyses cannot unambiguously determine the direction of causation between land rights and resource use decisions.

While most studies focus on the potential gains from more secure land rights, production and investment effects strongly depend on the effectiveness of market linkages. Some research is devoted to the specification of the market conditions that could make property rights a feasible instrument for land use planning. Others remain, however, rather sceptical about the disappearance of market failures and pay attention to farmers' motives for maintaining some collective property or fragmented holdings as a risk diversification device (Carter, 1984; Blarel *et al.*, 1992; Otsuka *et al.*, 1997). While land tenure insecurity may have adverse effects on factor use and investments, it could still be relatively efficient from the perspective of farm household risk.

Finally, results are clearly influenced by the existing heterogeneity among farm households, the relationships that each of these farmers maintains with factor and commodity markets, as well as by the degree of development of these markets. Tenure security appears to yield stronger effects on input use and land-saving investments in situations of severe land scarcity. Fixed investment effects related to tenure security are most relevant for medium-size households that only face supply-side credit constraints. However, when labour markets are well developed, farmers tend to prefer investment of resources in labour-saving technologies in order to enable their engagement in off-farm employment. Hiring-out of land to households with a labour surplus could then lead to factor price equalisation. Income effects depend on the available input substitution opportunities and differences in supervision costs.

18.6 Discussion

We discussed in this paper major differences between land property and transfer rights and their implications for production and investment behaviour. Land rights influence a wide range of farm household decisions on input use, labour allocation and investment choice. Therefore, an analytical framework is required that accounts for these simultaneous effects and permits their systematic disentanglement.

For a thorough analysis of the dynamic relationship between property rights and farm household behaviour, we can build on a variety of partial approaches that address the following effects of tenure regimes:

1) the function of property rights as a collateral for formal lending;
2) the role of transfer rights for enhancing the use of off-farm income for investments in landed property;
3) the function of recognised land use and transfer rights for decisions with respect to hiring-in or hiring-out land.

These effects permit analysis of the implications of land rights for the choice of the land use pattern, the impact of land rights on farmers' willingness to invest in attached and non-attached capital, and the indirect effect of established land rights on farm household risk behaviour.

We can conclude that the impact of secure property, use and transfer rights on input use and investment behaviour is basically dependent on farmers' access to factor markets for land, labour and capital. Since more secure land rights tend to reduce transaction costs for market exchange, farmers' willingness to invest is usually improved when risks decrease.

Finally, some implications of these findings can be derived for the design of appropriate land policies in developing countries for improving resource management regimes. Reform of property rights is now firmly on the policy agenda in a large number of countries, including China, large parts of sub-Saharan Africa and many countries of Latin America. There is also growing understanding of the importance of well-defined land rights for natural resource management regimes (see Hanna & Munasinghe, 1995 for an overview). Major attention is given to the design of appropriate governance systems that guarantee farmers access to credit and labour markets (Eggertsson, 1990; Bowles *et al.*, 1993). At the demand side, the short time horizon typical for small farmers and the behavioural uncertainty related to shared land rights have often been considered major impediments for sustainability[4]. Recent studies indicate, however, that sustainable resource management is not depending on a particular property regime, but rather on well-specified and generally accepted rules that are congruent with the ecological and social context (Jodha, 1992; Ostrom, 1990).

Policy measures aiming at titling of property rights or legalisation of transfer rights cannot yield any tangible results as long as access to factor markets remains constrained. Since farmers' decisions regarding input use, labour allocation and fixed investment are strongly dependent on opportunity costs and risks, an integrated 'package' of market and institutional reforms is required to enable simultaneous improvement of farmer's welfare and sustainability of land use systems (Ehui & Spencer, 1993).

18.7 References

Atwood, D.A. (1990): Land registration in Africa: the impact on agricultural production. World Development 18, 659-671

Basu, K. (1986): The market for land: an analysis of interim transactions. Journal of Development Economics 20, 163-177

Benda-Beckman, F. von (2000): Between free riders and free raiders: Property rights and soil degradation in context. This volume

Besley, T. (1995): Property rights and investment incentives: theory and evidence from Ghana. Journal of Political Economy 103, 903-937

Binswanger, H.P., K. Deininger & G. Feder (1993): Power, distortions, revolt and reform in agricultural land relations. In: J. Behrman & T.N. Srinivasan (Eds.) Handbook of development economics, Vol. 3B, Elsevier, Amsterdam, 2659-2772

Blarel, B., P. Hazell, F. Place & J. Quiggin (1992): The economics of farm fragmentation: evidence from Ghana and Rwanda. World Bank Economic Review 6, 233-254

Bowles, S., H. Gintis & B. Gustafsson (1993): Markets and democracy: participation, accountability and efficiency. Cambridge University Press, Cambridge

Bromley, D.W. (1991): Environment and economy: property rights and public policy. Basil Blackwell, Oxford

Bruce, J.W. & S.E. Migot-Adhollah (1993): Searching for land tenure security in Africa. Kendall-Hunt Publishing Co, Dubuqe, Iowa

[4] Sustainable resource management mostly requires some collective action given the spatial and intertemporal implications of natural resource management systems. Since land markets do not automatically reflect environmental amenities, additional instruments are required to account for externalities (Larson & Bromley, 1990).

Byerlee, D. & A. Valdez (1996): Land tenure insecurity and farm productivity in Latin America: overview and policy perspectives. World Bank Paper, Washington

Carter, M.R. (1984): Identification of the inverse relationship between farm size and productivity: an empirical analysis of peasant agricultural production. Oxford Economic Papers 36, 131-145

Carter, M.R. (1999): Old questions and new realities: contemporary land and land policy research in Latin America. Paper presented at CEDLA/CERES/WAU workshop Land in Latin America, Amsterdam, May 1999

Carter, M.R. & P. Olinto (1998): Getting institutions right for whom: the wealth-differentiated impact of property rights reform. Paper prepared for XXI LASA Congress. Chicago, September 1998

Carter, M.R. & Y. Yao (1999) Specialization without regret: transfer rights and agricultural investment in an industrializing economy. Working Paper. University of Wisconsin, Madison

Carter, M.R. & F. Zimmerman (1999): The dynamic cost and persistence of asset inequity in an agrarian economy. Working Paper. University of Wisconsin, Madison

Eggertsson, T. (1990): Economic behaviour and institutions. Cambridge University Press, Cambridge

Ehui, S.K. & D.S.C. Spencer (1993): Measuring the sustainability and economic viability of tropical farming systems: a model for sub-Saharan Africa. Agricultural Economics 9, 179-296

Faruque, R. & K. Carey (1997): Research on land markets in south Asia: What have we learned? World Bank Policy Research Working Paper No. 1754, Washington

Feder, G., T. Oncham, Y. Chalamwoond & D. Hongladaron (1988): Land policies and farm productivity in Thailand. Johns Hopkins University Press, Baltimore

Feder, G., L.J. Lau, L.Y. Lin & Y. Luo (1990): The relationship between credit and productivity in Chinese agriculture: A microeconomic model of disequilibrium. American Journal of Agricultural Economics 72, 1151-1157

Feder, G., L.J. Lau, L.Y. Lin & Y. Luo (1992): The determinants of farm investment and residential construction in post-reform China. Economic Development and Cultural Change 40, 1-26

Gavian, S. & S. Ehui (1999): Measuring the production efficiency of alternative land tenure contracts in a mixed crop-livestock system in Ethiopia. Agricultural Economics 20, 37-49

Gavian, S. & M. Fafchamps (1996): Land tenure and allocative efficiency in Niger. American Journal of Agricultural Economics 78, 460-71

Hanna, S. & M. Munasinghe (1995): Property rights and the environment. The Beijer International Institute of Ecological Economics and The World Bank, Washington

Hayami, Y. & V.W. Ruttan (1985): Agricultural development: an international perspective. The Johns Hopkins University Press, Baltimore

Hayes, J., M. Roth & L. Zepeda (1997): Tenure security, investment and productivity in Gambian agriculture: a generalized probit analysis. American Journal of Agricultural Economics 79, 369-382

Jodha, N.S. (1992): Common property resources: a missing dimension of development strategies. World Bank Discussion Paper No. 169, Washington

Larson, B.A. & D.W. Bromley (1990): Property rights, externalities and resource degradation. Journal of Development Studies 33, 235-262

Lawrey, S.W. (1993): Transactions in cropland held under customary tenure in Lesotho. In: T.J. Bassett & D.E. Crummy (Eds.) Land in African agrarian systems. University of Wisconsin Press, Madison, 57-74

Li, G. S. Rozelle & L. Brandt (1998): Tenure, land rights, and farmer investment incentives in China. Agricultural Economics 19, 63-71

Liu, S., M.R. Carter & Y.Yao (1998): Dimensions and diversity of property rights in rural China: Dilemmas on the road to further reform. World Development 26, 1789-1806

Lopez, R. (1996): Land titles and farm productivity in Honduras. University of Maryland. Department of Agricultural and Resource Economics, Maryland

Melmed-Sanjak, J.S. & M.R. Carter (1990): The economic viability and stability of 'capitalised family farming': An analysis of agricultural decollectivisation in Peru. Journal of Development Studies 27, 190-210

Ostrom, E. (1990): Governing the commons: the evolution of institutions for collective action. Cambridge University Press, Cambridge

Otsuka, K., S. Suyanto & T.P. Tomich (1997): Does land tenure insecurity discourage tree planting? Evolution of customary land tenure and agroforestry management in Sumatra. IFPRI EPTD Discussion Paper No. 31, Washington

Pickney, T.C. & P.K. Kimayu (1994): Land tenure reform in East Africa: good, bad or unimportant? Journal of African Economies 3, 1-28

Place, F. & P. Hazell (1993): Productivity effects of indigenous tenure systems in sub-Saharan Africa. American Journal of Agricultural Economics 75, 10-19

Place, F. & K. Otsuka (1997): Population pressure, land tenure, and tree resource management in Uganda. IFPRI EPTD Discussion Paper No. 24, Washington

Platteau, J.P. (1996): The evolutionary theory of land rights as applied to sub-Saharan Africa: a critical assessment. Development and Change 27, 29-86

Shively, G.E. (1997): Assets, attitudes, beliefs and behaviours: explaining patterns of soil conservation adoption among low-income farmers. Internal Paper Dept. Agricultural Economics, Purdue University, Purdue

19 Assessing Social Factors in Sustainable Land-Use Management: Social Capital and Common Land Development in Rajasthan, India

Anirudh Krishna & Norman Uphoff

Cornell International Institute for Food, Agriculture and Development (CIIFAD), Cornell University, Ithaca, U.S.A.

Abstract

Performance with respect to common land development is compared among 64 villages in north India, all of which participated in a government-funded program, beginning in 1991. Results vary considerably even though the same level of technical and financial support was provided to all participating villages, and even though these villages share similar agro-ecological condition. Three social factors, literacy, extent of political competition, and level of social capital – measured with the help of a locally-relevant scale – are all significantly associated with program performance. Purposive action to enhance the levels of these three variables can go a long way toward improving performance in common land development and other collective enterprises.

19.1 Social Factors in Soil Conservation

Technical and financial aspects are usually given overriding importance in soil conservation and land-use management plans. It is recognised that social and institutional factors also matter, but less attention is given to these aspects, which are regarded most often as additionally desirable elements, things to be clubbed on to the end of technical reports and financial planning exercises, but still not considered a central part.

One reason for this disregard has to do with lack of quantitative evaluations that provide evidence for the value actually added by specific social and institutional features. By how much has productivity risen in areas where local communities participated actively in program activities? What forms of local participation are most closely associated with gains in production? How much longer have benefits been sustained in areas where participation was relatively higher? These and other related questions are answered most often in terms of narrative or circumstantial evidence, which many technical specialists find hard to accept.

This paper seeks to fill this gap between proposition and proof. By comparing development and conservation results in 64 Indian villages that have all participated in a common program of watershed development and soil conservation, we can demonstrate the contribution that social factors have made to program results.

We compared program results by assessing how well common lands have been developed in each of the 64 selected villages. Nearly half or more of the land area in each village is not privately owned. Though ownership of village common lands vests in the state or local government, villagers usually have open access to these lands, which they use to graze their animals and to collect firewood. The critical importance of common land development for villagers in Rajasthan is examined in Section 19.2. Section 19.3 presents the scale of measurement, the Common Land Development Index (CLDI), that we used to compare program performance among participating villages.

We next assessed a host of social and administrative factors to see which among them might be salient for explaining inter-village differences. Since low and declining productivity of common lands is a problem faced by practically all village inhabitants, and since they have developed various collective responses for addressing this problem, differences in program performance can be explained in terms of the factors that assist or inhibit collective action.

We consider social capital among the several factors that are important for explaining performance results. Defining social capital as aspects of community organisation that contribute to mutually beneficial collective action, we devised a scale for measuring this attribute at the level of individual villages that is presented in Section 19.4.

Section 19.5 presents the results of this analysis. Social context does make a difference for program results, our analysis indicates, and three aspects of social context – levels of social capital, political competition and literacy – are particularly relevant for sustainable land-use management. Section 19.6 concludes the paper and presents some policy recommendations that emerge from this analysis.

19.2 Setting the Stage: Location and Context

The 64 villages comprising our data base have all been participating in the Rajasthan state government's programs for integrated watershed development, funded by the World Bank and the Government of India (GOI)[1]. These programs were launched in each participating village at about the same time, nearly seven years ago. This provides enough time to investigate the causes and correlates of differential performance, including an assessment of administrative and social factors.

More than 70% of Rajasthan villagers are dependent on agriculture and animal husbandry for their livelihoods. Animal husbandry serves both as a source of supplementary income and as insurance against the failure of uncertain and risk-prone cropping operations. On average, each household owns eight farm animals, mostly cows, sheep and goats. Though richer households own more cows and buffaloes than poorer ones, most households own at least some farm animals, and even the poorest households draw support from the small numbers of sheep and goats they own. All households are dependent to quite a significant extent on village common lands for obtaining fodder for their animals (Jodha, 1990; Brara, 1991; Krishna, 1997). Nearly half of all land of each village is not privately owned but vests, instead, with the state government or with village panchayats, units of local government with jurisdiction over one to four villages depending on size. In addition to serving as a source of fodder, such common lands are also the most important source of domestic energy in the form of fuelwood.

The preservation and development of village common lands is an important collective need among villagers of Rajasthan[2]. Many of these common land areas were on the verge of becoming – or had already become – wastelands. Abysmally low growth rates of biomass have restricted quite sharply the total off-take of fodder and fuelwood[3].

The program of watershed development addresses these issues centrally; and the largest share of program funds is provided for developing village common lands. Between five and eight residents in each participating village were elected by fellow villagers to a User Committee (UC). The program expected this group to organise and manage, on behalf of the village, a variety of soil and water conservation works on common lands, such as planting trees and grasses, enforcing rules for protection and extraction, and fencing these lands against stray

[1] To be eligible for program benefits, a village had to have less than 30% of its cropped area served by irrigation facilities. In addition, villagers had to be willing to contribute 10% of program resources in the form of labour or locally available materials. Little administrative discretion existed, thus, in selecting villages for coverage by the program.

[2] Of 2,397 respondents interviewed for this investigation, 2,123 (89%) included 'improvement of common lands' among the three most important development needs they identified, along with drinking water, schools, and improved (metalled or paved) roads.

[3] Before program implementation, common lands yielded on average less than 75 kilograms of dry grass per hectare. *More than ten times this amount* was observed being harvested after program implementation. *Baseline Survey Report*, College of Technology and Agricultural Engineering, Rajasthan Agricultural University, 1993.

cattle and human encroachment. With project advice and assistance, these activities were expected to sustainably enhance fodder and fuelwood yields. The Committee worked with government staff to draw up and implement a development plan for the common lands of its village. Its members were wholly responsible for formulating and implementing local rules that would govern the sharing of costs and the distribution of benefits.

Although common land development (CLD) represents a collectively organised response to a need felt acutely by most villagers[4], program results have varied considerably among villages. In terms of objective indicators (survival rates of trees, grass harvests per capita, etc.) as well as subjective evaluations (levels of individual satisfaction), high-performing villages stand out quite sharply from low-performing ones.

Data were collected from 64 program villages, 16 in each of four districts of Rajasthan – Ajmer, Bhilwara, Rajsamand and Udaipur. To ensure spread and range for the dependent variable, villages were selected to reflect high, medium and low levels of program performance, judged on the basis of information available prior to field investigations. The numbers of high-, medium- and low-performing villages were distributed practically evenly across the four districts.

The following sources of data were compiled or consulted:

(a) *Interviews with 2,397 individuals.* Households were selected on the basis of simple random sampling in each village. To have a gender-balanced sample, one man was selected at random from all odd-numbered and one woman from all even-numbered households in the sample. The mostly poor, uneducated, agriculture-dependent nature of the region's population is represented quite well by the sample, which has the following characteristics:

Education	No Education	= 72%
	Primary	= 22%
	Secondary	= 3%
	High School or College	= 3%
Caste	Upper Caste	= 34%
	Middle Caste	= 37%
	Scheduled Caste	= 13%
	Scheduled Tribe	= 16%
Primary occupation	Agriculturist	= 73%
	Agricultural labour	= 16%
	Service	= 11%

4 88% of all respondents stated that they had themselves, or through their representatives, participated in decision-making, and 87% felt that the program had delivered tangible benefits to the community, a remarkably positive evaluation for any public-sector program in India.

Landholding	Less than 0.5 hectares	= 22%
	0.5 to 1 hectare	= 33%
	1 to 2.5 hectares	= 21%
	2.5 to 5 hectares	= 16%
	More than 5 hectares	= 8%
Farm animals owned	None	= 10%
	1-2	= 19%
	3-5	= 32%
	6-10	= 23%
	11-20	= 9%
	21-40	= 4%
Type of dwelling	Kaccha (mud)	= 54%
	Pukka (brick or stone masonry)	= 46%

(b) *Focus group interviews with village leaders.* A standard instrument was used to interview User Committees members in each village. Field investigators also measured average fodder yields and plant survival rates on-site in each village after drawing a sample from the areas planted under program auspices.

(c) *Data obtained from official sources.* The Census of India and records from the state government's departments of agriculture, watershed development, elections, local self-government, land records, co-operatives, and animal husbandry were drawn on for additional data.

19.3 Evaluating Program Performance

An overall Common Land Development Index (CLDI) was constructed out of six individual elements. The six different criteria of measuring performance in watershed development are:

1. *Quantity of work.* Percentage of a village's total common land area that has been developed so far under the program.
2. *Quality of protection.* Survival rates for trees and shrubs planted under program auspices.
3. *Productivity.* Quantity of fodder and fuelwood harvested in the previous year, measured in headloads per family.
4. *Voluntary action.* How often in the last five years the User Committee voluntarily replanted and filled gaps in its forested areas.
5. *Diversification of activities.* How many times in the past five years the User Committee also took up other community development activities not directly related to watershed development.
6. *Support for local contributions.* If watershed development were to start afresh in the village, what percentage of total costs would villagers be willing to bear. While this represents a more subjective assessment than the other measures, it reflects villagers' willingness to mobilise local resources, an important factor for sustainable land-use management.

These six measures are quite highly correlated with one another, with all correlation coefficients significant at a 0.05 level or higher. Not surprisingly, factor analysis points to one common factor on which all six of these elements load quite highly.

Table 19.1: Common land development (factor pattern).

Quantity of work	0.59601
Quality of protection	0.69309
Productivity	0.82895
Voluntary action	0.81953
Diversification of activities	0.85734
Support for local contributions	0.65181

This single common factor has a communality of 3.514, which would account for almost 60% of the combined variance of the six individual elements. A scree plot shows a sharp discontinuity, or 'elbow', between this factor and a second one, reinforcing our conclusion that these six variables together represent a single common factor[5].

The *Common Land Development Index* (CLDI) was constructed by taking a simple sum of scores over these six items, after first standardising them by re-scoring the values so that each had a maximum range of one. Each variable thus had an equal weight within the index[6]. With six items combined to form the CLDI, a village could score a maximum of six points only if it ranked in the top bracket in each of the six individual categories. CLDI correlates quite highly with each of its six components[7].

Once this index had been constructed, we proceeded to test it as a dependent variable against a range of variables, representing alternative hypotheses that could explain collective action. Before doing so, however, and to facilitate the interpretation of results, we transformed this variable to have a range from zero to hundred. The variable resulting from this transformation is termed CLDINDEX.

---

5 The conclusion about a single common factor is given added weight by observing that root mean square off-diagonal residuals = 0.107, which is well within the acceptable limit of 0.126 suggested by Harman's criterion for a sample size of 64. Kaiser's Measure of Sampling Adequacy: over-all MSA = 0.81.

6 An alternative index which weighted the individual items by their factor scores was highly correlated (correlation coefficient = 0.97) with CLDI, indicating that the index is robust against alternative weighting schemes.

7 Cronbach's Alpha coefficient = 0.89. Each of the individual correlations among index components is significantly positive at the 0.0001 level, with a value of 0.60 or higher.

19.4 Scaling Social Capital

To be relevant to development efforts in Rajasthan, indicators of social capital need to correspond with the pattern of life in this largely agrarian setting. Measures of social capital have to relate to the types of networks or mutual support organisations that are found in this state[8]. We started by considering the types of activities in which people of this area are commonly engaged.

Not all activities observed in this area can automatically be considered as locally appropriate measures of social capital. Some activities are strongly individualised in their performance, while others are more typically undertaken together with other persons. Agreeing with Coleman (1988, 100-101), we think that social capital 'exists in the *relations* among persons'. Only those activities that are not mostly or usually undertaken in isolation from other people can be considered valid ground for finding components of social capital.

In the questionnaire we devised for this study, we included questions related to a large number of locally relevant activities. Activities for which more than 80% of respondents showed an individualistic inclination were excluded from the scope of our social capital inquiry.

Responses to six separate questions have been combined to construct an index of social capital which appears on conceptual as well as on empirical grounds to be valid for this region[9]. Given that formal associations are virtually absent in these villages, informal networks and established roles were assessed. This resulted in the following six questions:

1. *If a crop disease were to affect the entire standing crop of this village, then who do you think would come forward to deal with this situation?* Responses could range from 'Everyone would deal with the problem individually', scored 1, to 'The entire village would act together', scored 5. The average response score over all 2,397 individuals was 2.4.
2. *Who in this village has historically looked after the common pasture lands?* Responses ranged from 'No one does anything', scored 1, to 'We all discuss

[8] More detailed conceptual treatments related to context-specific measurement of social capital are presented in Uphoff (2000) and Krishna (2000a). Krishna & Shrader (2000) present a methodology and a set of empirically-validated tools that can be utilised for scaling social capital in specific contexts.

[9] Responses to a range of different questions indicated that social capital is a single underlying quality that manifests itself in multiple different behaviours and expectations. Some other questions related to other aspects of trust, reciprocity and solidarity among villagers could also have been combined within the index of social capital. However, village rankings do not change regardless of which sets of questions are selected for inclusion. For a further discussion on conceptual and empirical aspects of measurement, see Krishna & Uphoff (1999) and Krishna (2000a), especially Chapter 4.

and jointly decide what is to be done', scored 5. Average of all responses had a value of 3.2.

3. *Suppose two people in this village had a dispute with each other. Who do you think would resolve this dispute?* Five alternative responses were offered: 'No one', scored 1, through 'Their neighbours', scored 3, to 'The entire village collectively', scored 5. The average response was 3.5.
4. *Suppose some children of the village tend to stray from the correct path; for example, they are disrespectful to elders, they disobey their parents, are mischievous, etc. Who in this village feels it right to correct other people's children?* Four alternatives were posed: 'No one', scored 1; 'Only close relatives' scored 2; 'Relatives and neighbours', scored 3; and 'Anyone from the village', scored 4. Average response was 2.5.
5. *Which among the following is the most important reason why people in this village plant and protect grasses and trees?* The four alternatives, ranging from a purely instrumental motive, were: 'Because fodder and fuelwood are in short supply', scored 1, to community-mindedness, 'Because this activity keeps the villagers united', scored 4. The average response was 2.0.
6. *Suppose a friend of yours in this village faced the following alternatives:*
   - *to own and farm 10 units of land entirely by themselves (scored 1), or*
   - *to own and farm 25 units of land jointly with one other person (scored 2),*

   then which one would he or she prefer? The average response had a score of 1.4.

This last item operationalised in practical terms the factor of *trust* that is commonly included in most treatments of social capital. The second alternative would give each person access to more land (12.5 units), but they would have to work and share produce interdependently. This question was worded so that the respondent was not making a statement about his or her own level of trust, but rather about how he or she thinks others in the village will assess this trade-off. So this item reflects a village-level characteristic rather than just individual attitudes.

Responses to these six questions are quite highly correlated among themselves, and they load commonly on a single factor (Table 19.2). The single common factor accounts for 3.28, or about 55% of the combined variance[10]. These six separate variables we have combined into a *Social Capital Index* (SKI)[11].

---

[10] Cronbach's Coefficient Alpha = 0.855. Individual correlations with the index are all 0.75 or higher.

[11] Each variable is given an equal weight within the index, which is obtained by summing across the scores after first dividing each variable by its range, so that each variable has a maximum range of one. A further transformation results in an index that has a range from zero to hundred. This latter transformation is useful at a later stage for interpreting regression results

Table 19.2: Social capital (factor pattern).

Dealing with crop disease	0.73052
Dealing with common pastures	0.64826
Settling disputes	0.73272
Dealing with errant children	0.72029
Value placed on unity	0.78680
Trust placed in others	0.66859

19.5 Examining the Impact of Social Capital on Program Performance

Apart from social capital, a number of other factors have been previously suggested to explain differences in collective action levels among groups and communities. To assess and compare the relevance of each of these factors, we viewed them alongside social capital in regression analysis undertaken on the dependent variable, CLDINDEX.

These hypotheses are often simply stated, but they represent complex bundles of causal factors. We tried to operationalise each with concrete measures that reflect the determinants proposed in the literature and are quite specific and quantifiable. Our study was not designed to examine these respective hypotheses in detail, so we are assessing what we considered to be the best available operational measures of these factors that could influence collective action – instead of, or in addition to, social capital.

1. *Relative Need.* Perhaps differences in collective action for watershed development are caused by the degree of deprivation which makes collective action beneficial for persons engaging in it (Wade, 1994). Two variables were analysed to assess relative need:
   - COWUNITS. This variable measured the pressure of animal population upon available grazing land. The total number of cow units in a village divided by the area of available grazing land gave a measure of COWUNITS for each village[12].
   - MOISTRES. This was assessed by considering the ratio of rainfed (unirrigated) cropped area to irrigated cropped area.

[12] Since animals of different sizes have different feed requirements, we standardised for size and thus for grazing pressure by converting all animals into 'cow units', using formulae that are widely used by animal husbandry experts.

2. *Government Staff Support.* Even though most program details, implementation, and management were provided by the elected User Committees, the largest share of program funds was provided by the government, and government staff were also expected to provide technical support to UCs. Differences in the motivation and competence levels of government staff assigned to work with different villages were assessed by considering two questions included within the focus group questionnaires:
   - How frequently did staff of the government's Watershed Development Department visit your village?
   - Suppose that 100% of all villagers recognise the person who is the headmaster of the village school. On such a scale, what percentage of village residents do you think would recognise the local staff of the Watershed Department?

   Responses to these two questions were averaged to arrive at the variable OFFINPUT.

3. *Modernisation.* The assumption is often made that communities that are in some sense more 'modern', *i.e.*, more closely linked to external markets, more advanced in terms of mechanisation of production, better served with infrastructure facilities, etc., will also be more progressive, and willing and able to engage in behaviour that improves people's condition (Lipset, 1960, 1994; Bollen & Jackman, 1985). Three different sets of variables were used to assess any association that 'modernisation' might have with our dependent variable, CLDINDEX.
   - DISTMKT. Distance to market town, measured in kilometres.
   - MECHZN represents the number of tractors and mechanical threshers in the village per unit of cropped area.
   - INFRASTR added up the scores that each village had for level of service related to educational facilities, health care facilities, communication facilities, transport facilities, and electricity provision.

4. *Literacy.* In Third World countries, especially India, it has been argued that education is by itself the single most important variable responsible for economic as well as civic growth (Almond & Verba, 1989; Dreze & Sen, 1995). LITERACY was measured in proportion to village population.

5. *Heterogeneity and Stratification.* According to some social science analyses, more homogeneous populations facilitate collective action, while populations divided by caste, class and faction deter it (*e.g.* Johnston & Clark, 1982). We focus on caste here, since it is widely regarded as the most divisive influence in Indian villages, as well as on economic differences. Two sets of variables were constructed to assess the divisive influence of caste:
   - NCASTES reflects the number of distinct caste groups residing in a village.
   - DIVISION. In order to assess the separate and combined influence of all sorts of divisive factors at village level, we asked respondents a series of five questions. To what extent, we asked, do differences such as the following

tend to divide people in your community: Differences in education? Differences in wealth? Differences in caste? Differences between young and old? Differences between old inhabitants and new settlers? The summed score on these five variables reflects the combined divisive influence of these five constituent variables in any particular village.

6. *Political Competition.* Political competition was included in our analysis following a round of field visits by Krishna in May to August 1997. Attending meetings of local governments and User Committees and speaking with leaders and other villagers, Krishna found high levels of debate and competition co-existing alongside high levels of solidarity in the high-performing villages. An index of political competition (POLCOMP) was constructed by analysing the number of votes polled by different candidates in the last-held elections for the local government[13].

7. *Population.* Village population was also considered as an independent variable to test Olson's (1965) hypothesis concerning differences in capacity for collective action between large- and small-sized groups.

After operationalising all the above variables, and calculating their values for the 64 villages comprised in our database, we next conducted regression analysis to examine which of these variables would be significant for explaining differences in CLDINDEX. Table 19.3 reports the results.

Neither MOISTRES, nor the other variable related to the relative need hypothesis, COWUNITS, turned up with a significant coefficient under any alternative specification of the model. Similarly, none of the modernisation variables were significantly associated with the dependent variable. Variables related to social heterogeneity were not significant for the analysis.

Only three variables are consistently significant. The coefficient for the Social Capital Index was robust, no matter what other variables were considered in addition. The coefficient for the Political Competition Index was also consistently significant and robust, no matter what the alternative specifications. Literacy was the third variable that achieved significance[14].

---

[13] We use the Herfindahl-Hirschmann 'concentration index' which relies upon the inverse of the sum of squares of vote shares obtained by each candidate to an election. For each ward, we take the share of total votes cast in favour of each candidate. These vote shares are then squared, and the squared shares are totalled across candidates. The inverse of this total constitutes an index of political competition, as described in detail by Tagepaara & Shugart (1989).

[14] The amount of staff input was moderately significant, at the 0.1 level. However, the coefficient for this variable is not as robust as those of the other three variable. The effects of staff inputs are evident, further, only when more of these are made in combination with high levels of social capital and political competition.

Table 19.3: OLS regressions on common land development with CLDINDEX as the dependent variable.

	Model 1	Model 2	Model 3	Model 4
Intercept	-18.90 (24.68)	-34.3* (18.04)	-44.239** (17.627)	-40.93** (16.89)
SKI	0.689** (0.276)	0.691** (0.272)	0.721** (0.288)	0.703** (0.261)
DIVISION	-0.001 (0.041)			
POLCOMP	0.42*** (0.121)	0.38** (0.118)	0.376** (0.120)	0.391*** (0.117)
MOISTRES		0.029 (0.164)		
OFFINPUT			0.216* (0.124)	0.210* (0.122)
MECHZN			0.205 (0.337)	
INFRASTR		-1.31 (0.884)		
LITERACY		0.732* (0.355)	0.471 (0.321)	0.596* (0.298)
NCASTES		0.123 (0.690)		
POPULATION		0.005 (0.003)	0.002 (0.003)	
N	64	64	62	64
R^2	0.277	0.354	0.331	0.315
Adj-R^2	0.226	0.283	0.271	0.286
F-ratio	5.453	5.018	5.536	8.91
F-probability	.0009	.0004	.0003	.0001

Note: standard errors are in parentheses. $^*p<=0.1$; $^{**}p<=0.05$: $^{***}p<=0.001$.

19.6 What Do These Results Mean?

Even though a similar set of technical interventions was supported across all of these 64 villages, which additionally shared similar agro-economic conditions, and even though the level of program funding was standardised and consistently maintained, substantial differences remained in performance levels across participating villages. Common lands have been developed in some villages to the extent that supplies of fodder and fuelwood are maintained at high levels even in years when rainfall is short. In other villages that participated in the program, however, hardly any permanent gains appear to have accrued.

Two significant conclusions emerge from these findings. First, technical and financial parameters are not the sole or even very strong determinants of program performance. The same set of technical interventions introduced similarly across villages sharing fairly similar agro-climatic conditions ended up producing quite dissimilar effects. Villages located a few kilometres from each other and receiving the same levels of technical and financial support have performed very differently from each other.

Social organisation makes a substantial difference to program performance. Among different features of social organisation, three in particular stand out: levels of social capital, political competition and literacy were consistently and significantly associated with program performance. Performance in sustainable land use programs can be improved substantially, this analysis suggests, if efforts are made to enhance the levels of these three social indicators.

While increasing literacy is an important part of the agenda of development, the other two factors are relatively less well recognised in the literature on the planning and evaluation of development. Social capital has an evidently important effect on performance. Villages that have stronger predispositions and capacities for collective action, captured here by the variable SKI, perform better compared to others where collective action is less well supported by norms and roles. Similarly, increasing political competition at the local level is also associated with improving development results. By decentralising more authority to local governments, and by encouraging the growth of active and representative political parties, governments of developing countries can indirectly contribute to capacities for sustainable development.

Very little is known about factors that assist with building up social capital. The constrained view presented by Putnam *et al.* (1993) is that social capital is mostly historically created and determined and can hardly be enhanced within the short term. This view is challenged by other visions of collective enterprise found in the literature. Ostrom (1990) emphasises that collective action becomes successful when individuals in a community commonly abide by rules that have been constructed collectively. Members' socialisation results in rule-compliant behaviour, particularly when members have previously participated in the process of rule-formation. Comparative studies of local institutional development (Korten, 1980; Esman & Uphoff, 1984; Uphoff, 1986; Rondinelli, 1993; Uphoff *et al.* 1998) affirm that participatory action is an important factor in successful development at local levels. Hirschman (1984) suggests from his study of local development behaviour in Latin America that participation in the past can help generate enhanced participation in the future. Prior experience with collective action can thus help energise new common endeavours.

In our ongoing work of investigating social capital in relation to sustainable development, we have found each of these views to have some merit. Our preliminary findings indicate that history does matter to some extent, but history hardly determines the level of social capital. Rules and participation in development programs are both equally relevant in adding to a community's stock of social capital. Additionally, social capital is consistently higher among communities that have access to more sources of information[15].

Communities can be assisted to institute local rules that are respected by their members; participation can and should be made a constant feature in development programs; and flows of information can and must be enhanced, especially in developing countries. Instituting local rules, encouraging participation, and enhancing information are desirable ends in and of themselves. Additionally, they may have the desirable by-product of increasing the stock of social capital[16]. The results of increasing stocks of social capital will be felt in terms of improved development performance, not least in those programs that are concerned with sustainable land-use management.

19.7 References

Almond, G. & S. Verba. (1989): The civic culture: political attitudes and democracy in five Nations. Sage Publications, London

Bollen, K. & R. Jackman (1985): Political democracy and the size distribution of income. American Sociological Review 48, 468-79

Brara, R. (1991): Are grazing lands 'Wastelands'? Some evidence from Rajasthan. Economic and Political Weekly, Bombay, February 22

Coleman, J.S. (1988): Social capital in the creation of human capital. American Journal of Sociology 94, 95-120

Dreze, J. & A. Sen (1995): India: Economic development and social opportunity. Oxford University Press, Delhi

Esman, M.J. & N. Uphoff (1984): Local organisations: intermediaries in rural development. Cornell University Press, Ithaca, NY

---

15 These 'significant associations' are presented in detail in Krishna & Uphoff (1999). We speak in terms of significant associations and not in cause-and-effect terms because our analysis like those of other analysts is limited by the availability of only cross-sectional data. Analysis comparing trends over time will be required before causes of high social capital can be established with higher confidence.

16 Additionally, the *flows* of benefits arising from a given *stock* of social capital can be enhanced through purposive action focusing upon these elements. The corresponding argument and some real-world examples are presented in Krishna (2000).

Hirschmann, Albert O. (1984): Getting Ahead Collectively: Grassroots Experiences in Latin America. New York: Pergamon Press.

Jodha, N.S. (1990): Common property resources: contributions and crisis. Economic and Political Weekly. Bombay, June 30.

Johnston, B.F. & W.C. Clark (1982): Redesigning rural development: a strategic perspective. Johns Hopkins University Press, Baltimore

Korten, D.C. (1980): Community organization and rural development: a learning process ppproach. Public Administration Review 40, 480-511

Krishna, A. (1997): Participatory watershed development and soil conservation in Rajasthan, India. In: A. Krishna, N.T. Uphoff & M.J. Esman (Eds.). Reasons for hope: instructive experiences in rural development. Kumarian Press, West Hartford

Krishna, A. (2000): Creating and harnessing social capital. In: P. Dasgupta & I. Serageldin (Eds.). Social capital: a multifaceted perspective. The World Bank, Washington, DC

Krishna, A. (2000a): Social capital, collective action, and the State: understanding economic development, community harmony, and democratic governance. In: Rural North India. Ph.D. dissertation, Department of Government, Cornell University, Ithaca

Krishna, A. & E. Shrader (2000): Cross-cultural measures of social capital: a tool and results from India and Panama. Social Capital Initiative Working Paper Series. The World Bank, Washington, D.C.

Krishna, A. & N. Uphoff (1999): Mapping and measuring social capital: a conceptual and empirical study of collective action for conserving and developing watersheds in Rajasthan, India. Social Capital Initiative Working Paper Series No. 13. The World Bank, Washington, DC

Lipset, S.M. (1994): The social requisites of democracy revisited: 1993 presidential address. American Sociological Review 59, 1-22.

Lipset, S.M. (1960): Political man: the social bases of politics. Doubleday, New York

Olson, M. (1965): The logic of collective action: public goods and the theory of groups. Harvard University Press, Cambridge, MA

Ostrom, E. (1990): Governing the commons: the evolution of institutions for collective action. Cambridge University Press, Cambridge, U.K.

Putnam, R.D., R. Leonardi & Raffaella Nanetti. (1993): Making democracy work: civic traditions in modern Italy. Princeton University Press, Princeton

Rondinelli, D.A. (1993): Development projects as policy experiments: an adaptive approach to development administration. Methuen, London

Tageepara, R. & M.S. Shugart (1989): Seats and votes: the effects and determinants of electoral systems. Yale University Press, New Haven.

Uphoff, N.T. (1986): Local institutional development: an analytical sourcebook with cases. Kumarian Press, West Hartford

Uphoff, N.T. (2000): Understanding social capital: learning from the analysis and experience of participation. In: P. Dasgupta & I. Serageldin (Eds.). Social capital: a multifaceted perspective. The World Bank, Washington, DC

Uphoff, N.T., M. Esman & A. Krishna. (1998): Reasons for success: learning from instructive experiences in rural development. Kumarian Press, West Hartford

Wade, R. (1994): Village republics: economic conditions for collective action in South India. Institute for Contemporary Studies, San Francisco

20 Modelling Economic Policy Reforms and Sustainable Land Use in Developing Countries: Key Issues

Marijke Kuiper[1], Nico Heerink[1] & Herman van Keulen[2]

[1] Development Economics Group, Department of Economics and Management Wageningen University and Research centre, Wageningen, the Netherlands

[2] Plant Research International, Wageningen University and Research centre, Wageningen, the Netherlands

Abstract

Four key issues emerge, from the models presented in this book, in quantifying the effects of economic policy reforms on sustainable land use in developing countries: (1) conceptualisation of household objectives, market imperfections, agricultural production relationships, and sustainability, (2) specification of interactions between agricultural inputs, agricultural activities, spatially dispersed land units, household objectives, and economic agents, (3) method of aggregating micro-relationships to regional or higher levels, and (4) incorporation of a time dimension in the relationships. The way in which the various models presented in this volume incorporate these four key issues is reviewed. Based on this review, it is recommended that future quantitative research should concentrate on (1) incorporating the institutional environment of rural households, (2) testing the validity of basic assumptions, such as separability of household decision making, and/or the sensitivity of model outcomes to changes in these assumptions, and (3) addressing the aggregation bias in regional or higher-level models resulting from non-linear micro relationships and from off-site effects of soil erosion.

20.1 Introduction

In the previous chapters, the effects of economic policies on agricultural resource dynamics have been examined from various angles and at different spatial levels. The aims of this final chapter are to reflect on the advances made in the modelling work included in this volume, to identify the strengths, weaknesses and complementarities of the various approaches, and to identify directions for future research.

Because of the apparent diversity of the approaches, we first define a set of criteria that can be used for their consistent comparison. We identify four sets of key issues that are encountered when quantifying effects of economic policy on (sustainable) land use. These key issues refer to fundamental choices in model development and applications. We will not give a comprehensive review of each of these issues, but limit the discussion to their relevance for the topic of this book. We then proceed by summarising the way in which the different approaches presented in the book deal with the four key issues, to arrive at a consistent view on their strengths and limitations. At the same time, points at which they are complementary, and areas where future research could extend existing approaches, can be identified.

The key issues are illustrated from both an economic and a biophysical perspective in Section 20.2. Section 20.3 examines the ways in which the different models address the key issues. In the last part we assess the contribution of different types of bio-economic models to quantification of the effects of economic policy reforms on sustainable agricultural land use. Here, we also identify issues that are underexposed in current modelling exercises and need to be addressed in future research.

20.2 Key Issues in Modelling Bio-Economic Interactions

Central in the analysis of the effects of economic policies on agricultural resource use is the behaviour of economic agents and its effect on natural resources. The response of economic agents to a policy has to be formulated first. Then the implications for the agricultural resource base can be identified. A third step is to describe feedback relations between changes in agricultural resources and economic behaviour. Although the approaches presented in this book operate in two different disciplines, four sets of key issues can be identified that apply to both the economic and the biophysical components. These issues can be captured under the headings of *conceptualisation*, *interactions*, *aggregation*, and *time dimension*. We will address these from both an economic and a biophysical perspective.

20.2.1 Conceptualisation

Only two of the models in this volume are strictly analytical (*i.e.* Antle & Stoorvogel and Albersen & Sun), in the sense of being abstract theoretical constructs without an application to empirical data (Kaimowitz & Angelsen, 1997). Use of an applied model requires conceptualisation of general theoretical constructs (*e.g.* utility) into a specific operational definition (*e.g.* full income). The assumptions that guide this step from theory to application, drive the outcome of the models, and therefore deserve considerable thought and explicit treatment. In this context we will discuss issues related to the conceptualisation of household behaviour, institutions, production sets and sustainability.

We limit the discussion on the behaviour of economic agents to rural households, since in most developing countries they take the vast majority of the land use decisions. If other economic agents, such as plantations or government estates, play an important role, modelling of their behaviour requires similar careful treatment. We furthermore start from the premise that the models are intended to support policymaking, with policies defined as the exogenous variables whose impact on land use is being addressed. Modelling of government behaviour is thus not treated.

A key issue in the literature on modelling of behaviour of rural households in developing countries, is the *separability* of production and consumption decisions. Production, labour supply and consumption decisions can be separated (and thus analysed recursively) if all markets function perfectly (Sadoulet & de Janvry, 1995; Singh *et al.*, 1986). In such a situation, the consumption and labour supply decisions of a household do not affect the household's production decisions. Perfect markets are a sufficient, but not necessary condition for separability: imperfect markets do not necessarily imply interlinked production, labour and consumption decisions. To illustrate this point, consider an output that has to be sold to a government board against a fixed price. Such a monopsony does not cause non-separability of household decisions, since the output price is still exogenous to the household. In most situations, the effect of market imperfections will not be as clear-cut as in this example. Therefore, the impact of market imperfections has to be carefully examined, before taking the decision of modelling household behaviour as either recursive or simultaneous decision-making, while taking into account that market imperfections might only affect part of the population (see for example Sadoulet *et al.*, 1996). In the case of separability, production and consumption decisions may be modelled as profit and consumption maximisation, respectively. In household models assuming non-separability, maximisation of full income (consumption goods and leisure) is generally taken as the objective. As discussed in Chapter 1, responses to price changes may be weaker or even reversed, if one switches from assuming separable to non-separable decision-making[1].

The above discussion follows the basic assumption in economic theory that human behaviour can be understood in terms of an optimisation problem. This point deserves some attention, since different assumptions on this issue distinguish

[1] See also Kaimowitz & Angelsen (1997) for a discussion of the impact of separability assumptions on the relation between higher agricultural prices and the level of deforestation.

dynamic simulation models from the other models. A distinction can be made between 'substantive rationality' (behaviour appropriate with achieving given goals under given constraints) and 'procedural rationality' (behaviour as outcome of appropriate deliberation) (Rubinstein, 1998). Economically rational behaviour is of the substantive type, *i.e.* 'any behaviour that solves some optimisation problem (...), independent of whether it comes about as the result of careful thought, or whether it comes about through tradition, routine, or in any other way' (Börgers, 1996, 3). Essential is that substantive rationality does not require economic agents to consciously solve an optimisation problem; it is a theoretical concept, also used in other empirical sciences. To illustrate this point, the shape of bubbles in soapy water can be modelled as the outcome of an optimisation problem, but obviously does not assume conscious deliberation by the molecules. In applied models, this concept of rational behaviour, combined with specific assumptions on the utility function, offers a framework for understanding and predicting the behaviour of economic agents (Börgers, 1996).

Empirical evidence suggests that economic agents display both rational and so-called irrational (*i.e.* non-optimising) behaviour. However, clear evidence is lacking on the circumstances under which certain behaviour dominates, and to what extent this could be accounted for in more sophisticated optimisation models (Börgers, 1996; Rubinstein, 1998). Theoretically, it would be possible to model objectives on the basis of non-optimising behaviour of households in all four model types, using, for example, bounded rationality (Rubinstein, 1998) or an evolutionary approach (Nelson & Winter, 1982) as a basis for household behaviour. Both approaches use rules of thumb, or routines, for modelling behaviour. Under specific conditions, these rules can be consistent with rational behaviour (see for example Blonski, 1999; Lettau & Uhlig, 1999). In the end, for applied models the same holds as for separability: the assumptions on behaviour, whether rational or irrational, should be checked empirically.

A second important aspect of household behaviour is associated with the issue of *risk*, especially in developing countries were low-income households generally lack the resources to deal with fluctuations in income. Modelling of risk requires conceptualisation of both the variability in prices and/or production, and the attitude of the households towards risk and their coping behaviour. Whereas variability can be expressed objectively, for example through including a measure of fluctuations in production levels in the model, the attitude of households towards such fluctuations is a subjective issue, that is not easily measured.

The importance of market functioning for the separability of household decisions has already been mentioned. Other *institutions*, such as property rights, may also have a strong impact on the behaviour of rural households. Institutions can be defined as 'rules that can constrain behaviour over a certain domain and give rise to behavioural regularities' (Lin & Nugent, 1995, 2307). Hence, they include markets as well as laws, contracts, norms and values. Social interactions, other than markets or other exchange relations, can be highly relevant for understanding land use decisions. Definition of the socio-economic environment of households in terms of institutions will make a limitation of the analysis to (the absence of) markets a more conscious decision, ideally supported by empirical facts.

A major reason for defining the institutional environment in terms of (formal) markets may be the lack of an operational theory that can be used to describe the role and development of other institutions (Lin & Nugent, 1995; Roumasset, 1995). This in contrast to a well-developed theory of markets, that forms the core of standard economic theory. As a result, most authors recognise the importance of institutions other than markets, but then tend to neglect them as soon as formalisation of the model starts (Kaimowitz & Angelsen, 1997). The major exception seems to be the issue of property rights in relation to land use and soil conservation investment (*e.g.* Larson & Bromley, 1990; Persson & Munasinghe, 1996; Unemo, 1996). The study of the effects of social capital by Krishna and Uphoff illustrates, that considering institutions that are traditionally not included in quantitative analyses, could improve the understanding of effects of policy reforms on land degradation.

Where at the economic side one has to grapple with objectives and institutions, at the biophysical side a set of equally complicated conceptualisation problems awaits: specification of a production set and defining sustainability. A *production set* specifies the relation between inputs and outputs. As for utility, specifying this relation in an applied model creates problems. In particular, one has to select the elements to be included in the production set. Inputs applied in the production process like labour, seeds, impact of weather, notably rainfall, are obvious elements to be considered, just as the output of crop or livestock products. Most production functions in economic models do not go beyond this. However, such a specification ignores the effects of soil characteristics on yields, as well as the possible effects (positive or negative) of agricultural production on soil quality. From a biophysical perspective, the contribution to, and effect on a range of characteristics such as soil depth, soil density, nutrient contents, acidity, *etc.* should be accounted for when estimating the production function. Unfortunately, complex interactions between bio-physical processes and agricultural production make it difficult, if not impossible, to generate such a complete specification (see Kruseman & Van Keulen).

These difficulties are directly related to the fourth conceptual issue, *sustainability*. An often cited definition of sustainable development is 'development that meets the needs of the present without compromising the ability of future generations to meet their own needs' (WCED, 1987, 43). Sustainability, thus, is related to balancing the needs of current and future generations. This temporal component in the definition of sustainability creates a number of controversies, due to uncertainty with respect to the needs of future generations, and their capacity to meet these, taking into account the possibilities of technological innovations (van Pelt, 1993).

Sustainability in the present context implies estimating the effects of current land use activities on the future productive capacity of the land. If current activities do not affect its productive capacity, impairing the production possibilities of future generations is not an issue. Following Pearce & Turner (1990) and Bouman *et al.* (2000), three different concepts of sustainability may be distinguished:

1. Maintaining the resource base, *i.e.* utilise renewable resources at rates lower than or equal to the natural rate at which they regenerate. With respect to soil quality, this concept may be quantified in terms of changes in soil nutrient (nitrogen (N), phosphorus (P), potassium (K)) and organic matter stocks.
2. Protecting the environment, *i.e.* keep waste flows to the environment at or below its assimilative capacity. This concept may be quantified in terms of nitrogen waste flows, biocide emissions, and so on.
3. Efficient use of non-renewable resources, *i.e.* optimise the efficiency with which non-renewable resources are used, subject to substitutability between resources and technical progress. This concept may be operationalised by including an economic objective function in models that explain choices between alternative land use options, taking into account (long-term) interactions between soil quality and agricultural production.

The spatial level at which sustainability is defined is also important, because of its strong impact on the restrictions imposed on resource use: the higher the spatial level, the less restrictive the sustainability requirements (van Pelt, 1993). For example, if erosion causes soil nutrient and productivity losses upstream, but, through deposition, improves soil fertility and productivity downstream, the sustainability of land use at the watershed level (measured either through the first or the third concept) may not be affected.

20.2.2 Interactions

A first issue in modelling biophysical interactions has already been touched upon in the discussion of the spatial level at which sustainability is defined. *Land units* can interact through flows of soil and water, transporting nutrients, organic matter, pesticide residues, *etc.* These flows can result in either negative (*e.g.* siltation of reservoirs and rivers) or positive (*e.g.* increased fertility through downstream sedimentation) external effects. The spatial scale can affect the possible impact of, for example, erosion to a large extent: a study in a basin in Wisconsin indicated, that only between 8 and 31% of the eroded soil actually left the watershed (Trimble, 1999). Such interactions between land units can only be accounted for by including spatial dimensions in the model.

As a second set of biophysical interactions, *viz.* interdependencies among different production activities, can be distinguished. Interactions occur first of all through competition for constrained inputs, such as lànd, labour and capital. These are included in all models describing more than one activity. We, therefore, focus here on *complementarities* among activities. These can be bio-physical, as for crop and livestock systems that are linked through the exchange of manure and fodder. Another example is the use of intermediate goods in production. Complementarities can also originate from socio-economic circumstances. For example, when access to fertiliser is limited to farmers growing a specific cash crop (*e.g.* cotton in some West-African countries), fertiliser availability for other crops will be linked to cash crop production.

A third set of biophysical interactions, those among different inputs for a single activity, can be distinguished at a lower level of aggregation. *Interactions among inputs* refer to both substitution (where inputs can compensate for each other) and synergy (where the effect of a particular input depends on the presence of other inputs). Substitution can exist between factors of production, a standard concept in economics, but also between external inputs (*e.g.* fertiliser) and soil characteristics (*e.g.* soil nutrient contents). Accurate quantitative specification of the latter type of substitution is important for assessing the sustainability of the system in terms of soil quality. If soil fertility (defined in physical and chemical characteristics) can at all times be restored through the use of external inputs, there is no risk of technically impairing the production possibilities of future generations. However, thresholds may exist, below which the decline in soil quality becomes irreversible (Rietkerk, 1998; Erenstein, 1999). This limits the scope for restoring soil fertility through external inputs and increases the need for more judicious behaviour.

Complex interactions exist between the physical, chemical and biological characteristics of the soil, complicating the definition of soil quality (Jaenicke & Lengnick, 1999), and thus the quantification of sustainability of land use, as indicated by Kruseman & Van Keulen. In addition, this complexity affects the specification of the input-output relations. For example, the response of crops to chemical fertiliser applications is positively related to the soil organic matter content (and hence to the use of animal manure and/or crop residues). Capturing the synergetic interactions between inputs implies abandoning the continuous concave production functions (as often used in economic models) in favour of *e.g.* discrete input-output coefficients. This, however, yields problems with scale and factor substitutions that both play a prominent role in economic analyses of land use (Ruben *et al.*, 1998). As synergistic relations between inputs create problems in linking biophysical and economic models, some important aspects of the production process have to be sacrificed at either the economic or the biophysical side of bio-economic models.

At the economic side, two more sets of interactions require attention. A single household might pursue multiple goals (*e.g.* income maximisation, food security, risk avoidance), that could conflict with each other. Intra-household differences in access to resources, for example based on gender, could also be a reason for conflicting goals within a single household. To quantify the behaviour of the household as a whole, some kind of *weighing of objectives* is necessary. The second set of economic interactions refers to those among households, or between households and other *economic agents*, like, for example, (local) government, enterprises or plantations. These can be addressed in a partial way, for example by incorporating share-cropping arrangements in farm household models, or through a general equilibrium approach, tracing all flows of goods and income. Recent analyses using village or micro economy-wide models have shown that income and expenditure linkages can play a major role even at the micro level. Modelling of these micro-linkages can result in policy-induced supply responses opposite to those derived from partial household models (Taylor & Adelman, 1996).

20.2.3 Aggregation

Policies are not targeted towards individual households or single plots. To assess whether they realise their objectives, some aggregation of their effects to regional or higher levels is required. Such a scaling-up of the analysis generates problems in (1) maintaining the assumption of exogeneity for some critical variables, (2) incorporating the diversity of the smaller units, and (3) dealing with non-linearities. We will discuss these issues from an economic and a biophysical perspective.

The most commonly treated aggregation problem in the economic literature on land use models, is the *endogeneity* of prices. Prices of inputs, outputs and production factors that are exogenous to farm households may become endogenous at the regional level, when these commodities or factors are traded within the region and costs of transactions with markets outside the region are high. These equilibrium effects can already occur at lower levels of aggregation, as the village-town model by Dyer *et al.* illustrates. Accounting for price and income effects in a general equilibrium setting has a strong policy implication. The Debreu-Mantel-Sonnenschein result shows that aggregating over consumers results in loss of general properties of the excess demand function (Ginsburgh & Keyzer, 1997, 13). As a result, no generalised predictions on the effects of economic policies are possible, not even under strict assumptions on production and consumption behaviour.

Interactions at higher levels of aggregation also complicate matters at the biophysical side. An agronomic example of endogenous variables at higher levels of analysis is the incidence of pests and diseases. At field or household level, the risk of pests and diseases can be considered exogenous, even though that risk can be affected by crop rotations. At the regional level, however, the cropping pattern can affect population dynamics of pests and diseases, making their impact an endogenous factor (Bouman *et al.*, 2000). Given a similarly high dependence on local conditions as above, no general recommendations can be derived.

A second problem in aggregation is how to deal with the *diversity* in micro-units. Responses to (policy-induced) price changes may differ substantially for different farm household types within the same region, depending on farm size, production system, and other farm characteristics. Similarly, crop responses to fertiliser or manure applications may vary considerably among plots, depending on soil type and quality, water availability, and other land characteristics. By specifying aggregate relationships for a whole region or watershed, essential information on differences in responses among micro-units may be lost. One way to deal with this problem is to distinguish groups of households (or plots or other micro-units) on the basis of a number of the relevant characteristics, and develop separate models for 'representative' households and/or plot types in each group. The higher the level of disaggregation, the smaller the aggregation bias is likely to be (Mohamed, 1999).

The use of representative households, plot types or other relevant micro-units does not solve the problem of aggregating non-linear relations (Antle and Stoorvogel). When a concave relationship exists between, for example, fertiliser price and use of fertiliser for all households in a certain group, or between nutrient availability and plant growth for all plots with certain characteristics, a negative aggregation bias will result, if the same concave specification is used to represent the aggregate relationship between average fertiliser use and average price, or between average nutrient availability and plant growth (de Wit & Van Keulen, 1989; Heerink, 1994; Heerink & Folmer, 1994). If the micro-relationship is convex, the aggregation bias is positive. When the specification of the functional relationship at the micro-level and the distribution of the explanatory variable(s) over the population of households or plot types are both (approximately) known, this type of aggregation bias can simply be corrected at the aggregate level (see Heerink, 1994; Heerink & Folmer, 1994)[2].

These examples of aggregation issues in the economic and biophysical realm point to the need for a careful balancing of aggregate and micro patterns. Where crucial information may be lost through aggregating over units, without accounting for their variability, 'overemphasising micro issues while ignoring broader trends, can be compared to rearranging deck chairs on the Titanic' (Harrington *et al.*, 1995, 25 (cited in Erenstein, 1999, 30)). The challenge is thus to minimise this *aggregation bias* when looking at aggregate characteristics of either households or biophysical units, without getting bogged down by too much detail. A minimum requirement from an economic point of view, however, seems that the aggregate model should reflect individual rational behaviour. If this condition is not met, realising simulated policy results can be expected to encounter significant problems.

20.2.4 Time Dimension

In the preceding section we have dealt with aggregation over spatially different units at the same point in time. In this section we shift our attention to aggregating

---

[2] For convex relationships, the aggregation bias is in fact a measure of the degree of inequality in the distribution of the explanatory variable over the relevant population. The shape of the micro-relationship determines the exact specification of the inequality measure that is needed to correct for the aggregation bias.

over different *time periods*, which we consider a last key issue in modelling economic policy reform and sustainable land use. A major reason is that negative impacts on the quality of the agricultural resource base generally only show up after a considerable period of time. Furthermore, there is uncertainty about the effect(s) of degradation on production, aggravated by the *cumulative impact* and possible synergy between different interventions in the resource base (van Pelt, 1993; Erenstein, 1999). To account for these phenomena, a long-term perspective is required, combined with a method to address the uncertainty surrounding the effects of degradation (Kruseman & van Keulen).

From an economic perspective, this creates a problem of intergenerational welfare comparison under uncertainty. In addition, the long-term perspective introduces the possibility of *technological change*. Changing production environments (both biophysical and socio-economic) can induce technological innovations. These can either mitigate or aggravate the impact of degradation on the welfare of future generations. Moreover, even without technological innovations, the impact of degradation, in economic terms, can change over time as a result of changing socio-economic conditions (van Pelt, 1993; Erenstein, 1999). For example, road construction may make cultivation of vegetables on previously isolated lands economically attractive. In this case, soil productivity losses in value terms will increase. A last complication is the possibility of irreversibility of degradation processes (Rietkerk, 1998). If threshold values for soil quality indicators exist, the valuation of the soil resource should account for both use value and option value, *i.e.* the value of keeping the option open of being able to use the resource (Dasgupta & Mäler, 1995).

Where uncertainty and irreversibility affect the value of the resources for each generation, discounting affects the balance between current and future values, adding two additional layers of (economic) complications to the analysis of resource use over time. First, the levels of the resources as such are not important, but the welfare that can be derived from them. Given the long-term nature of degradation processes, this requires definition of an aggregate *welfare* function, spanning multiple generations. Such a welfare function requires definition of both current and future utility functions, which is problematic in itself. As the welfare of different generations is of interest, there is no *a priori* reason to impose constraints on the use of resource stocks. Instead, the focus should be on guaranteeing a certain level of welfare for future generations. Such a constraint on future welfare could then, in a second step, lead to a *derived* constraint on resource levels (Dasgupta & Mäler, 1995).

The second complication arises from balancing the welfare of different generations in the aggregate social welfare function, *i.e.* selecting a discount rate. This total discount rate can be divided in two parts. One part accounts for the expected increase in future consumption, for example due to productivity-increasing technological change. The second part is the relative weight attached to future generations' welfare. In economic literature, this distinction between consumption and well-being components in the discount rate is often not made explicitly. If an increase in consumption is expected, applying a positive consumption discount rate only serves to partition this welfare gain (possibly generated through investments by earlier generations) equally between generations. This rate for distributing consumption gains among generations can be positive, even if the well-being discount rate is zero, *i.e.* if welfare of different generations receives equal weight (Dasgupta & Mäler, 1995). The uncertainties surrounding both, the impact of degradation and technological change, complicate the determination of an appropriate discount rate, affecting both, the elements of consumption and of well-being.

Further pursuing the idea that the welfare of different generations is the characteristic to be balanced over time, a discount rate has to be established that guarantees a minimum level of welfare, for example the current level, for all generations. The concept of balance suggests that neither the present generation, nor future generations, should fully determine the optimal path. The use of a positive well-being discount rate does not satisfy this condition, since it ignores the welfare of generations in the distant future. But, if a measure of the welfare of future generations (such as average welfare over all generations) is added to the sum of total discounted welfare, a balanced measure of aggregate welfare results, doing justice to the interests of both current and future generations (Dasgupta & Mäler, 1995; Chichilnisky, 1997).

In summary, the important elements that have to be addressed to account for the long-term impact of degradation, are: a long-term perspective on soil quality accounting for technological change and cumulative effects of soil degradation, and specification of a balanced measure of aggregate welfare, doing justice to both current and future generations' interests.

20.3 Addressing The Key Issues: Comparing Approaches

The models presented in the previous chapters have been classified in four groups, according to the modelling technique used: econometric models, partial programming models, general equilibrium models and dynamic simulation models. In econometric models, relationships are quantified by applying statistical methods to empirical data (Kaimowitz & Angelsen, 1997). Partial programming models use linear or non-linear programming techniques on both empirical data and data from other sources, such as expert knowledge or stylised facts. General equilibrium models can also use programming techniques and data from other sources, but aim at modelling an entire economic system (Ginsburgh & Keyzer, 1997), where partial programming models deal with only part of an economy. The last group of models differs from the previous ones in that it does not rely on optimisation (*i.e.* rational behaviour of economic agents), nor on a need to reach equilibrium, but models relationships on the basis of decision rules or empirical regularities. Dynamic simulation models can, such as partial programming and general equilibrium models, include other types of data, like expert knowledge, in addition to empirically observed data.

Tables 20.1 to 20.4 summarise the way in which the different models deal with the four key issues. To keep the tables as concise as possible, models are identified by the name of the first author only, with the chapter number in superscript. Below, we shortly discuss some general patterns for each of the key issues, based on the summary descriptions in the tables. Only those models are included that explicitly describe the behaviour of economic agents in relation to their biophysical environment. The chapters omitted from the tables do not fit into this framework, either because they do not discuss a specific model (the first two stage-setting chapters and the chapter by von Benda-Beckmann), or because the model lacks description of behaviour (Keyzer & Sonneveld and Krishna & Uphoff). The chapter by Albersen & Sun is also left out for that latter reason. Although they refer to a social welfare optimisation framework in which their transformation function can be incorporated, the main body of the chapter focuses on the modelling of degradation without addressing the behaviour of economic agents. Where relevant, contributions from the chapters not included in the tables will be incorporated in the discussions below.

20.3.1 Conceptualisation

The results of the modelling exercises are driven by the assumptions on the objectives pursued by the agents. In all models summarised in the tables, rational or maximising behaviour of economic agents is assumed. The only exception is the model of Struif Bontkes, which relies on decision rules, postulated without theoretical or empirical justification. The model by Abiye & Aune focuses on the dynamics of soil nitrogen stocks and is therefore included under the heading of dynamic simulation models. Although reference is made to net present values of income streams, their model does not incorporate the decision-making behaviour of the households and is thus unable to choose between scenarios.

With respect to the link between production and consumption decisions, all groups of models comprise examples of separable and non-separable household models, as indicated by objectives in Table 20.1. Only two (econometric) models, by de la Brière and Romano, test the separability assumption. More generally, whereas general equilibrium models are calibrated to reproduce base-year observations organised in a social accounting matrix, and econometric models rely on statistical tests, no standard procedures exist for testing the empirical relevance for partial programming and dynamic simulation models. As a result, information on their accuracy and appropriateness of assumptions tends to be limited.

Table 20.1: Conceptualisation.

Model [Chapter]	Household behaviour: Objectives	Household behaviour: Risk	Market imperfections
Econometric models			
Antle [10]	- profit	-	-
Brière [6]	- food, non-food cons.	-	- land, food
Romano [7]	- value of production	-	- labour
Ruben [18]	- consumption	- uncertain land rights & non-farm jobs	- credit
Partial programming models			
Barbier [12]	- net income, value livestock & forest, leisure	-	- land, output, migration
Deybe [14]	- net individual income / wealth & revenue common resources, food preferences	- deviation from minimum income / variability of revenue	- credit, imperfect price information, minimum consumption requirement
Jansen [13] (S)*	- economic surplus	-	- land
Jansen [13] (U)*	- production, leisure, purchased products / long-term income / profit	-	- land, credit
Jansen [13] (O)*	- profit	-	-
Kruseman [8]	- consumption, income adjusted for soil quality	- weather risk	- land, food crops, inputs, labour, capital
Shiferaw [5]	- net income, value of leisure	-	- land, credit, labour, food, feed
General equilibrium models			
Dyer [11]	- consumption, savings	-	- village markets for land, capital, labour, staples
Glomsrød [15]	- profit & utility, subsistence production	- incomplete adjustment to (expected) wage differentials	- credit, frontier production
Dynamic simulation models			
Abiye [3]	- net present value income streams	-	-
Struif Bontkes [9]	- food & cash requirements, luxury items, livestock	- risk in cash crop production	- cash crops, inputs, hired labour

Table 20.1: Conceptualisation (continued).

Model [Chapter]	Production set		Sustainability indicator
	Outputs	Inputs	
Econometric models			
Antle [10]	- crops	- variable & fixed factors, environm. characteristics	- trade-off curve
Brière [6]	- food crops	- area, soil fertility, labour, household characteristics	- adoption of conservation
Romano [7]	- farm production	- area, soil quality, family labour, capital, farm household characteristics	- adoption of conservation
Ruben [18]	- crop production	- land, capital, variable inputs	- land improvements
Partial programming models			
Barbier [12]	- crops	- potential yield, fertiliser, erosion, soil depth	- soil depth
Deybe [14]	- crops, livestock	- land, labour, credit, manure, fertiliser, fodder	- value of production of reference crop
Jansen [13] (S)*	- crops, livestock	- land, labour, input costs	- N, P, K stocks, biocides, N loss, greenhouse gasses
Jansen [13] (U)*	- crops, livestock	- land, labour, fixed & variable inputs, capital,	- N, P, K stocks
Jansen [13] (O)*	- livestock	- fertiliser, feed, stocking rate, production & maintenance costs	- N stocks
Kruseman [8]	- crops, livestock, forestry	- land, seed, labour, nutrients	- SOM, nutrient stocks
Shiferaw [5]	- crops, livestock	- land, labour, fertiliser, traction, soil depth, N, P	- N stock, rooting depth
General equilibrium models			
Dyer [11]	- basic grains, other crops, livestock, non-ag. products	- land, labour, animal & physical capital	- staple productivity, land area
Glomsrød [15]	- agr. products (sectors by crops), non-ag. products	- land, labour, fertiliser, capital, intermediates, N, pesticides	- N stock, deforested area
Dynamic simulation models			
Abiye [3]	- crops, livestock	- land, fertiliser	- N stock
Struif Bontkes [9]	- crops, livestock	- potential prod., labour, N, P, water, pests & diseases	- N, P stocks

* S = SOLUS, U = UNA-DLV, O = Optimal control

Another important observation on the behaviour of economic agents is the absence of risk considerations in most models, despite their focus on developing countries (see Table 20.1). The types of uncertainties that are dealt with refer to the short run, and none of the models addresses the uncertainty with respect to soil quality.

In Table 20.1 reference is made to markets only, since none of the models explicitly addresses other institutions. An exception is the study on the impact of uncertain property rights in the model by Ruben *et al.*, which is captured under the heading of risk. Most models incorporate various types of market imperfections, varying from complete absence of markets to limits on the total amount that can be demanded or supplied. We also consider the existence of village markets as a market imperfection, since perfect markets would imply full integration of village and town economies in the national economy, eliminating village equilibrium effects.

Important to note is that in a number of models a fixed quantity of land available for agricultural production is assumed. This might be justified at for example the regional level, where it could be assumed that no land from other regions can be hired and/or bought. However, in most areas it is possible for households to at least rent additional land (or hire out land). In these cases, the assumption of failing land markets might be artificial, biasing model outcomes. Interesting is that this assumption is often not explicitly treated, when discussing the market imperfections addressed by the model, but only becomes evident after examining the model constraints.

We now turn to the biophysical part of the conceptualisation issues, starting with the outputs and inputs of the production set. As expected, given the topic we are dealing with, most models focus on agricultural production. Only the two general equilibrium models by Dyer *et al.* and Glomsrød take into account non-agricultural production. The model by Deybe incorporates a link with urban consumption, but assumes fixed food budgets. The second point to note is that the econometric models do not incorporate both crops and livestock, where most other model types do. At the input side, some models take into account the impact of soil quality characteristics on yield, by specifying a potential yield corrected for differences in soil characteristics. This entails an implicit assumption of fixed labour and other material inputs (at the levels required to realise potential production), eliminating the response of these factors to relative price changes.

The last column in Table 20.1 shows the way in which sustainability of land use is quantified. Nutrient stocks are most often used as an indicator of soil quality. The models that include a feedback loop by incorporating land quality indicators as inputs in the production set can quantify the effect of current land use on future production possibilities. These models use the third concept of sustainability specified above. In the models with feedback, nutrient stocks (especially nitrogen) and soil depth are used as indicators of soil quality. Changes in these characteristics feed back to production possibilities in future periods. Selection of the appropriate indicators of soil quality depends on the conditions perceived as most limiting in a case study area. Models that do not incorporate such feedback effects implicitly adopt the first sustainability concept given above. The only exception is the use of a reference crop in the model by Deybe, that enforces consideration of long-term future impacts, while avoiding specification of activity-specific feedback loops. The second sustainability concept, referring to waste flows to the environment, is not included in any of the models.

The level or tightness of the sustainability criterion depends on the level of analysis and the interactions between different land units (both discussed in Subsection 20.2.2). Here, it should be noted that the calculation of the stocks in the feedback loops generally takes place at the level of land types. This implies that crops cultivated on the same land type can compensate for each other's depletion of stocks, which alleviates the constraints on production. Consideration of within-site variability, as in the model by Antle & Stoorvogel, suggests that such use of representative land types may lead to biases in the assessment of a system's sustainability, if no proper correction is made for prevailing non-linearities.

20.3.2 Interactions

The model by Antle & Stoorvogel considers within-site spatial variability in erosion, but not the interactions among land units. Only the model by Barbier accounts for the interactions of land units, through water availability. External effects of land degradation are not incorporated in any of the models.

Complementary relations between crop and livestock activities are not included in any of the econometric models. These models also use continuous production functions, as do the general equilibrium models. Assuming a concave continuous production function implies that it is increasing in all inputs, *i.e.* other inputs (like fertiliser or labour) can compensate for loss of soil fertility. Therefore, synergetic effects between inputs cannot be accounted for in these specifications. Threshold effects of water, nitrogen and phosphorus are included only in the model by Struif Bontkes.

Table 20.2 shows two major ways of weighing different elements in the objective function. Objectives are converted in either monetary values or in utility levels. This conversion to a single unit of measurement avoids conflicts between different elements in the objective function. For the dynamic simulation model, which does not include a single objective function, it is unclear how possibly conflicting outcomes of different decision rules are dealt with.

Finally, interactions between agents are modelled in both the partial programming and general equilibrium models. Technically speaking, each transaction of inputs or outputs is an interaction with another agent, but the entries in Table 20.2 refer to those cases in which the behaviour of both the supply and the demand side are modelled. Thus, that excludes the transactions with fixed prices and/or exogenously imposed quantity constraints. As expected, the two general equilibrium models explicitly address the interactions at different markets. Interactions in a single market are also modelled in a partial programming and dynamic simulation setting. The latter interactions only deal with price formation. Second-round income and expenditure linkages between groups of economic agents are included in the general equilibrium models only.

20.3.3 Aggregation

The aggregation problems encountered depend on the level of analysis. Table 20.3 therefore starts by indicating the levels distinguished in each model. Most of the models include the household level. For the model by Antle & Stoorvogel, no specification is given on the agent (farmers, a central planner) performing the profit maximisation. In the model of Glomsrød, aggregate demand and supply at sector level are derived from individual consumer and producer behaviour.

The levels incorporated directly indicate which models may have to address endogenous prices. Despite the aggregation to the watershed level in the model by Barbier and to the policy-relevant (presumably more aggregate level) in the model by Antle & Stoorvogel, these two models do not account for endogeneity of prices. Endogeneity of biophysical processes at higher levels of analysis is not addressed in any of the models. More importantly, as mentioned before, none of the models includes off-site effects of degradation. Such off-site effects can have a significant impact, positive or negative, on agricultural production at a higher level of aggregation.

Spatial heterogeneity is addressed in several models by distinguishing different household types or land use types. The aggregation bias resulting from non-linearities is explicitly addressed only by Antle & Stoorvogel, comparing within-site variability to a representative soil description. Although non-linearities are characteristic of many biophysical relationships (*e.g.* nutrient availability and plant growth, rainfall and yields) and economic relationships (*e.g.* convexities in production, Engel curves in consumption), the aggregation bias resulting from such non-linearities is not corrected in any of the models.

In three models (Barbier, Deybe and the SOLUS model in Jansen *et al.*), the objective function is defined at the watershed or regional level. This implies that household-level rationality of the model outcomes is not assured. Such a 'central planner' approach allows increases in aggregate production resulting from reductions in household production for some households (and increases for others), as has been shown in a similar regional model by Jansen & Stoorvogel (1996).

Table 20.2: Interactions.

Model [Chapter]	Land units	Complementarity	Inputs[#]	Weighing of objectives	Economic agents
Econometric models					
Antle [10]	-	-	- p.f.	- monetary	-
Brière [6]	-	-	- concave p.f.	- utility	-
Romano [7]	-	-	- concave p.f	- monetary	-
Ruben [18]	-	-	- concave p.f	- utility	-
Partial programming models					
Barbier [12]	- water	-	- linearly approximated p.f.	- monetary	- watershed labour market
Deybe [14]	-	- crop-livestock	- technical coeff.	- monetary, utility	- transfer of resources, common resources
Jansen [13] (S)*	-	-	- technical coeff.	- monetary	- labour & output market
Jansen [13] (U)*	-	-	- technical coeff.	- mixed utility & monetary	- output market
Jansen [13] (O)*	-	-	- p.f.	- monetary values	-
Kruseman [8]	-	- crop-livestock	- technical coeff.	- mixed utility & monetary	- regional factor & commodity market
Shiferaw [5]	-	- crop-livestock	- p.f.	- monetary values	-
General equilibrium models					
Dyer [11]	-	- crop-livestock	- concave p.f.	- utility levels	- land, capital, labour & staple markets
Glomsrød [15]	-	- intermediates	- concave p.f.	- utility & monetary values	- markets for labour, land, savings & commodities
Dynamic simulation models					
Abiye [3]	-	-	- p.f.	- monetary values	-
Struif Bontkes [9]	-	- crop-livestock	- minimum of yield potentials	- decision rules	- cereal market

[#] p.f .= production function; * S = SOLUS, U = UNA-DLV, O = Optimal control

Table 20.3: Aggregation.

Model Chapter	Levels & diversity	Endogenous prices	Aggregation bias
Econometric models			
Antle [10]	- land units, scale relevant for policies	-	- site specific vs. representative site
Brière [6]	- 2 household types	-	-
Romano [7]	- 2 household types	-	-
Ruben [18]	- household	-	-
Partial programming models			
Barbier[12]	- 18 land units, 6 farm types, watershed	-	- maximisation of watershed utility
Deybe [14]	- farm types, regions, urban consumer types, sectors	- endogenous consumer prices	- maximisation of regional net income
Jansen [13] (S)*	- 3 soil groups, 2 sub-regions, region, nation	- labour, output	- maximisation of regional economic surplus
Jansen [13] (U)*	- 3 soil groups, 4 farm types, region	- output	-
Jansen [13] (O)*	- plot	-	-
Kruseman [8]	- land units, household types, region	- factors, commodities	-
Shiferaw [5]	- 8 land units, 4 household types	-	-
General equilibrium models			
Dyer [11]	- 2 households, villages, town	- capital, land, labour, staples	-
Glomsrød [15]	- sectors	- land, consumer goods, domestic producer prices	-
Dynamic simulation models			
Abiye [3]	- 3 household types	-	-
Struif Bontkes [9]	- 3 soil types, 4 farm types, region	- cereal prices	-

* S = SOLUS, U = UNA-DLV, O = Optimal control

20.3.4 Time Dimension

Table 20.4 summarises various time-related issues. All models, except the model by Dyer *et al.* and the SOLUS model in Jansen *et al.,* have a multi-period structure. Several of the models include alternative or potential technologies, for example introduction of conservation measures. The models including such alternative technologies, offer specified alternative options, leaving the adoption decisions to the households. This implies exogenous technological change, since the development of alternatives is not induced by changes in the bio-physical and/or socio-economic conditions. A number of models keeps track of the cumulative aspects of degradation through changes in soil characteristics indicative of soil quality, such as nutrient stocks and soil depth.

Finally, most models select the intertemporal land use pattern in a deterministic way, by optimising over the whole planning period at its start. The objective function flows in these models are discounted (expect in the model by Ruben *et al.*), thus giving precedence to current generations' welfare. Four models take a different approach. The model by Antle & Stoorvogel appears to be recursive, since it only describes period-wise profit maximisation. Two models are essentially static, with long-term impacts being anticipated through a reference crop (Deybe) or adjustments of present income (Kruseman). The models by Glomsrød and Struif Bontkes do not allow any anticipation of the effects of degradation due to their recursive nature: households can only act in response to the cumulative effects of past actions.

20.4 Concluding Remarks

Different models serve different purposes. Hence, it is not possible to select the 'best' type of model for addressing the impact of economic policies on sustainable land use. As discussed in Chapter 1, the impact of macroeconomic and agricultural sector policies on the socio-economic environment of households can be analysed by macro general equilibrium models, that provide insights in economic incentives originating outside the agricultural sector and in feedback mechanisms between agricultural and non-agricultural sectors through production and income linkages. The impact of the socio-economic environment (prices, institutions, transaction costs) on farm household decision making in developing countries can be analysed by farm household and village economy models that take into account the impact of market imperfections on household behaviour, and the possibility of feedback mechanisms (through income and expenditure effects, and local price formation processes) at the village level.

Table 20.4: Time dimension.

Model [Chapter]	Time period	Technological change	Cumulative effects	Balancing welfare
Econometric models				
Antle [10]	- multiple periods	-	- soil quality depends on history of land use	-
Brière [6]	- 3 periods	-	- soil fertility	- discounting
Romano [7]	- infinite	-	- soil quality index	- discounting
Ruben [18]	- 2 periods	-	-	- summing of utilities
Partial programming models				
Barbier [12]	- multiple years	-	- soil depth	- discounting
Deybe [14]	- multiple periods & long run impact on reference crop	- potential technologies	- reference crop	- recursive, discounting of reference crop
Jansen [13] (S)*	- 1 year	- alternative technologies	-	- discounting income flows of perennials
Jansen [13] (U)*	- single & multiple periods	- alternative technologies	-	- discounting
Jansen [13] (O)*	- infinite	- alternative technologies	- N stock	- discounting
Kruseman [8]	- multiple periods & long run impact of soil quality on present income	- alternative technologies	-	- recursive, discounting of long run impact
Shiferaw [5]	- 10 seasons, 10 years	- conservation technologies	- nutrient stock, rooting depth	- discounting
General equilibrium models				
Dyer [11]	- static	-	-	-
Glomsrød [15]	- multiple years	-	- N stock, forest area	- recursive
Dynamic simulation models				
Abiye [3]	- multiple years	-	- N stock	- discounting
Struif Bontkes [9]	- 45 years	-	- N, P stocks	- recursive

* S = SOLUS, U = UNA-DLV, O = Optimal control

Bio-economic models are especially suited for analysing the links between socio-economic and bio-physical processes at the household level. These models try to integrate the complexity of bio-physical processes, including feedback effects on agricultural productivity, with socio-economic analyses of farm household decision making.

20.4.1 Complementary Advantages of Different Approaches

In addition, a number of strong and weak points can be identified, related to the type of modelling technique that is used. Starting with the econometric models, their strongest point is testing of the economic theory on which they are based. Their reliance on statistical methods represents at the same time their weakness, since it limits their use to extrapolations of past trends. This implies that they cannot be used, for example, to analyse possible impacts of technological alternatives still in the research and development pipeline. The strong point of the partial programming and general equilibrium models is their capacity to deal with changes in the socio-economic and biophysical environment, including technologies not yet practiced. Overall, the econometric, partial and general equilibrium models offer complementary advantages. Econometric models can quantify economic and biophysical components, partial programming models are well suited for dealing with the interactions between economic and biophysical processes, while the general equilibrium models offer a theoretically consistent way of modelling interactions between economic agents.

The major difference of dynamic simulation models from the other three model types is their origin in the (bio)physical science, as opposed to the economic origin of the others. Economic theory can be incorporated in these models by providing the basis for behavioural descriptions of the agents. In a similar way, (results of) crop simulation models are used in the partial programming models to provide the link between production and soil quality, as, for example, the reference crop in the model by Deybe. It should, however, be noted that a stream of dynamic simulation models has been developed within an evolutionary economics paradigm, criticising the reductionist and equilibrium assumptions of conventional economic theory on which the models in this volume are based. Discussion of these issues is beyond the scope of this book (useful references in this context are Costanza *et al.* (1993) and Hodgson (1997)).

20.4.2 Directions for Future Research

This brings us to considering in what way future work could build on current achievements, using the four key issues as a guideline. Examining the different tables, it follows that more explicitly addressing the institutional environment of the rural households would be one direction. Major work in this area is required, since the formal analysis of institutional issues largely is still an uncharted area. The study of Krishna and Uphoff provides a first step in this direction by defining and estimating different elements of social capital. Their results show that such a type of analysis may deepen the understanding of success or failure of policies. Extension of the models with descriptions of institutions, and their responses to policies, does not have to start from scratch, but can build on work from other social disciplines. As the chapter by von Benda-Beckmann shows, property rights represent a much more complex issue than can be captured by changing the planning horizon, or assigning a probability to loosing land.

Richer specifications of the institutional environment may also provide a handle for modelling coping with risk, for example through social relations as insurance mechanisms. More generally, issues of both economic and biophysical uncertainties represent a second important direction for future research. Especially for low-income households in fragile tropical environments, they form major constraints. A first step could be to systematically explore how different assumptions on the effects of degradation would affect model outcomes, *i.e.* to critically reflect on the uncertainties surrounding the feedback between land use and degradation. Although it can be argued, following Keyzer & Sonneveld, that these uncertainties are so large that functional descriptions cannot be formulated, a more pragmatic stand is needed. To analyse the effects of (new) economic policies on sustainability, descriptions of the degradation processes and their consequences for production are required. However, the uncertainties should be acknowledged, and the implications of necessary simplifying and incomplete assumptions be examined through sensitivity analyses. That might provide indications of the robustness of model outcomes for these (as well as other) key assumptions, hopefully enlarging the scope for applying the results in policy formulations.

A more general case can be made for empirical justification of key assumptions and documentation of sensitivity analyses. Given the impact of, for example, the assumptions on separability of household decisions, empirical justification for these assumptions and/or tests of the sensitivity of model outcomes to changes in these assumptions would be of great value. Reporting the results of sensitivity analyses not only increases the credibility of modelling results. It also serves to identify critical assumptions and parameters. This can provide a direction for future research efforts into areas where it would be most useful for increasing our understanding of the impact of economic policies on sustainable land use.

A third direction for future research would be finding a balanced way of aggregating micro-level studies to levels relevant for policy-makers, accounting for micro-level variability and rationality, while focusing on general patterns. Understanding behavioural responses to policy instruments at those levels (particularly farm households) where land use decisions are actually taken, is crucial for successful implementation of any policy aimed at, either directly or indirectly, promoting sustainable land use. Since the success or failure of policies is measured at regional or higher levels, issues of interactions between economic agents and land units, as well as spatial heterogeneity need to be addressed. Two important issues in this respect are largely neglected in current model approaches.

One is the impact of positive or negative off-site effects of soil erosion on agricultural productivity elsewhere within (and outside) the same region. The second is the bias introduced by aggregating non-linear relationships. Although many biophysical, econometric and general equilibrium models derive their specifications from observed (or hypothesised) non-linear relationships at the micro level, the aggregation bias resulting from application of these specifications at higher levels of analysis is neglected almost entirely.

The above (incomplete) list for extending current bio-economic modelling of land use involves several theoretical and practical pitfalls. On the other hand, the proposed extensions would expand both theoretical and empirical boundaries of bio-economic modelling, which would benefit both scientific fields. Moreover, analysis of interactions between policies and sustainable land use in developing countries is an issue of direct relevance for large parts of the current and future world population. Such a combination of theoretical challenge and social relevance can do nothing else but inspire more innovative research in the near future.

20.5 References

Blonski, M. (1999): When is rational behaviour consistent with rules of thumb? A link between evolutionary terminology and neo-classical methodology. Journal of Mathematical Economics 32, 131-144

Börgers, T. (1996): On the relevance of learning and evolution to economic theory. Centre on Economics Learning and Social Evolution working paper, Department of Economics, University College London, London

Bouman, B.A.M., H.G.P. Jansen, R.A. Schipper, J. Bouma, A. Kuyvenhoven & M.K. van Ittersum (2000): A toolbox for land use analysis. In: B.A.M. Bouman H.G.P. Jansen, R.A. Schipper, H. Hengsdijk & A. Nieuwenhuyse (Eds.) Tools for land use analysis at different scales – with case studies from Costa Rica. Kluwer, Boston

Chichilnisky, G. (1997): What is sustainable development? Land Economics 72, 467-491

Costanza, R., L. Wainger, C. Folke & K.-G. Mäler (1993): Modelling complex ecological economic systems. Bioscience 43, 545-555

Dasgupta, P. & K. Mäler (1995): Poverty, institutions, and the environmental resource base. In J. Behrman & T. N. Srinivasan (Eds.) Handbook of development economics. Elsevier Science B.V., Amsterdam, *etc.*

Erenstein, O.C.A. (1999): The economics of soil conservation in developing countries: the case of crop residue mulching. Ph.D. Thesis, Wageningen Agricultural University, Wageningen

Ginsburgh, V. & M. Keyzer (1997): The structure of applied general equilibrium models. The MIT press, Cambridge (Massachusetts) & London.

Harrington, L., P. Jones and M. Winograd (1995): Operacionalizacion del concepto de sostenibilidad: un metodo basado en la productividad total. In: J. Berdegue and E. Ramirez (Eds.) Operacionalizacion del concepto de sistemas de produccion sostenibles. RIMISP, Santiago de Chile

Heerink, N. (1994): Population growth, income distribution, and economic development - theory, methodology, and empirical results. Springer, Berlin

Heerink, N. & H. Folmer (1994): Income distribution and the fulfilment of basic needs: Theory and empirical evidence. Journal of Policy Modelling 16, 625-652

Hodgson, G.M. (1997): Economics and evolution and the evolution of economics. In: J. Reijnders (Ed.) Economics and evolution. Edward Elgar, Cheltenham (UK), Lyme (N.H.)

Jaenicke, E.C. & L.L. Lengnick (1999): A soil quality index and its relationship to efficiency and productivity growth measures: two decompositions. American Journal of Agricultural Economics 81, 881-893

Jansen, H.G.P. & J. Stoorvogel (1996): Incorporating the farm level in regional agricultural land use models: quantification of different sources of aggregation bias. Paper presented at 'Workshop on sustainable land use: issues and models at different scale levels', held at Wageningen Agricultural University

Kaimowitz, D. & A. Angelsen (1997): A guide to economic models of tropical deforestation. Centre for International Forestry Research, Jakarta

Larson, B.A. & D.W. Bromley (1990): Property rights, externalities, and resource degradation - locating the tragedy. Journal of Development Economics 33, 235-262

Lettau, M. & H. Uhlig (1999): Rules of thumb versus dynamic programming. The American Economic Review 89, 148-174

Lin, J.Y. & J.B. Nugent (1995): Institutions and economic development. In: J. Behrman & T.N. Srinivasan (Eds.) Handbook of development economics. Elsevier Science B.V., Amsterdam, *etc.*

Löfgren, H. & S. Robinson (1999): To trade or not to trade: non-separable farm household models in partial and general equilibrium. TMD Discussion Paper 37, Trade and Macroeconomics Division, IFPRI, Washington D.C.

Mohamed, A.M. (1999): An integrated agro-ecological and agro-economic framework for land use planning and policy analysis. Ph.D. Thesis, Wageningen University, Wageningen

Nelson, R.R. & S.G. Winter (1982): An evolutionary theory of economic change. The Belknap Press of Harvard University Press, Cambridge (MA), London

Pearce, D.W. & R.K. Turner (1990): Economics of natural resources and the environment. Harvester Wheatsheaf, New York

Pelt, M.J.F. van (1993): Ecological sustainability and project appraisal: case studies in developing countries. Avebury Aldershot, *etc.*

Persson, A. & M. Munasinghe (1996): Economywide policies and deforestation: The case of Costa Rica. In: M. Munasinghe (Ed.) Environmental impacts of macroeconomic and sectoral policies. World Bank, Washington, D.C.

Rietkerk, M. (1998): Catastrophic vegetation dynamics and soil degradation in semi-arid grazing systems. Ph.D. Thesis, Wageningen Agricultural University, Wageningen

Roumasset, J.A. (1995): Induced institutional change: a neo-classical synthesis. In: B.M. Koppel (Ed.) Induced innovation theory and international agricultural development. The Johns Hopkins University Press, Baltimore and London

Ruben, R., H. Moll & A. Kuyvenhoven (1998): Integrating agricultural research and policy analysis: analytical framework and policy applications for bio-economic modelling. Agricultural Systems 58, 331-349

Rubinstein, A. (1998): Modelling bounded rationality. The MIT Press, Cambridge (MA), London

Sadoulet, E. & A. de Janvry (1995): Quantitative development policy analysis. The Johns Hopkins University Press, Baltimore, London

Sadoulet, E., A. de Janvry & C. Benjamin (1996): Household behaviour with imperfect labour markets. Working Paper 786, Department of Agricultural and Resource Economics, University of California, Berkeley

Singh, I., L. Squire & J. Strauss (1986): Agricultural household models. Extensions, applications and policy. John Hopkins University Press, Baltimore and London

Taylor, J.E. & I. Adelman (1996): Village economies: the design estimation and use of village-wide economic models. Cambridge University Press, Cambridge, *etc.*

Trimble, S.W. (1999): Decreased rates of alluvial sediment storages in the Coon Creek Basin, Wisconsin, 1975-93. Science 285, 1244-1246

Unemo, L. (1996): Environmental impact of governmental policies and external shocks in Botswana: A CGE modeling approach. In: M. Munasinghe (Ed.) Environmental impacts of macro-economic and sectoral policies. World Bank, Washington, D.C.

WCED. (1987). Our common future. Oxford University Press, Oxford.

Wit, C.T. de & H. van Keulen (1989): Modelling production of field crops and its requirements. Geoderma 40, 253-265

Contributions to Economics

Simon Duindam
Military Conscription
1999. ISBN 3-7908-1203-X

Bruno Jeitziner
Political Economy of the Swiss National Bank
1999. ISBN 3-7908-1209-9

Irene Ring et al. (Eds.)
Regional Sustainability
1999. ISBN 3-7908-1233-1

Katharina Müller/Andreas Ryll/
Hans-Jürgen Wagener (Eds.)
Transformation of Social Security: Pensions in Central-Eastern Europe
1999. ISBN 3-7908-1210-2

Stefan Traub
Framing Effects in Taxation
1999. ISBN 3-7908-1240-4

Pablo Coto-Millán
Utility and Production
1999. ISBN 3-7908-1153-X

Frank Riedel
Imperfect Information and Investor Heterogeneity in the Bond Market
2000. ISBN 3-7908-1247-1

Kirsten Ralf
Business Cycles
2000. ISBN 3-7908-1245-5

Michele Bagella/
Leonardo Becchetti (Eds.)
The Competitive Advantage of Industrial Districts
2000. ISBN 3-7908-1254-4

Frank Bohn
Monetary Union and Fiscal Stability
2000. ISBN 3-7908-1266-8

Michael Malakellis
Integrated Macro-Micro-Modelling Under Rational Expectations
2000. ISBN 3-7908-1266-8

Stefan Baumgärtner
Ambivalent Joint Production and the Natural Environment
2000. ISBN 3-7908-1290-0

Henri Capron, Wim Meeusen (Eds.)
The National Innovation System of Belgium
2000. ISBN 3-7908-1308-7

Tobias Miarka
Financial Intermediation and Deregulation
2000. ISBN 3-7908-1307-9

Chisato Yoshida
Illegal Immigration and Economic Welfare
2000. ISBN 3-7908-1315-X

Nikolaus Thumm
Intellectual Property Rights
2001. ISBN 3-7908-1329-X

Johannes Bröcker/Hayo Herrmann (Eds.)
Spatial Change and Interregional Flows in the Integrating Europe
2001. ISBN 3-7908-1352-4

Kirstin Hubrich
Cointegration Analysis in a German Monetary System
2001. ISBN 3-7908-1352-4

GPSR Compliance

The European Union's (EU) General Product Safety Regulation (GPSR) is a set of rules that requires consumer products to be safe and our obligations to ensure this.

If you have any concerns about our products, you can contact us on ProductSafety@springernature.com

In case Publisher is established outside the EU, the EU authorized representative is:

Springer Nature Customer Service Center GmbH
Europaplatz 3
69115 Heidelberg, Germany

Batch number: 10725973

Printed by Printforce, the Netherlands